Nutraceutical Fatty Acids from Oleaginous Microalgae

Scrivener Publishing
100 Cummings Center, Suite 541J
Beverly, MA 01915-6106

Publishers at Scrivener
Martin Scrivener (martin@scrivenerpublishing.com)
Phillip Carmical (pcarmical@scrivenerpublishing.com)

Nutraceutical Fatty Acids from Oleaginous Microalgae

A Human Health Perspective

Edited by
Alok Kumar Patel
and Leonidas Matsakas

Scrivener Publishing

WILEY

This edition first published 2020 by John Wiley & Sons, Inc., 111 River Street, Hoboken, NJ 07030, USA and Scrivener Publishing LLC, 100 Cummings Center, Suite 541J, Beverly, MA 01915, USA

For more information about Scrivener publications please visit www.scrivenerpublishing.com.

Wiley Global Headquarters
111 River Street, Hoboken, NJ 07030, USA

For details of our global editorial offices, customer services, and more information about Wiley products visit us at www.wiley.com.

Library of Congress Cataloging-in-Publication Data

ISBN 9781119631712

Cover image: Alok Kumar Patel
Cover design by Kris Hackerott

Set in size of 11pt and Minion Pro by Manila Typesetting Company, Makati, Philippines

10 9 8 7 6 5 4 3 2 1

Contents

1

Introduction to Essential Fatty Acids

Alok Patel*, Ulrika Rova, Paul Christakopoulos and Leonidas Matsakas

Biochemical Process Engineering, Division of Chemical Engineering, Department of Civil, Environmental, and Natural Resources Engineering, Luleå University of Technology, Luleå, Sweden

Abstract

Certain omega-3 fatty acids, such as α-linolenic acid (ALA), and omega-6 fatty acids, such as linoleic acid (LA), cannot be synthesized in the human body and are recognized as essential fatty acids. While some long-chain polyunsaturated fatty acids (LC-PUFA) such as eicosapentaenoic acid (EPA) and docosahexaenoic acid (DHA) can be synthesized from the parent omega-3 fatty acids (ALA), this is done at a very low conversion rate, hence it must be taken through diet to fulfill the daily intake requirement. Both EPA and DHA have several vital activities in the human body, such as anti-inflammatory effects and being the structural component of the cell membrane. The fatty acids DHA, arachidonic acid (AA), and LA accumulate most usually in tissues, whereas DHA mostly accumulates in retina and brain gray matters and it is important for proper visual and neurological development during gestation period and postnatal period. Replacement of saturated fatty acids with omega-3 and omega-6 fatty acids in daily diet reduces the risk of cardiovascular disease and prevents diseases such as Alzheimer's, bipolar disorder, and schizophrenia. Proper EPA and DHA content also help individuals with type 2 diabetes to reduce the elevated serum triacylglycerides. It also facilitates infants to reduce the risks of fatal myocardial infarction and other cardiovascular diseases. Hence, as recommended by the American Heart Association, it is necessary to consume fish, and especially oily fish at least twice per week as it is an excellent source of these fatty acids. Marine fishes of Salmonidae, Scombridae, and Clupeidae families are important sources of omega-3 fatty acids but due to the increasing demand of PUFA and diminishing aquatic ecosystem, fishes are not a sustainable source to serve as a long-term feedstock for omega-3. Plants can synthesize some of PUFA such as oleic acid, LA, GLA (γ-linolenic acid), ALA, and octadecatetraenoic acid but due to lacking some essential

**Corresponding author*: alok.kumar.patel@ltu.se

Alok Kumar Patel and Leonidas Matsakas (eds.) Nutraceutical Fatty Acids from Oleaginous Microalgae: A Human Health Perspective, (1–22) © 2020 Scrivener Publishing LLC

enzymes for PUFA synthesis such as desaturase and elongases, they are incapable of synthesizing EPA and DHA. Oleaginous microalgae and thraustochytrids could be a sustainable option to produce microbial EPA and DHA.

Keywords: Oleaginous microorganisms, lipid accumulation, fatty acid profile, microalgae, nutraceuticals, omega-3 fatty acid, human health

1.1 Introduction

Fatty acids or lipids serve in diverse metabolic functions related to growth and maintenance of cells and tissues, and act as caloric energy molecules involved in various cellular signaling events that accompany several physiological processes after metabolism [1, 2]. Lipids are usually originated from the acetate route and derivatives of polyketides [3]. Chemically lipids are hydrocarbons of C6 to C32 long-chain, containing hydrophilic carboxyl group at one end and methyl group at the terminal end. These hydrocarbon chains are made up of an even number of carbon atoms in naturally occurring fatty acids; however, they can be branched or cyclic that is present in some bacterial strain [3]. The hydrocarbon chain can be saturated, monounsaturated or polyunsaturated depending on the presence and numbers of double bond [4, 5]. From the different fatty acid types, polyunsaturated fatty acids (PUFAs) have greater physiological importance because of their medicinal properties [6]. Omega-3 and omega-6 fatty acids belong to a category of PUFAs where the first double bond is located between the 3rd and 4th carbon atom near the methyl end (or omega end) of omega-3 fatty acids (n-3) whereas the first double bond is situated between the 6th and 7th carbon in the case of omega-6 fatty acids (Figure 1.1).

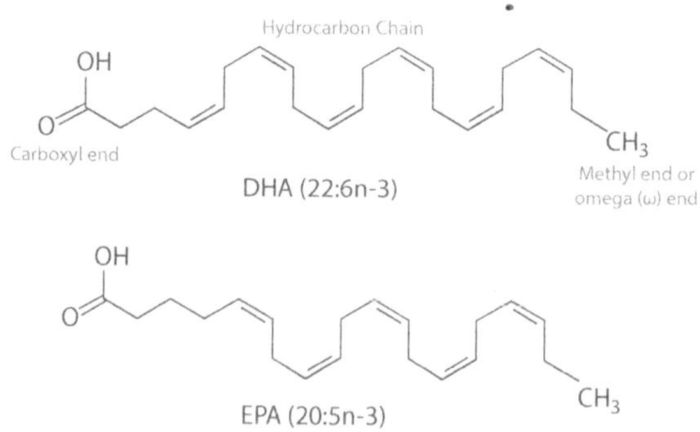

Figure 1.1 The chemical structure of DHA and EPA, representative omega-3 and omega-6 PUFAs.

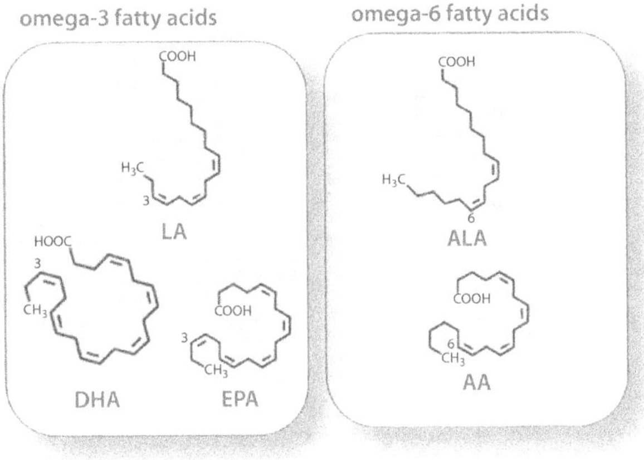

Figure 1.2 Molecular structure of omega-3 fatty acids and omega-6 fatty acids.

These fatty acids have unique structures due to the presence of the double bond as it introduces the bends in the hydrocarbon chain which affect its physical properties (Figure 1.2) [7]. Mammals can synthesize saturated and unsaturated fatty acids from carbon present in carbohydrates and proteins while they show an inability to synthesize PUFAs due to the lack of certain enzymes that are necessary to introduce the cis double bonds at n-3 and n-6 position [5]. These omega-3 and omega-6 fatty acids are considered as essential fatty acids for human beings and must be taken through diet. α-linoleic acid (ALA, 18:3n-3) and linoleic acid (LA, 18:2n-6) are the parent fatty acids of omega-3 and omega-6 fatty acid series, respectively [4].

Humans can synthesize long-chain PUFA (LC-PUFA) such as eicosapentaenoic acid (EPA; 20:5n-3) and docosahexaenoic acid (DHA; 22:6n-3), from ALA and dihomo-γ-linolenic acid (DGLA; 20:3n-6) and AA (20:4n-6), from LA. Some investigators showed that only 2 to 10% of ALA can be converted into EPA and DHA [8], while others suggested that 7% of ALA is converted into EPA while low conversion rate (0.013%) is reported for DHA [9]. Hussein *et al.* (2005) suggested that only 0.3% of EPA and <0.01% of DHA can be converted from ALA [10]. The conversion of ALA to EPA and DHA is too low and cannot meet the daily intake requirement. These fatty acids have several health benefits as they are incorporated in various parts of the body including cell membrane and play a vital role in anti-inflammatory process [11, 12]. Both EPA and DHA are essential for proper aging and fetal development [13], DHA is mainly incorporated in the retina and brain. They are also used to treat and prevent several viral diseases.

1.2 Biosynthesis of PUFAs

Fatty acids are absorbed in intestines after hydrolysis from dietary fats (triacylglycerols and phospholipids) by pancreatic enzymes [14]. Micelles are formed in intestines after mixing fatty acids and other fat digestion products in the presence of bile salts. Fats are absorbed to an extent of 85-90% throughout the small intestine from these mixed micelles [14]. Omega-3 fatty acids are considered as essential fatty acids as they cannot be synthesised by humans and animals, due to lack of the Δ-12 and Δ-15 desaturase enzymes [4, 5]. As humans and animals are unable to produce a *de novo* synthesis of omega-3 fatty acids, these must be taken through their diet from other sources [15]. However, they have the capability to synthesise small amounts of DHA and some intermediate products such as EPA from the parent omega-3 fatty acids such as ALA, and the rest of the required amount is fulfilled from direct consumption of DHA and EPA [16]. The parent ALA intake is done also through our diet from plant sources. The metabolism starts by the action of Δ-6 desaturase on ALA, for the unsaturation in aliphatic carbon chain, followed by the addition of two carbon atoms in the aliphatic carbon chain by the action of elongases. This structure of fatty acids allows the action of Δ-5 desaturase and finally formation of EPA [16]. Some previous studies showed that a desaturation process occurs in endoplasmic reticulum [16]. However, the role of Δ-4 desaturase was established from the microbial pathway for DHA synthesis (22:6ω-3) from 22:5ω-3. This part of the pathway was interpreted after the discovery of 24:5ω-3 from the elongation of 22 carbon chain and further desaturated into 24:6ω-3 in the mammals [17] and there is a reverse pathway of partial oxidation of 24:6ω-3 to synthesize DHA in peroxisomes [18]. This is also evidenced by the treatment of peroxisomal disorders, such as Zellweger syndrome, in infants by dietary supplementation of DHA [19]. The unsaturation and elongation process during synthesis of fatty acids is presented in Figure 1.3.

There is another pathway of fatty acid synthesis that was discovered in the thraustochytrids strain *Schizochytrium* sp. and is known as polyketide synthesis pathway (PKS). This pathway is distinguished from the conventional Fatty acid synthesis (FAS) pathway in terms of oxygen requirements, as it does not require molecular oxygen for the synthesis of PUFAs [23]. In this microorganism, C14 and C16 fatty acid synthesis follows the FAS system, while the synthesis of LC-PUFA follows the PKS pathway [24]. A major difference between the FAS and PKS pathways is that the elongation of fatty acids is done through dehydration and isomerization of fatty acyl intermediates in the PKS pathway, whereas in the FAS pathway it occurs

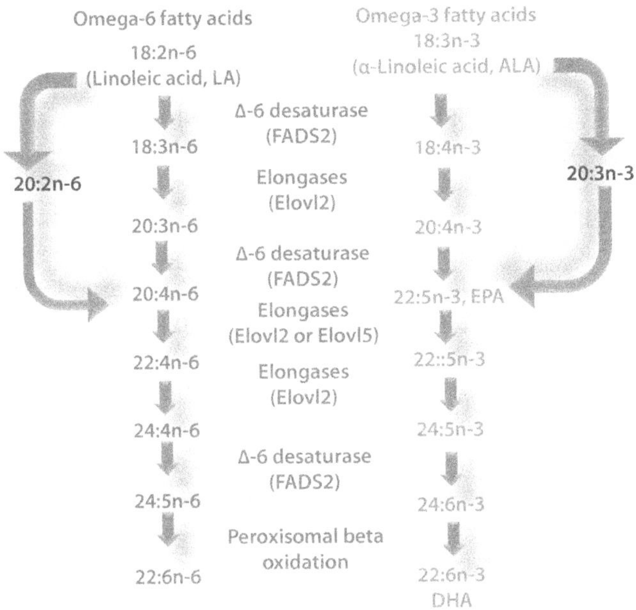

Figure 1.3 Synthesis of EPA and DHA from ALA through a series of desaturation and elongation reactions [3, 20–22].

through elongases and desaturases [25]. Although LC-PUFA synthesis in thraustochytrids is not fully understood yet, the identification of a Δ-4 fatty acid desaturase in thraustochytrids indicated that DHA synthesis might be a result of the FAS pathway [23]. In FAS route of PUFA synthesis, a series of desaturases and elongases works on either the C16:0 or C18:0 to form unsaturated fatty acids and DHA is synthesized from 22:5 by delta-4 desaturase [24]. It has been proved by genomc annotation results of thraustochytrids *S. limacinum* that this microorganism does not have delta-4 desaturase as reported in FAS route of DHA synthesis. DHA is usually synthesized by the PKS pathway present in this microorganism [23, 26, 27] which require a different series of enzymes such as 3-ketoacyl synthase, 3-ketoacyl-ACP reductase, enoyl reductase, and dehydrase/isomerase [28].

1.3 Sources of Essential Fatty Acids and Daily Intake Requirement

The daily requirement of the parent omega-3 fatty acid LA is 17-20 g for men and 12-13 g for women and can be obtained from vegetable oils such as soybean, corn oil or seeds of various plants [29]. The daily intake of ALA is 1.8-2 g for men and 1.4-1.5 g for women and can be taken through, for example, flaxseeds and canola oil [30]. Animals have the capacity to produce AA from LA, so certain animal products such as meat and eggs are an excellent source of AA [31]. Evening primrose oil and borage seed oil are good sources of γ-linolenic acid (GLA) [32].

The daily intake requirements of EPA and DHA for adults are 200 mg to 1000 mg according to various agencies [30]. The major dietary sources of EPA and DHA are fishes [33] (see Table 1.1). Cod liver oil, for example, is not only enriched with EPA and DHA but also contains vitamins such as vitamin A [34]. Krill oil may be an alternative to fish oils, as it a good source of omega-3 fatty acids and has a high content of EPA and DHA [35]. Nowadays, due to the increasing interest in vegetarian foods, people avoid consuming non-vegetarian sources of DHA and EPA. Moreover, a major concern with consuming certain fishes is the possibility of contamination with several harmful elements such as arsenic, lead and mercury [34]. Some plant and vegetable oils are good source of polyunsaturated fatty acids and are listed in Table 1.2. The only sustainable replacement of non-vegetarian sources of PUFAs is the oils obtained from microbial sources. Microalgae, fungi and bacteria can be sources of microbial oils. These microorganisms are known to be natural producers and the original sources of omega-3 PUFAs [36]. A list of thraustochytrids and microalgae for their omega-3 fatty acid production capabilities are presented in Table 1.3 and Table 1.4, respectively. Despite the high production cost, microorganisms offer certain advantages for the production of omega-3 PUFAs, such as high growth rates, simple nutrient input requirement, controllable culture conditions, easy genetic manipulation and well-annotated genomes and metabolic

Table 1.1 PUFA content of some fish species [39].

Fish species	g DHA per serving	g EPA per serving
Salmon, Atlantic, farmed cooked, 3 ounces	1.24	0.59
Salmon, Atlantic, wild, cooked, 3 ounces	1.24	0.59
Herring, Atlantic, cooked, 3 ounces	0.94	0.77
Tuna, light, canned in water, drained, 3 ounces	0.17	0.02

Table 1.2 Omega-3 and omega-6 fatty acids content (% w/w) of plant and vegetable oils.

Plant or vegetable seeds	PUFA content	Omega-3 fatty acid content	Omega-6 fatty acid content	References
Chia	80.4	NA	NA	[40]
Flax	73.6	NA	NA	[40]
Hemp	62.8	0.4	62.4	[41]
Olive	18.0	1.6	16.4	
Sunflower	62.4	0.4	62.2	
Sesame	41.2	0.2	40.9	
Rapeseed	20.9	1.2	19.6	
Canola	30.7	NA	NA	[42]
Walnut	65.0	NA	NA	[43]
Coconut	1.6	0.0	1.6	[41]
Perilla	75.9	NA	NA	[40]
Peanut	18.2	0.0	18.2	[41]

NA, not available.

pathways [37]. Microbial oils usually contain a significant amount of natural antioxidants such as carotenoids and tocopherols, which play a role in protecting omega-3 PUFAs from oxidation. The first microbial oil containing GLA that was produced was from a filamentous fungus, *Mucor circinelloides* and its commercial production lasted from 1985 to 1990 [38].

1.4 Biological Role of Essential Fatty Acids

1.4.1 Effect on Cell Membrane Structure

Omega-3 and omega-6 fatty acids are important cell membrane structural elements. They affect cellular properties such as membrane fluidity and permeability, and the activities of phospholipids bounded enzymes [65]. Cell membrane composition and molecular structure are totally dependent on

Table 1.3 PUFA content of Thraustochytrids species.

Microorganisms	Cell dry weight (g/L)	DHA concentration (%, total lipid)	References
Aurantiochytrium sp. ATCC PRA-276	17.0	12.5	[44]
	11.24	35.76	[45]
Aurantiochytrium sp. KRS101	4.40	14.31	[46]
	5.50	14.18	
Aurantiochytrium sp. KRS101	9.00	19.88	[47]
Schizochytrium limacinum SR 21	20.10	18.38	[48]
Aurantiochytrium 4W-1b	9.01 ± 0.62	27.9	[49]
Aurantiochytrium SW1	19.0	25	[50]
Aurantiochytrium sp. T66	11.24	35.76	[45]
Schizochytrium SR21	21.0	34.9	[51]
S. limacinum SR21	33.24	23.48	[48]
A. limacinum (ATCC MYA-1381)	8.86	15.01	[52]
S. limacicum SR21	10.90 ± 0.30	34	[53]
Schizochytrium sp.	71	48.95	[54]

the dietary fatty acid consumption as well as endogenous metabolisms and it was evidenced by increased concentration of omega-3 in immune cells, cardiac tissue and other cells throughout the body after intake of omega-3 fatty acids [5, 65–67].

Table 1.4 EPA and DHA content of microalgae.

Microalgal strains	EPA (%)	DHA (%)	References
Phaeodactylum tricornutum	19.87	4.89	[55]
P. tricornutum	29	NA	[56]
Nannochloropsis sp.	26.7		[57]
Nannochloropsis oceanica	23.4	NA	[58]
Nannochloropsis salina	28	NA	[59]
Pinguiococcus pyrenoidosus	22.03		[60]
Chlorella minutissima	39.9	NA	[61]
Pavlova viridis	36		[62]
Pavlova lutheri	27.7		[63]
Isocrysis galbana	28		[64]
Pavlova lutheri	41.5		[63]

1.4.2 Impact on Vision

DHA is predominantly integrated into the membranes of retinal cells. The retina stores and recycles DHA even if the consumption of omega-3 fatty acids is limited [68–70]. Studies on DHA showed that it regenerates rhodopsin pigments in retina that usually converts the light from retina to a visual image in the brain [70, 71].

1.4.3 Brain Function

DHA is a structural component of postsynaptic neuronal cell membranes, with the highest amount of DHA and AA to be found in the phospholipids of brain gray matter, representing their importance in central nervous system regulation [72]. DHA starts accumulating in the fetal brain tissue during the last trimester and persists at very high rates until the end of the second year of life [7, 72]. For a proper brain function, DHA should be taken through food, since its endogenous formation is low. Its deficiency showed malfunction of brain, posing a risk of creating a problem of learning deficits [73].

1.4.4 Biosynthesis of Lipid Mediators

Eicosanoids are usually 20 carbon long chain of hydrocarbons and act as bioactive lipid mediators that mainly play an important role in inflammatory and immune response [6]. DGLA, AA and EPA are usually secreted from the cellular membranes after activation by hormones, cytokines and other signals and become substrates to produce eicosanoids [74]. It has been suggested that AA derived eicosanoids are more potent inducers of inflammatory response than the EPA derived eicosanoids [2, 75]. Isoprostanes are produced from free radicals induced oxidation of PUFAs; therefore, they act as markers for oxidative stress and have both pro- and anti-inflammatory effects [74–76].

1.4.5 Effect of Omega Fatty Acids on the Regulation of Gene Expression

Omega-3 and omega-6 fatty acids induce a large number of gene expressions that are involved in inflammation and other immune responses by interrelating through specific transcription factors, such as peroxisome proliferator-activated receptors (PPARs) [77]. In several instances, PUFAs act as steroid hormones to the gene expression by binding with receptors like PPARs and further bind with the promoters of genes that can regulate transcription. They can act as regulators of transcription factors such as NFκB and Sterol Regulatory Element–binding Proteins (SREBP-1) inside the cell's nucleus [78]. For example, NFκB is suppressed by omega-3 fatty acid to inhibit the production of multiple genes involved in inflammation like eicosanoids and cytokines. SREBP-1 regulates the transcription factor that involves the *de novo* synthesis of fatty acids, and is also affected by dietary omega-3 fatty acids that suppress the SREBP-1 to further synthesis of omega-3 fatty acids [79].

1.5 Effect of Essential Fatty Acid on Human Health (Disease Prevention and Treatment)

1.5.1 Neonatal Development

DHA is mostly accumulated in the brain and retina during the last trimester of pregnancy and two years from birth [80]. Human breast milk contains approximate 12% of omega-6 and 1.3% of omega-3 fatty acids [81]. Before 2001, infant milk formula contained only ALA as nutritional supplements; however, due to the low conversion of ALA to DHA and EPA, this was not sufficient [82]. Studies on randomized controlled trials testing the visual and neurological development in infants (19 trials, 1,949 infants) with and without DHA supplementation, showed the beneficial effect of DHA supplementation on visual acuity from one month to 12 months after birth [83]. Epidemiological studies showed that omega-3 and omega-6 fatty acid intake from fish and seafood during pregnancy has a positive response on the growth of offspring [84].

1.5.2 Gestation and Pregnancy

The results of randomized controlled trials of omega-3 fatty acids intake during low-risk pregnancy found no positive effect on gestational diabetes, preeclampsia and hypertension but due to high consumption of omega fatty acids, the duration of pregnancy can be increased modestly because of unbalanced prostaglandins production involved in parturition. [85–87]. The fish oil supplementation during last trimester increases the time of pregnancy without affecting foetus growth and on the course of labour [86]. However, according to a meta-analysis of six randomized controlled trials, the gestation period can be increased by 1.6 day in low-risk pregnancies with omega-3 supplementation [88]. According to a 2007 meta-analysis of randomized controlled trials it was evidenced that omega-3 fatty acids do not affect the gestation period in high-risk pregnancies but can reduce early premature birth [89]. It is highly recommended by experts in the US that pregnant and lactating women take at least 200 mg DHA daily for proper development of the foetus and newborn baby [89, 90]. According to the European Food and Safety Authority (EFSA) recommendations, the amount of DHA daily intake should be 100-200 mg along with 250 mg/day EPA as for normal adults.

1.5.3 Cardiovascular Disease

Both omega-3 and omega-6 fatty acids intake are directly corelated with the reduction in risk of cardiovascular diseases [91]. As per observation from randomized controlled trials, the American Heart Association suggested that high intake of omega-6 fatty acids such as LA can reduce the risk of coronary heart disease [91, 92]. The outcomes of 11 cohort studies on 344,696 individuals for the replacement of 5% of energy intake from SFA to PUFAs showed almost 26% less coronary deaths [91].

1.5.4 Cancer Inhibition

Some studies showed that intake of omega-3 fatty acids reduce the risk of cancer due to their potent anti-inflammatory effects and inhibitory effect on cell growth factors [93–96]. Recent studies suggested that increased intake of omega-3 fatty acids or increased blood levels of omega-3 fatty acids reduced the risk of breast and colorectal cancers [97–99]. Opposite to these findings, some studies suggested that high omega-3 intake increases the risk of prostate cancer; however, some scientists have doubts on this findings [100].

1.5.5 Rheumatoid Arthritis

Several metanalysis and randomized controlled trials suggested that intake of PUFAs has positive effects on reducing the risk of rheumatoid arthritis, an autoimmune inflammatory disease of multiple joints that is associated with cardiovascular diseases [101–105]. Intake of PUFAs reduced the leukotriene B4 which is a proinflammatory marker and also has a positive effect on blood profile levels, with the triacylglycerol level found to be reduced [106].

1.5.6 Effect on Suicide Risk in Mood Disorders

The deficiency of PUFAs and the ratio of omega-3 to omega-6 fatty acids leads to unipolar and dipolar disorders and is further associated with suicidal behaviour that was evidenced after recent epidemiological studies [107, 108]. After treating patients with mental disorders by PUFAs intake, the results showed reduced depression but no statistically significant changes were observed on omega-3 fatty acids depletion in both control and the patient with mental illness or with suicidal behaviour [109].

1.6 Concluding Remarks

The long-chain omega-3 fatty acids EPA and DHA can be synthesized from ALA, but it is recommended that both should be derived through diet. The most common PUFAs that accumulate in tissues are LA, AA and DHA. Fish oil alone cannot satisfy the ever-increasing global demands for omega-3 omega-6 PUFAs due to declining fish stocks and marine ecosystem pollution, which has led to increased interest in sustainable alternative sources. Two possible alternatives to fish oil are vegetable oils from genetically engineered plant oils and microbial oils, with the omega-3 PUFAs content to be higher in the latter. Several examples of genetically engineered plant species with the ability to produce omega-3 PUFAs are *Brassica juncea, Arabidopsis thaliana*, and *Camelina sativa*. Although there are numerous advantages for transgenic plants, their production depends on seasonal and climatic conditions and arable land availability. However, microorganisms are natural sources of microbial oils and can serve as an excellent source of omega-3 and omega-6 PUFAs of nutritional significance.

References

1. R. Zárate, N. el Jaber-Vazdekis, N. Tejera, J.A. Pérez, C. Rodríguez, Significance of long chain polyunsaturated fatty acids in human health, *Clin. Transl. Med.* 6 (2017) 25. doi:10.1186/s40169-017-0153-6.
2. A.J. de Jong, M. Kloppenburg, R.E.M. Toes, A. Ioan-Facsinay, Fatty acids, lipid mediators, and T-cell function, *Front. Immunol.* 5 (2014) 3–9. doi:10.3389/fimmu.2014.00483.
3. P.M. Dewick, The Acetate Pathway: Fatty Acids and Polyketides, in: *Med. Nat. Prod. A Biosynthetic Approach*, John Wiley & Sons, Ltd, Chichester, UK, 2009: pp. 39–135. doi:10.1002/9780470742761.ch3.
4. R.K. Saini, Y.S. Keum, Omega-3 and omega-6 polyunsaturated fatty acids: Dietary sources, metabolism, and significance—A review, *Life Sci.* 203 (2018) 255–267. doi:10.1016/j.lfs.2018.04.049.
5. G. Kaur, D. Cameron-Smith, M. Garg, A.J. Sinclair, Docosapentaenoic acid (22:5n-3): A review of its biological effects, *Prog. Lipid Res.* 50 (2011) 28–34. doi:10.1016/j.plipres.2010.07.004.
6. T. Ishihara, M. Yoshida, M. Arita, Omega-3 fatty acid-derived mediators that control inflammation and tissue homeostasis, *Int. Immunol.* 31 (2019) 559–567. doi:10.1093/intimm/dxz001.
7. S.M. Innis, Dietary omega 3 fatty acids and the developing brain, *Brain Res.* 1237 (2008) 35–43. doi:10.1016/j.brainres.2008.08.078.
8. C.C. Chiu, K.P. Su, T.C. Cheng, H.C. Liu, C.J. Chang, M.E. Dewey, R. Stewart, S.Y. Huang, The effects of omega-3 fatty acids monotherapy in Alzheimer's disease and mild cognitive impairment: A preliminary randomized double-blind placebo-controlled study, *Prog. Neuro-Psychopharmacology Biol. Psychiatry.* 32 2008) 1538–1544. doi:10.1016/j.pnpbp.2008.05.015.
9. P.L.L. Goyens, M.E. Spilker, P.L. Zock, M.B. Katan, R.P. Mensink, Compartmental modeling to quantify α-linolenic acid conversion after longer term intake of multiple tracer boluses, *J. Lipid Res.* 46 (2005) 1474–1483. doi:10.1194/jlr.M400514-JLR200.
10. N. Hussein, E. Ah-Sing, P. Wilkinson, C. Leach, B.A. Griffin, D.J. Millward, Long-chain conversion of [13C]linoleic acid and α-linolenic acid in response

to marked changes in their dietary intake in men, *J. Lipid Res.* 46 (2005) 269–280. doi:10.1194/jlr.M400225-JLR200.

11. N. Lazzarin, E. Vaquero, C. Exacoustos, E. Bertonotti, M.E. Romanini, D. Arduini, Low-dose aspirin and omega-3 fatty acids improve uterine artery blood flow velocity in women with recurrent miscarriage due to impaired uterine perfusion, *Fertil. Steril.* 92 (2009) 296–300. doi:10.1016/j.fertnstert.2008.05.045.
12. J.K. Kiecolt-Glaser, M.A. Belury, R. Andridge, W.B. Malarkey, B.S. Hwang, R. Glaser, Omega-3 supplementation lowers inflammation in healthy middle-aged and older adults: A randomized controlled trial, *Brain. Behav. Immun.* 26 (2012) 988–995. doi:10.1016/j.bbi.2012.05.011.
13. S. Krauss-Etschmann, R. Shadid, C. Campoy, E. Hoster, H. Demmelmair, M. Jiménez, A. Gil, M. Rivero, B. Veszprémi, T. Decsi, B. V. Koletzko, Effects of fish-oil and folate supplementation of pregnant women on maternal and fetal plasma concentrations of docosahexaenoic acid and eicosapentaenoic acid: A European randomized multicenter trial, *Am. J. Clin. Nutr.* 85 (2007) 1392–1400. doi:10.1093/ajcn/85.5.1392.
14. M.H. Davidson, Omega-3 fatty acids: New insights into the pharmacology and biology of docosahexaenoic acid, docosapentaenoic acid, and eicosapentaenoic acid, *Curr. Opin. Lipidol.* 24 (2013) 467–474. doi:10.1097/MOL.0000000000000019.
15. D.R. Tocher, M.B. Betancor, M. Sprague, R.E. Olsen, J.A. Napier, Omega-3 long-chain polyunsaturated fatty acids, EPA and DHA: Bridging the gap between supply and demand, *Nutrients.* 11 (2019) 1–20. doi:10.3390/nu11010089.
16. S.M. Innis, Perinatal biochemistry and physiology of long-chain polyunsaturated fatty acids, *J. Pediatr.* 143 (2003) 1–8. doi:10.1067/s0022-3476(03)00396-2.
17. H. Sprecher, Q. Chen, F.Q. Yin, Regulation of the biosynthesis of 22:5n-6 and 22:6n-3: A complex intracellular process, *Lipids.* 34 (1999) S153–S156. doi:10.1007/bf02562271.
18. S. Ferdinandusse, S. Denis, P.A.W. Mooijer, Z. Zhang, J.K. Reddy, A.A. Spector, R.J.A. Wanders, Identification of the peroxisomal β-oxidation enzymes involved in the biosynthesis of docosahexaenoic acid, *J. Lipid Res.* 42 (2001) 1987–1995.
19. A.M. Paker, J.S. Sunness, N.H. Brereton, L.J. Speedie, L. Albanna, S. Dharmaraj, A.B. Moser, R.O. Jones, G. V. Raymond, Docosahexaenoic acid therapy in peroxisomal disorders: Results of a double blind randomized clinical trial, *Neurology.* 70 (2008) A122-123, Abstract.
20. M.H. Liang, J.G. Jiang, Advancing oleaginous microorganisms to produce lipid via metabolic engineering technology, *Prog. Lipid Res.* 52 (2013) 395–408. doi:10.1016/j.plipres.2013.05.002.
21. Y. Xin, C. Shen, Y. She, H. Chen, C. Wang, L. Wei, K. Yoon, D. Han, Q. Hu, J. Xu, Biosynthesis of Triacylglycerol Molecules with a Tailored PUFA

Profile in Industrial Microalgae, *Mol. Plant.* 12 (2019) 474–488. doi:10.1016/j.molp.2018.12.007.

22. M.L. Hamilton, R.P. Haslam, J.A. Napier, O. Sayanova, Metabolic engineering of Phaeodactylum tricornutum for the enhanced accumulation of omega-3 long chain polyunsaturated fatty acids, *Metab. Eng.* 22 (2014) 3–9. doi:10.1016/j.ymben.2013.12.003.
23. J.G. Metz, P. Roessler, D. Facciotti, C. Levering, F. Dittrich, M. Lassner, R. Valentine, K. Lardizabal, F. Domergue, A. Yamada, K. Yazawa, V. Knauf, J. Browse, Production of polyunsaturated fatty acids by potyketide synthases in both prokaryotes and eukaryotes, *Science* (80-.). 293 (2001) 290–293. doi:10.1126/science.1059593.
24. C. Ratledge, Fatty acid biosynthesis in microorganisms being used for Single Cell Oil production., *Biochimie.* 86 (2004) 807–15. doi:10.1016/j.biochi.2004.09.017.
25. A.H. Metherel, R.P. Bazinet, Updates to the n-3 polyunsaturated fatty acid biosynthesis pathway: DHA synthesis rates, tetracosahexaenoic acid and (minimal) retroconversion, *Prog. Lipid Res.* 76 (2019) 101008. doi:10.1016/j.plipres.2019.101008.
26. J.G. Wallis, J.L. Watts, J. Browse, Polyunsaturated fatty acid synthesis: What will they think of next?, *Trends Biochem. Sci.* 27 (2002) 467–473. doi:10.1016/S0968-0004(02)02168-0.
27. X. Qiu, Biosynthesis of docosahexaenoic acid (DHA, 22:6-4, 7,10,13,16,19): Two distinct pathways, *Prostaglandins Leukot. Essent. Fat. Acids.* 68 (2003) 181–186. doi:10.1016/S0952-3278(02)00268-5.
28. C. Ye, W. Qiao, X. Yu, X. Ji, H. Huang, J.L. Collier, L. Liu, Reconstruction and analysis of the genome-scale metabolic model of schizochytrium limacinum SR21 for docosahexaenoic acid production, *BMC Genomics.* 16 (2015) 1–11. doi:10.1186/s12864-015-2042-y.
29. M.M. Manore, S.L. Meacham, New dietary reference intakes set for energy, carbohydrates, fiber, fat, fatty adds, cholesterol, proteins, and amino acids, 2003.
30. W.S. Harris, International recommendations for consumption of long-chain omega-3 fatty acids, *J. Cardiovasc. Med.* 8 (2007). doi:10.2459/01.JCM.0000289274.64933.45.
31. P. Howe, B. Meyer, S. Record, K. Baghurst, Dietary intake of long-chain ω-3 polyunsaturated fatty acids: Contribution of meat sources, *Nutrition.* 22 (2006) 47–53. doi:10.1016/j.nut.2005.05.009.
32. C. Gómez Candela, L.M. Bermejo López, V. Loria Kohen, Importancia del equilibrio del índice omega-6/omega-3 en el mantenimiento de un buen estado de salud recomendaciones nutricionales, *Nutr. Hosp.* 26 (2011) 323–329. doi:10.3305/nh.2011.26.2.5117.
33. D. Dave, W. Routray, Current scenario of Canadian fishery and corresponding underutilized species and fishery byproducts: A potential source

of omega-3 fatty acids, *J. Clean. Prod.* 180 (2018) 617–641. doi:10.1016/j.jclepro.2018.01.091.

34. J.L. Domingo, A. Bocio, G. Falcó, J.M. Llobet, Benefits and risks of fish consumption. Part I. A quantitative analysis of the intake of omega-3 fatty acids and chemical contaminants, *Toxicology.* 230 (2007) 219–226. doi:10.1016/j.tox.2006.11.054.
35. S.M. Ulven, B. Kirkhus, A. Lamglait, S. Basu, E. Elind, T. Haider, K. Berge, H. Vik, J.I. Pedersen, Metabolic effects of krill oil are essentially similar to those of fish oil but at lower dose of EPA and DHA, in healthy volunteers, *Lipids.* 46 (2011) 37–46. doi:10.1007/s11745-010-3490-4.
36. A. Beopoulos, J. Cescut, R. Haddouche, J.-L. Uribelarrea, C. Molina-Jouve, J.-M. Nicaud, Yarrowia lipolytica as a model for bio-oil production, *Prog. Lipid Res.* 48 (2009) 375–387. doi:10.1016/j.plipres.2009.08.005.
37. A. Beopoulos, J.M. Nicaud, Yeast: A new oil producer?, *OCL - Ol. Corps Gras Lipides.* 19 (2012) 22–28. doi:10.1684/ocl.2012.0426.
38. C. Ratledge, Microbial oils: an introductory overview of current status and future prospects, *OCL.* 20 (2013) D602. doi:10.1051/ocl/2013029.
39. U.S. Department of Agriculture, *Agric. Res. Serv.* (n.d.). https://fdc.nal.usda.gov/ (accessed December 24, 2019).
40. O.N. Ciftci, R. Przybylski, M. Rudzińska, Lipid components of flax, perilla, and chia seeds, *Eur. J. Lipid Sci. Technol.* 114 (2012) 794–800. doi:10.1002/ejlt.201100207.
41. J. Orsavova, L. Misurcova, J. Vavra Ambrozova, R. Vicha, J. Mlcek, Fatty acids composition of vegetable oils and its contribution to dietary energy intake and dependence of cardiovascular mortality on dietary intake of fatty acids, *Int. J. Mol. Sci.* 16 (2015) 12871–12890. doi:10.3390/ijms160612871.
42. V. Kostik, S. Memeti, B. Bauer, Fatty acid composition of edible oils and fats, *J. Hyg. Eng. Des.* 4 (2013) 112–116.
43. M. Dogan, A. Akgul, Fatty acid composition of some walnut (Juglans regia L.) cultivars from east Anatolia, *Grasas y Aceites.* 56 (2005) 328–331. doi:10.3989/gya.2005.v56.i4.101.
44. V.J.M. Furlan, V. Maus, I. Batista, N.M. Bandarra, Production of docosahexaenoic acid by Aurantiochytrium sp. ATCC PRA-276, *Brazilian J. Microbiol.* 48 (2017) 359–365. doi:10.1016/j.bjm.2017.01.001.
45. A. Patel, U. Rova, P. Christakopoulos, L. Matsakas, Simultaneous production of DHA and squalene from Aurantiochytrium sp. grown on forest biomass hydrolysates, *Biotechnol. Biofuels.* 12 (2019) 1–12. doi:10.1186/s13068-019-1593-6.
46. W.K. Park, M. Moon, S.E. Shin, J.M. Cho, W.I. Suh, Y.K. Chang, B. Lee, Economical DHA (Docosahexaenoic acid) production from Aurantiochytrium sp. KRS101 using orange peel extract and low cost nitrogen sources, *Algal Res.* 29 (2018) 71–79. doi:10.1016/j.algal.2017.11.017.
47. W.K. Hong, D. Rairakhwada, P.S. Seo, S.Y. Park, B.K. Hur, C.H. Kim, J.W. Seo, Production of lipids containing high levels of docosahexaenoic acid by

a newly isolated microalga, Aurantiochytrium sp. KRS101, *Appl. Biochem. Biotechnol.* 164 (2011) 1468–1480. doi:10.1007/s12010-011-9227-x.
48. K.P. Patil, P.R. Gogate, Improved synthesis of docosahexaenoic acid (DHA) using Schizochytrium limacinum SR21 and sustainable media, *Chem. Eng. J.* 268 (2015) 187–196. doi:10.1016/j.cej.2015.01.050.
49. A. Nakazawa, H. Matsuura, R. Kose, K. Ito, M. Ueda, D. Honda, I. Inouye, K. Kaya, M.M. Watanabe, Optimization of Biomass and Fatty Acid Production by Aurantiochytrium sp. Strain 4W-1b, *Procedia Environ. Sci.* 15 (2012) 27–33. doi:10.1016/j.proenv.2012.05.006.
50. Y. Nazir, S. Shuib, M.S. Kalil, Y. Song, A.A. Hamid, Optimization of Culture Conditions for Enhanced Growth, Lipid and Docosahexaenoic Acid (DHA) Production of Aurantiochytrium SW1 by Response Surface Methodology, *Sci. Rep.* 8 (2018) 1–12. doi:10.1038/s41598-018-27309-0.
51. T. Nakahara, T. Yokochi, T. Higashihara, S. Tanaka, T. Yaguchi, D. Honda, Production of docosahexaenoic and docosapentaenoic acids by Schizochytrium sp. isolated from yap islands, JAOCS, *J. Am. Oil Chem. Soc.* 73 (1996) 1421–1426. doi:10.1007/BF02523506.
52. S. Abad, X. Turon, Biotechnological production of docosahexaenoic acid using aurantiochytrium limacinum: Carbon sources comparison and growth characterization, *Mar. Drugs.* 13 (2015) 7275–7284. doi:10.3390/md13127064.
53. Y. Liang, N. Sarkany, Y. Cui, J. Yesuf, J. Trushenski, J.W. Blackburn, Use of sweet sorghum juice for lipid production by Schizochytrium limacinum SR21, *Bioresour. Technol.* 101 (2010) 3623–3627. doi:10.1016/j.biortech.2009.12.087.
54. L.J. Ren, X.J. Ji, H. Huang, L. Qu, Y. Feng, Q.Q. Tong, P.K. Ouyang, Development of a stepwise aeration control strategy for efficient docosahexaenoic acid production by Schizochytrium sp., *Appl. Microbiol. Biotechnol.* 87 (2010) 1649–1656. doi:10.1007/s00253-010-2639-7.
55. A. Patel, L. Matsakas, K. Hrůzová, U. Rova, P. Christakopoulos, Biosynthesis of Nutraceutical Fatty Acids by the Oleaginous Marine Microalgae Phaeodactylum tricornutum Utilizing Hydrolysates from Organosolv-Pretreated Birch and Spruce Biomass, *Mar. Drugs.* 17 (2019) 119. doi:10.3390/md17020119.
56. W. Yongmanitchai, O.P. Ward, Growth of and omega-3 fatty acid production by Phaeodactylum tricornutum under different culture conditions, *Appl. Environ. Microbiol.* 57 (1991) 419–425. doi:10.1121/1.3458814.
57. H. Hu, K. Gao, Optimization of growth and fatty acid composition of a unicellular marine picoplankton, Nannochloropsis sp., with enriched carbon sources, *Biotechnol. Lett.* 25 (2003) 421–425. doi:10.1023/A:1022489108980.
58. V. Patil, T. Källqvist, E. Olsen, G. Vogt, H.R. Gislerød, Fatty acid composition of 12 microalgae for possible use in aquaculture feed, *Aquac. Int.* 15 (2007) 1–9. doi:10.1007/s10499-006-9060-3.

59. J. Van Wagenen, T.W. Miller, S. Hobbs, P. Hook, B. Crowe, M. Huesemann, Effects of light and temperature on fatty acid production in Nannochloropsis salina, *Energies.* 5 (2012) 731–740. doi:10.3390/en5030731.
60. M. Sang, M. Wang, J. Liu, C. Zhang, A. Li, Effects of temperature, salinity, light intensity, and pH on the eicosapentaenoic acid production of Pinguiococcus pyrenoidosus, *J. Ocean Univ. China.* 11 (2012) 181–186. doi:10.1007/s11802-012-1868-z.
61. A. Seto, H.L. Wang, C.W. Hesseltine, Culture conditions affect eicosapentaenoic acid content of Chlorella minutissima, *J. Am. Oil Chem. Soc.* 61 (1984) 892–894. doi:10.1007/BF02542159.
62. C. Hu, M. Li, J. Li, Q. Zhu, Z. Liu, Variation of lipid and fatty acid compositions of the marine microalga Pavlova viridis (Prymnesiophyceae) under laboratory and outdoor culture conditions, *World J. Microbiol. Biotechnol.* 24 (2008) 1209–1214. doi:10.1007/s11274-007-9595-0.
63. A.P. Carvalho, F.X. Malcata, Optimization of ω-3 fatty acid production by microalgae: Crossover effects of CO2 and light intensity under batch and continuous cultivation modes, in: *Mar. Biotechnol.* 2005: pp. 381–388. doi:10.1007/s10126-004-4047-4.
64. T. Yago, H. Arakawa, T. Morinaga, M. Yoshioka, Global Change: Mankind-Marine Environment Interactions, in: S.G. Ceccaldi H.J., Dekeyser I., Girault M. (Ed.), *Glob. Chang. Mankind-Marine Environ. Interact.*, Springer, Dordrecht, 2011: pp. 43–45. doi:10.1007/978-90-481-8630-3.
65. R.C. Valentine, D.L. Valentine, Omega-3 fatty acids in cellular membranes: A unified concept, *Prog. Lipid Res.* 43 (2004) 383–402. doi:10.1016/j.plipres.2004.05.004.
66. P.C. Calder, N-3 Fatty acids, inflammation and immunity: New mechanisms to explain old actions, *Proc. Nutr. Soc.* 72 (2013) 326–336. doi:10.1017/S0029665113001031.
67. P.C. Calder, Polyunsaturated fatty acids and inflammatory processes: New twists in an old tale, *Biochimie.* 91 (2009) 791–795. doi:10.1016/j.biochi.2009.01.008.
68. M. Tanito, R.E. Anderson, Dual roles of polyunsaturated fatty acids in retinal physiology and pathophysiology associated with retinal degeneration, *Clin. Lipidol.* 4 (2009) 821–827. doi:10.2217/clp.09.65.
69. H. Shindou, H. Koso, J. Sasaki, H. Nakanishi, H. Sagara, K.M. Nakagawa, Y. Takahashi, D. Hishikawa, Y. Iizuka-Hishikawa, F. Tokumasu, H. Noguchi, S. Watanabe, T. Sasaki, T. Shimizu, Docosahexaenoic acid preserves visual function by maintaining correct disc morphology in retinal photoreceptor cells, *J. Biol. Chem.* 292 (2017) 12054–12064. doi:10.1074/jbc.M117.790568.
70. B.G. Jeffrey, H.S. Weisinger, M. Neuringer, D.C. Mitchell, The role of docosahexaenoic acid in retinal function, *Lipids.* 36 (2001) 859–871. doi:10.1007/s11745-001-0796-3.

71. J.P. SanGiovanni, E.Y. Chew, The role of omega-3 long-chain polyunsaturated fatty acids in health and disease of the retina, *Prog. Retin. Eye Res.* 24 (2005) 87–138. doi:10.1016/j.preteyeres.2004.06.002.
72. L. Lauritzen, P. Brambilla, A. Mazzocchi, L.B.S. Harsløf, V. Ciappolino, C. Agostoni, DHA effects in brain development and function, *Nutrients.* 8 (2016) 1–17. doi:10.3390/nu8010006.
73. S.C. Dyall, Long-chain omega-3 fatty acids and the brain: A review of the independent and shared effects of EPA, DPA and DHA, *Front. Aging Neurosci.* 7 (2015) 1–15. doi:10.3389/fnagi.2015.00052.
74. A. Molfino, M.I. Amabile, M. Monti, M. Muscaritoli, Omega-3 Polyunsaturated Fatty Acids in Critical Illness: Anti-Inflammatory, Proresolving, or Both?, *Oxid. Med. Cell. Longev.* 2017 (2017) 1–6. doi:10.1155/2017/5987082.
75. M.R. Flock, W.S. Harris, P.M. Kris-Etherton, Long-chain omega-3 fatty acids: Time to establish a dietary reference intake, *Nutr. Rev.* 71 (2013) 692–707. doi:10.1111/nure.12071.
76. G. Bannenberg, C.N. Serhan, Specialized pro-resolving lipid mediators in the inflammatory response: An update, *Biochim. Biophys. Acta - Mol. Cell Biol. Lipids.* 1801 (2010) 1260–1273. doi:10.1016/j.bbalip.2010.08.002.
77. P.T. Price, C.M. Nelson, S.D. Clarke, Omega-3 polyunsaturated fatty acid regulation of gene expression, *Curr. Opin. Lipidol.* 11 (2000) 3–7. doi:10.1097/00041433-200002000-00002.
78. H. Sampath, J.M. Ntambi, Polyunsaturated fatty acid regulation of gene expression, *Nutr. Rev.* 62 (2004) 333–339. doi:10.1301/nr.2004.sept.333-339.
79. M.S. Brown, J.L. Goldstein, The SREBP pathway: Regulation of cholesterol metabolism by proteolysis of a membrane-bound transcription factor, *Cell.* 89 (1997) 331–340. doi:10.1016/S0092-8674(00)80213-5.
80. P. Guesnet, J.M. Alessandri, Docosahexaenoic acid (DHA) and the developing central nervous system (CNS) - Implications for dietary recommendations, *Biochimie.* 93 (2011) 7–12. doi:10.1016/j.biochi.2010.05.005.
81. K. Moon, S.C. Rao, S.M. Schulzke, S.K. Patole, K. Simmer, Longchain polyunsaturated fatty acid supplementation in preterm infants, *Cochrane Database Syst. Rev.* 2016 (2016). doi:10.1002/14651858.CD000375.pub5.
82. E. Larque, H. Demmelmair, B. Koletzko, Perinatal Supply and Metabolism of Long-Chain Polyunsaturated Fatty Acids, *Ann. N. Y. Acad. Sci.* 967 (2006) 299–310. doi:10.1111/j.1749-6632.2002.tb04285.x.
83. A. Qawasmi, A. Landeros-Weisenberger, M.H. Bloch, Meta-analysis of LCPUFA supplementation of infant formula and visual acuity, *Pediatrics.* 131 (2013). doi:10.1542/peds.2012-0517.
84. A. Qawasmi, A. Landeros-Weisenberger, J.F. Leckman, M.H. Bloch, Meta-analysis of long-chain polyunsaturated fatty acid supplementation of formula and infant cognition, *Pediatrics.* 129 (2012) 1141–1149. doi:10.1542/peds.2011-2127.

85. C.M. Smuts, M. Huang, D. Mundy, T. Plasse, S. Major, S.E. Carlson, A randomized trial of docosahexaenoic acid supplementation during the third trimester of pregnancy, *Obstet. Gynecol.* 101 (2003) 469–479. doi:10.1016/S0029-7844(02)02585-1.
86. S.F. Olsen, J. Dalby S.O. rensen, N.J. Secher, M. Hedegaard, T. Brink Henriksen, H.S. Hansen, A. Grant, Randomised controlled trial of effect of fish-oil supplementation on pregnancy duration, *Lancet.* 339 (1992) 1003–1007. doi:10.1016/0140-6736(92)90533-9.
87. J.L. Onwude, R.J. Lilford, H. Hjartardottir, A. Staines, D. Tuffnell, A randomised double blind placebo controlled trial of fish oil in high risk pregnancy, *BJOG An Int. J. Obstet. Gynaecol.* 102 (1995) 95–100. doi:10.1111/j.1471-0528.1995.tb09059.x.
88. H. Szajewska, A. Horvath, B. Koletzko, Effect of n−3 long-chain polyunsaturated fatty acid supplementation of women with low-risk pregnancies on pregnancy outcomes and growth measures at birth: a meta-analysis of randomized controlled trials, *Am. J. Clin. Nutr.* 83 (2006) 1337–1344. doi:10.1093/ajcn/83.6.1337.
89. A. Horvath, B. Koletzko, H. Szajewska, Effect of supplementation of women in high-risk pregnancies with long-chain polyunsaturated fatty acids on pregnancy outcomes and growth measures at birth: A meta-analysis of randomized controlled trials, *Br. J. Nutr.* 98 (2007) 253–259. doi:10.1017/S0007114507709078.
90. E. Larqué, A. Gil-Sánchez, M.T. Prieto-Sánchez, B. Koletzko, Omega 3 fatty acids, gestation and pregnancy outcomes, *Br. J. Nutr.* 107 (2012). doi:10.1017/S0007114512001481.
91. W.S. Harris, D. Mozaffarian, E. Rimm, P. Kris-Etherton, L.L. Rudel, L.J. Appel, M.M. Engler, M.B. Engler, F. Sacks, Omega-6 fatty acids and risk for cardiovascular disease: A science advisory from the American Heart Association nutrition subcommittee of the council on nutrition, physical activity, and metabolism; council on cardiovascular nursing; and council on epidem, *Circulation.* 119 (2009) 902–907. doi:10.1161/CIRCULATIONAHA.108.191627.
92. P. Kris-Etherton, J. Fleming, W.S. Harris, The Debate about n-6 Polyunsaturated Fatty Acid Recommendations for Cardiovascular Health, *J. Am. Diet. Assoc.* 110 (2010) 201–204. doi:10.1016/j.jada.2009.12.006.
93. D. Gaillard, R. Negrel, M. Lagarde, G. Ailhaud, Requirement and role of arachidonic acid in the differentiation of pre-adipose cells, *Biochem. J.* 257 (1989) 389–397. doi:10.1042/bj2570389.
94. B. Mirnikjoo, S.E. Brown, H.F.S. Kim, L.B. Marangell, J.D. Sweatt, E.J. Weeber, Protein Kinase Inhibition by ω-3 Fatty Acid, *J. Biol. Chem.* 276 (2001) 10888–10896. doi:10.1074/jbc.M008150200.
95. C.N. Serhan, N. Chiang, T.E. Van Dyke, Resolving inflammation: Dual anti-inflammatory and pro-resolution lipid mediators, *Nat. Rev. Immunol.* 8 (2008) 349–361. doi:10.1038/nri2294.

96. S. Abad, X. Turon, Valorization of biodiesel derived glycerol as a carbon source to obtain added-value metabolites : Focus on polyunsaturated fatty acids, *Biotechnol. Adv.* 30 (2012) 733–741. doi:10.1016/j.biotechadv.2012.01.002.
97. S. Wu, B. Feng, K. Li, X. Zhu, S. Liang, X. Liu, S. Han, B. Wang, K. Wu, D. Miao, J. Liang, D. Fan, Fish consumption and colorectal cancer risk in humans: A systematic review and meta-analysis, *Am. J. Med.* 125 (2012). doi:10.1016/j.amjmed.2012.01.022.
98. J.S. Zheng, X.J. Hu, Y.M. Zhao, J. Yang, D. Li, Intake of fish and marine n-3 polyunsaturated fatty acids and risk of breast cancer: Meta-analysis of data from 21 independent prospective cohort studies, *BMJ.* 347 (2013) 1–10. doi:10.1136/bmj.f3706.
99. C.H. MacLean, S.J. Newberry, W.A. Mojica, P. Khanna, A.M. Issa, M.J. Suttorp, Y.-W. Lim, S.B. Traina, L. Hilton, R. Garland, S.C. Morton, Effects of Omega-3 Fatty Acids on Cancer Risk, *JAMA.* 295 (2006) 403. doi:10.1001/jama.295.4.403.
100. W. Alexander, Prostate cancer risk and omega-3 fatty acid intake from fish oil: A closer look at media messages versus research findings, *P. T.* 38 (2013) 561–564.
101. R.W. Gan, M.K. Demoruelle, K.D. Deane, M.H. Weisman, J.H. Buckner, P.K. Gregersen, T.R. Mikuls, J.R. O'Dell, R.M. Keating, T.E. Fingerlin, G.O. Zerbe, M.J. Clare-Salzler, V.M. Holers, J.M. Norris, Omega-3 fatty acids are associated with a lower prevalence of autoantibodies in shared epitopepositive subjects at risk for rheumatoid arthritis, *Ann. Rheum. Dis.* 76 (2017) 147–152. doi:10.1136/annrheumdis-2016-209154.
102. E. Rajaei, K. Mowla, A. Ghorbani, S. Bahadoram, M. Bahadoram, M. Dargahi-Malamir, The Effect of Omega-3 Fatty Acids in Patients With Active Rheumatoid Arthritis Receiving DMARDs Therapy: Double-Blind Randomized Controlled Trial, *Glob. J. Health Sci.* 8 (2015) 18–25. doi:10.5539/gjhs.v8n7p18.
103. L. Navarini, A. Afeltra, G. Gallo Afflitto, D.P.E. Margiotta, Polyunsaturated fatty acids: Any role in rheumatoid arthritis?, *Lipids Health Dis.* 16 (2017) 1–15. doi:10.1186/s12944 017-0586-3.
104. X. Li, X. Bi, S. Wang, Z. Zhang, F. Li, A.Z. Zhao, Therapeutic potential of ω-3 polyunsaturated fatty acids in human autoimmune diseases, *Front. Immunol.* 10 (2019) 1–14. doi:10.3389/fimmu.2019.02241.
105. U. Akbar, M. Yang, D. Kurian, C. Mohan, Omega-3 fatty acids in rheumatic diseases a critical review, *J. Clin. Rheumatol.* 23 (2017) 330–339. doi:10.1097/RHU.0000000000000563.
106. A. Gioxari, A.C. Kaliora, F. Marantidou, D.P. Panagiotakos, Intake of ω-3 polyunsaturated fatty acids in patients with rheumatoid arthritis: A systematic review and meta-analysis, *Nutrition.* 45 (2018) 114-124.e4. doi:10.1016/j.nut.2017.06.023.
107. M.D. Lewis, J.R. Hibbeln, J.E. Johnson, Y.H. Lin, D.Y. Hyun, J.D. Loewke, Suicide Deaths of Active-Duty US Military and Omega-3 Fatty-Acid Status, *J. Clin. Psychiatry.* 72 (2011) 1585–1590. doi:10.4088/JCP.11m06879.

108. M.E. Sublette, J.R. Hibbeln, H. Galfalvy, M.A. Oquendo, J.J. Mann, Omega-3 polyunsaturated essential fatty acid status as a predictor of future suicide risk, *Am. J. Psychiatry.* 163 (2006) 1100–1102. doi:10.1176/ajp.2006.163.6.1100.
109. M. Pompili, L. Longo, G. Dominici, G. Serafini, D.A. Lamis, J. Sarris, M. Amore, P. Girardi, Polyunsaturated fatty acids and suicide risk in mood disorders: A systematic review, *Prog. Neuro-Psychopharmacology Biol. Psychiatry.* 74 (2017) 43–56. doi:10.1016/j.pnpbp.2016.11.007.

2

Nutraceutical Fatty Acid Production in Marine Microalgae and Cyanobacteria

Anders K. Nilsson[1,2], Carlos Jiménez[3] and Angela Wulff[1*]

[1]*Department of Biological and Environmental Sciences, University of Gothenburg, Göteborg, Sweden*
[2]*Section for Ophthalmology, Department of Clinical Neuroscience, Institute of Neuroscience and Physiology, Sahlgrenska Academy, University of Gothenburg, Göteborg, Sweden*
[3]*Department of Ecology, Faculty of Sciences, University of Málaga, Bulevar Louis Pasteur, Málaga, Spain*

Abstract

In light of the potential negative impacts that fisheries and fish aquaculture have on marine ecosystems, alternative sources for long-chain polyunsaturated fatty acids (LC-PUFAs) for human consumption are warranted. *n*-3 and *n*-6 PUFAs have documented beneficial effects on human health and wellbeing. Of particular interest for their nutraceutical properties are the C_{20} and C_{22} LC-PUFAs, represented by arachidonic acid (ARA) of the *n*-6 series and eicosapentaenoic- and docosahexaenoic acid (EPA and DHA, respectively) of the *n*-3 series. LC-PUFAs can be endogenously synthesized from shorter-chain precursor molecules by humans. However, this synthesis is limited to why it is often promoted that preformed LC-PUFAs should be included in the diet, particularly EPA and DHA. Fatty fish and other seafood accumulate *n*-3 LC-PUFAs produced *de novo* at lower trophic levels, i.e., by marine microalgae and cyanobacteria, and are the main source of dietary EPA and DHA. In this chapter, we review LC-PUFA biosynthesis pathways and the current use of photosynthetic marine microbes for nutraceutical oil production. Different cultivation techniques, as well as optimal growth conditions, are discussed in the context of future possibilities of exploring microalgae and cyanobacteria for large-scale LC-PUFA production.

Keywords: Omega-3, omega-6, long-chain polyunsaturated fatty acids, arachidonic acid, eicosapentaenoic acid, docosahexaenoic acid

**Corresponding author*: angela.wulff@bioenv.gu.se

Alok Kumar Patel and Leonidas Matsakas (eds.) Nutraceutical Fatty Acids from Oleaginous Microalgae: A Human Health Perspective, (23–76) © 2020 Scrivener Publishing LLC

2.1 Introduction

The term nutraceutical stems from *Nutrition* and *Pharmaceutical* and was coined by the American physician Stephen L. DeFelice in 1989 [1]. Although nutraceuticals lack a strict definition, it typically refers to food-derived products with a nutritive value that can provide medical or health benefits, and be used in the prevention or treatment of diseases. Nutraceuticals supply bioactive compounds in a concentrated form, allowing intake in "dosages that exceed those that could be obtained from normal food" [2]. Long-chain polyunsaturated fatty acids (LC-PUFAs, here defined as ≥20 carbons in length) of the omega 3 and 6 families (*n*-3 and *n*-6, respectively) are a class of biochemicals that fall well into the category of nutraceuticals, with recognized roles in human health, including reducing risk of coronary disease [3, 4], anti-inflammatory actions [5], and neuroprotective properties [6, 7]. *n*-3 and *n*-6 fatty acids (FAs) are essential nutrients for humans and other animals, making them absolute dietary requirements. Eicosapentaenoic- and docosahexaenoic acid (EPA and DHA, respectively), both belonging to the *n*-3 family, are two LC-PUFAs that have gained special attention over the last decades for their health-promoting properties. The shorter chain counterpart of EPA and DHA, α-linoleic acid (ALA), is found in relatively high abundance in certain vegetable oils and can be metabolically converted to EPA and DHA in the human body. However, the conversion rate is low, particularly that of ALA to DHA, and only trace levels are endogenously synthesized. EPA and DHA are therefore sometimes considered as conditionally essential nutrients [8]. Various authorities and health organizations recommend a daily intake of preformed EPA + DHA at 250-1000 mg to obtain the health benefits associated with these FAs [9]. The average consumption of EPA + DHA is however well below these recommendations in many populations. For example, the mean intake of EPA + DHA in North America is estimated at around 100 mg/day [10].

Currently, nutritional EPA and DHA are principally derived from marine fish and seafood, either from wild capture fisheries or aquaculture farms. These are resource-consuming or limed sources in need of alternatives for long-term sustainable production. If the world population of 8 billion people was to adhere to recommendations and consume 500 mg of EPA + DHA per day and person, the annual consumption would add up to 1.5 million metric tons [11]. Total available EPA + DHA for human consumption is estimated at around 0.2 metric tons, leaving a production deficit of roughly 1.3 million metric tons [11]. Further exploration of natural

LC-PUFAs resources simply cannot fill this gap between availability and demand. Moreover, global warming is predicted to reduce DHA availability in fish by 10-58% by 2100 [12].

Many marine microalgal species produce substantial quantities of LC-PUFAs making microalgal oils a new promising environmentally friendly alternative to fish and fish oils. Microalgae and cyanobacteria are the primary producers of *n*-3 and *n*-6 FAs in the marine environment, continuously supplying higher trophic levels of the aquatic food web with precursors for, and preformed LC-PUFAs, such as EPA and DHA. Over 70% of total FAs may be constituted of EPA and DHA in certain microalgae. Also, other FAs with promising nutraceutical properties are abundantly produced by microalgae and cyanobacteria (e.g., the *n*-6 PUFAs γ-linolenic acid and arachidonic acid), as will be discussed in this chapter. From a sustainable perspective, using marine over freshwater algae has the advantage that only seawater is required, restricting the use of freshwater, another limited resource. Moreover, marine or salt-tolerant algae generally produce EPA and DHA to a greater extent than do freshwater green algae [13]. For these reasons, this chapter will solely focus on marine species of microalgae for FA production.

Microalgae display a remarkable diversity in size, morphology, metabolism and life cycles, and are found as single cells or in colonies forming chains or clusters. To date, there are over 150,000 algal species entries in the AlgaeBase database (https://www.algaebase.org/), with a majority of species being unicellular. Despite this great diversity, only a limited number of species are currently commercially grown for nutraceutical oil production. Marine microalgae are a diverse group of organisms where most species belong to the non-systematic group protists, including diatoms and dinoflagellates. Cyanobacteria are per definition not microalgae; algae are eukaryotes while cyanobacteria belong to the kingdom of prokaryotes. However, cyanobacteria and microalgae are evolutionary and metabolically related as algal plastids are of prokaryotic origin, derived from endosymbiosis. Furthermore, as oil production from microalgae and cyanobacteria can use similar manufacturing pipelines, we have chosen to consider both these organism groups in this chapter, although the emphasis will be on microalgae.

The human diet in the Western world has changed considerably over the last century. Among these changes, there has been a dramatic increase in the consumption of vegetable and seed oils rich in the *n*-6 FA linoleic acid (LA, 18:2 *n*-6) [14]. This has led to a shift in the dietary ratio of LA to ALA, likely resulting in lower tissue concentrations of EPA and DHA, and may be involved in the etiology of many diseases [14]. There is evidence that this trend is still ongoing in the 21st century [15]. Animal studies and clinical trials have unambiguously demonstrated that dietary enrichment of LC-PUFAs such as EPA and DHA increases their levels in circulating lipoproteins, adipose tissue and cellular membranes in various tissues. Many intervention studies support that algal oils and fish oils have similar effects on blood lipids: increasing levels of EPA and DHA and decreasing the fraction of *n*-6 FAs [16–24].

Production costs for microalgae oils are still high, limiting their potential use and commercial-scale production. Implementing results from basic science to industrial-scale production and further research are needed to cut costs and promote microalgae oils as a realistic challenger to today's LC-PUFA sources. Factors that need attention for exploration of marine microalgae for production of nutraceutical FAs include 1) bioprospecting – searching the seas for species with interesting FA profiles; 2) isolation and selection – identification of species or strains that perform well under laboratory conditions; 3) cultivation optimization – development of high-yield cultivation systems that rely on low-energy input; 4) extraction and refining – development of efficient and sustainable harvesting methods of desired FAs (Figure 2.1). In addition to these aspects, genetic engineering to increase yield and tailoring of FA profiles, which could not be achieved through conventional selection, should be considered.

2.2 Fatty Acid Synthesis

FAs are aliphatic carboxylic acids, normally in the range of 6-32 carbons in length. Most naturally occurring FAs are linear with an even number of carbon atoms, although certain organisms, particularly bacteria, also produce larger quantities of odd-numbered, branched and cyclic FAs. FAs are classified according to the number and positions of double bonds; saturated FAs (SFA) lack double-bound while mono- (MUFA) and polyunsaturated FAs (PUFA) contain one or more double bonds, respectively. Most double bonds in FAs found in nature are in *cis* configuration and multiple double bonds are generally interrupted with one methyl group. The position of the double bond closest to the terminal end (i.e., the omega carbon) of the acyl chain defines the FA family. Accordingly, an omega 3 (also *n*-3, where *n* denotes the carbon chain length) FA carries a double bond at the third carbon from the methyl end while an omega 6 (*n*-6) has its first double bond at the sixth carbon atom (Figure 2.2a-c) [25]. Double bond positions may also be indicated using Δ-numbering; in this case, the position is counted from the carboxyl end of the FA.

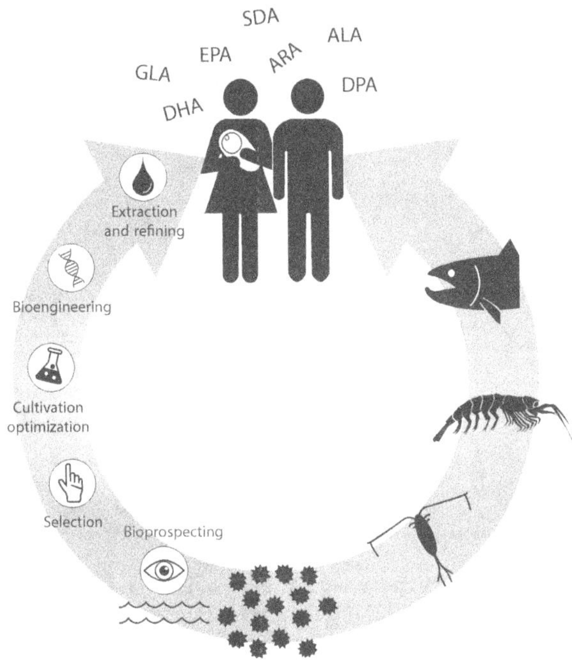

Figure 2.1 Overview of strategy for exploring microalgae for the production of nutraceutical FAs (green arrow) compared with a marine food chain where LC-PUFAs produced by microalgae are transferred up trophic levels before reaching the human end consumers (red arrow).

The following section will give a brief overview of FA synthesis and acyl lipid metabolism in microalgae; readers are referred elsewhere for excellent reviews on the topic [13, 26–30]. Much of our current knowledge on lipid metabolism and the functional characterization of enzymes involved come from studies of land plants, fungi, and animals. By gene sequence similarities, this knowledge has been inferred to algae and partly confirmed and complemented by experimental work, particularly in studies on the model green algae *Chlamydomonas reinhardtii* [31, 32]. However, as there is still limited biochemical data beyond what has been documented for green algae (and to some extent for other algae groups, such as diatoms [33] and red algae [34]), one can expect to find some variations to the pathways described here as other algal groups are investigated.

As in higher plants, *de novo* FA synthesis in algae is initiated in the stroma within chloroplasts. Fundamentally, sugars derived from photosynthetic CO_2 fixation provide the starting material for the production of acetyl-CoA, which is the carbon donor for FA synthesis. ATP-dependent

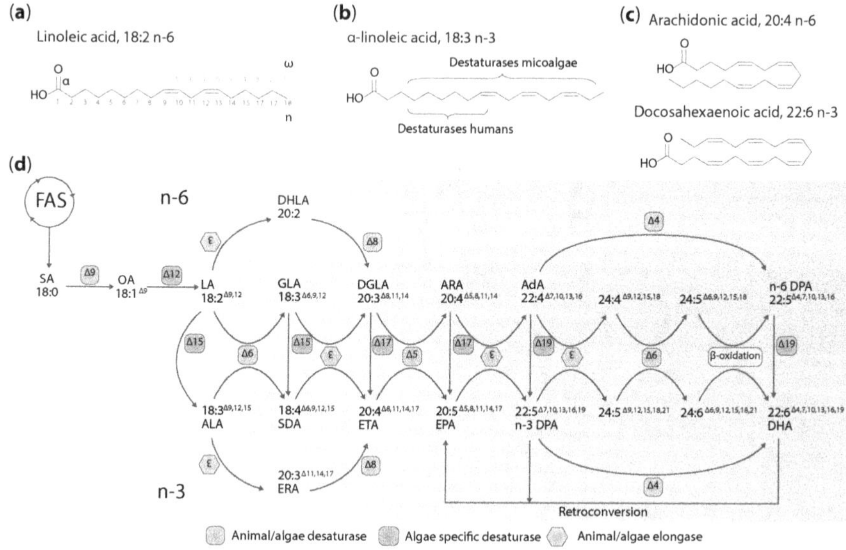

Figure 2.2 Nomenclature, structure, and synthesis of FAs. (a) Chemical structure of linoleic acid and numbering of carbon atoms according to systematic names (starting from the carboxyl end, α) and omega numbering (starting from methyl end, ω). (b) Structure of α-linolenic acid and indicated in what ranges of the fatty acyl humans and microalgae can introduce double bonds. (c) Structure of arachidonic- and docosahexaenoic acid. (d) FA biosynthesis pathway. AdA, adrenic acid; ALA, α-linolenic acid; ARA, arachidonic acid; DGLA, dihomo-γ-linolenic acid; DHA, docosahexaenoic acid; DHLA, dihomo-linoleic acid; DPA, docosapentaenoic acid; EPA, eicosapentaenoic acid; ERA, eicosatrienoic acid; ETA, eicosatetraenoic acid; FAS, FA synthase; GLA, γ-linolenic acid; LA, linoleic acid; OA, oleic acid; SA, stearic acid; SDA, stearidonic acid.

carboxylation of acetyl-CoA (catalyzed by acetyl CoA carboxylase, ACCase) generates malonyl-CoA that is enzymatically converted to malonyl-acyl carrier protein (ACP) (catalyzed by malonyl-CoA:acyl carrier protein malonyltransferase). In a cycle of condensation reactions, acetyl-CoA is extended by two carbon atoms, gained from malonyl-ACP, inside the multi-subunit complex FA synthase (FAS) to finally yield 16:0-ACP. The 16:0-ACP molecule can be directly acylated to chloroplast membrane lipids, released as non-esterified FAs (NEFA), or elongated to 18:0-ACP. Further metabolic reactions that include oxygenic FA desaturation and carbon chain elongation steps usually occur in the cytosolic endoplasmic reticulum (ER), but plastids may also host these processes. 18:0-ACP can be desaturated by a soluble Δ9 desaturase before or after export out of the chloroplast. LC-PUFA synthesis takes place while the FA is esterified on complex lipid substrates, such as phosphatidylcholine (PC), or alternatively, in the acyl-CoA pool. The 18:1 *n*-9 FA (oleic acid) may enter the *n*-6 pathway by the introduction of a double bond at C_{12} (Δ12 desaturase) yielding linoleic acid (LA, 18:2 *n*-6), or the *n*-3 pathway by an additional round of desaturation (Δ15) yielding α-linolenic acid (ALA. 18:3 *n*-3) (Figure 2.2d). From this point, two alternative pathways exist; elongation to a C_{20} molecule and desaturation by a Δ8 desaturase, or Δ6 desaturation to produce γ-linolenic acid (GLA, 18:3 *n*-6)/stearidonic acid (SDA, 18:4 *n*-3). Both these routes have been characterized in microalgae as well as in humans [13]. However, direct desaturation by Δ6 appears to be the main fate for LA and ALA in humans and most microalgae. Further elongation and desaturation reactions are similar in mammalian and algal cells, although certain crucial differences exist. Firstly, in mammals, DHA and *n*-6 docosapentaenoic acid (DPA) are mainly produced by two carbon chain shortening of 24:5 *n*-3 and 24:4 *n*-6, respectively, by peroxisomal beta-oxidation (in the so-called "Sprecher's shunt"), while in algae direct synthesis of DHA and *n*-6 DPA occurs by a Δ4 desaturase from *n*-3 DPA and adrenic acid (AdA, 22:4 *n*-6), respectively. The latter pathway is also present in humans, although it is thought to contribute to DHA formation to a lesser extent than the Sprecher pathway [35, 36]. Secondly, and importantly, in marine microalgae FAs of the *n*-6 series can be converted to *n*-3 FAs at several independent steps along the biosynthetic pathway by *n*-3 desaturases [37]. In addition to the described pathways of oxygenic desaturation and elongation, an anaerobic PUFA synthesis pathway has been described in both marine prokaryotes and eukaryotic microorganisms [38]. In this pathway, PUFA synthesis is catalyzed by the so-called PUFA synthase, a large polyketide synthase (PKS)-like multimeric enzyme complex [39]. The oleaginous marine heterokont *Schizochytrium* sp. uses a PKS-based pathway for DHA synthesis [38, 40]. Although often referred to as microalgae, species of the genera *Schizochytrium* belong to the family of *Thraustochytriaceae*, heterotrophic, fungus-like *Stramenopiles* [41].

Unlike microalgae, animals cannot introduce double bonds beyond carbon 9 of fatty acyl chains (Figure 2.2b and d). Humans are therefore unable to synthesize *n*-3 and *n*-6 FAs *de novo*. Instead, these FAs must be provided via diet. The need for the inclusion of LA in the diet to prevent symptoms of deficiency was demonstrated in the pioneering work of Burr and Burr nearly a century ago [42, 43]. Later, also ALA was recognized as an essential FA (EFA) that must be acquired through diet [44, 45]. Animal cells can elongate and desaturate LA and ALA to produce their longer and more unsaturated counterparts, primarily the LC-PUFAs ARA and EPA/DHA (Figure 2.2c and d). Note that this chain extension occurs at the carboxyl end of the FA with two carbon units being introduced per elongation cycle. The position of the terminal doubled bond relative to the methyl end is thus maintained while the chain length is increased and further desaturated [46]. The enzymes responsible for this metabolic conversion are shared, and in competition, between the substrates of the *n*-3 and *n*-6 pathways [47]. Consequently, a high intake of LA over ALA, as is common in the Western diet, pushes the production in favor of ARA over EPA and DHA [48]. This has been proposed to lead to a more pro-inflammatory physiological state, although there is some evidence that contradicts this hypothesis [49, 50]. In addition to dietary intake, the endogenous conversion of LA and ALA to LC-PUFAs is influenced by genetic polymorphisms that affect FA desaturation and elongation [51]. In particular, single nucleotide polymorphisms (SNPs) in a gene cluster that covers the Δ5 and Δ6-desaturases *FADS1* and *FADS2* are important regulators of LC-PUFA status [52–54].

2.3 Glycerolipid Synthesis and Lipid Accumulation

FAs are only present in small quantities in their free non-esterified form; instead, they are found as esters of glycerol, i.e., acyl lipids. Due to the phylogenetic diversity of marine microalgae, there is considerable variability in their acyl lipid composition, and generalizations are made here. Polar glycerolipids are the main lipid constituent of biological membranes and in algae include the phosphoglycerides (phosphatidylcholine [PC], phosphatidylethanolamine [PE], and phosphatidylglycerol [PG]) and the glycosylglycerides (monogalactosyldiacylglycerol [MGDG], digalactosyldiacylglycerol [DGDG] and sulphoquinovosyldiacylglycerol [SQDG]) (Figure 2.3). MGDG and DGDG are the dominant lipids in photosynthetic tissue, including those from plants, algae, and cyanobacteria, together with SQDG and PG found in lower quantities. As photosynthetic membranes are abundant in algae, MGDGs often show the greatest percentage contribution to total membrane lipids and contain the highest fraction of PUFAs [55]. In certain marine algae, betaine lipids (e.g., diacylglyceryl-trimethylhomoserine [DGTS]) are important membrane components that can replace common phosphoglycerides, and typically represent 10-20% of total membrane lipids [56, 57]. Non-polar glycerides include diacylglycerides (DAG) and triglycerides (TAG). TAG is the principal storage lipid and hence the bulk carrier of FAs under conditions when this lipid class prevails. Microalgae may accumulate as much as 20-50% of their dry weight in the form of TAGs under stress conditions such as N or P limitation, elevated salinity, high temperature or excess light. TAG is the desired

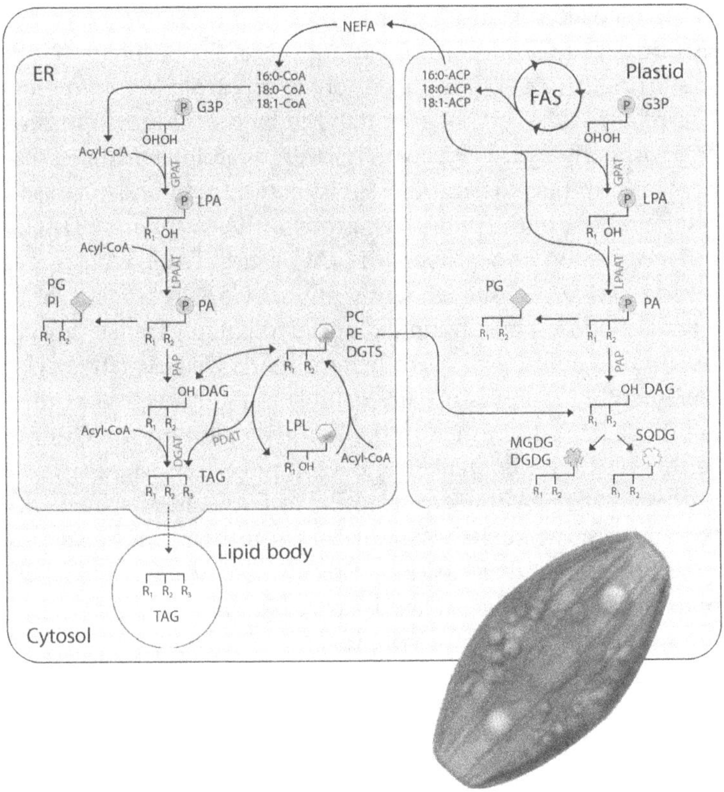

Figure 2.3 A generalized and simplified scheme of acyl lipid synthesis in microalgae. The insert shows an *Amphora* sp. cell where lipid bodies are seen. ACP, acyl carrier protein; DAG, diacylglycerol; DGAT, acyl-CoA:diacylglycerol acyltransferase; DGDG, digalactosyldiacylglycerol; DGTS, diacylglyceryl N,N,N-trimethylhomoserine; ER, endoplasmic reticulum; FAS, FA synthase; G3P, glycerol-3-phosphate; GPAT, acyl-CoA:sn-glycerol-3-phosphate acyltransferase; LPA, lysophosphatidic acid; LPAAT, acyl-CoA:lysophosphatidic acid acyltransferase; LPL, lysophospholipid; NEFA, Non-esterified FA; MAG, monoacylglycerol; MGDG, monogalactosyldiacylglycerol; PA, phosphatidic acid; PAP, PA phosphohydrolase; PC, Phosphatidylcholine; PDAT, phospholipid: diacylglyceryl acyltransferase; PE, phosphatidylethanolamine; PG, phosphatidylglycerol; PI, phosphatidylinositol; SQDG, sulfoquinovosyldiacylglycerol.

lipid fraction by the industry for commercial FA production, and TAG metabolism has therefore attracted considerable interest.

There are two parallel routes for glycerolipid synthesis in algae: one in the chloroplast and one in the ER, sometimes referred to as the "prokaryotic" and "eukaryotic" pathway, respectively, based on their evolutionary origin (Figure 2.3a). The ER pathway is also known as the "Kennedy pathway", named after its discoverer Eugene Kennedy [58]. FAs (acyl-CoA in the ER or acyl-ACP in plastids) and glycerol-3-phosphate (G3P) provide the starting material in the assembly of glycerolipids. Two sequential acylations of acyl-CoA to G3P are catalyzed by specific acyltransferases to produce phosphatidic acid (PA) with lyso-PA as an intermediate, reactions that occur both in plastids and ER. Phosphatidic acid phosphatase (PAP) catalyzes the release of the phosphate group of PA to generate DAG. In the ER, further acylation of acyl-CoA to DAG yields TAG. TAG can also be synthesized from DAG independently of acyl-CoA by transfer of an acyl moiety (catalyzed by phospholipid:diacylglycerol acyltransferase, PDAT) from membranes lipids, e.g., PC or betaine lipids [31]. As fatty acyls in PC and other membrane lipids usually exhibit a higher degree of unsaturation than those in DAG formed in the Kennedy pathway, this transfer can increase the LC-PUFA content in the TAG pool. The lyso-PC molecules that are formed after FA transfer from PC to DAG are esterified with new FAs, derived from the acyl-CoA pool, which in turn can be subjected to further desaturation and elongation. This cycle of deacylation-reacylation in the phospholipids, known as Lands' cycle [59], allows global remodeling of the fatty acyls within the glycerolipids.

DAG formed in the ER serves as the substrate for zwitterionic phosphoglyceride (PC and PE) synthesis and plastidial DAG is the substrate for galactocylglyceride (MGDG and DGDG) synthesis. PA is also used in the formation of anionic phosphoglycerides (PG) with cytidine diphosphate-DAG (CDP-DAG) as an intermediate, both in plastids and the ER. In some microalgae species, a plastidial TAG synthesis pathway appears to contribute to TAG accumulation [60]. Excessive synthesis of TAG in the ER initiates the formation of oil droplets that pinch off the ER when a critical size is reached (Figure 2.3b). These droplets, also called oil bodies or lipid bodies, are organelles consisting of a core of TAGs coated with a monolayer of phospholipids and proteins, derived from the outer ER leaflet, which facilitates their dispersion in the cytosol. Induction of TAG synthesis, and thereby oil droplet formation, is a key step in the production of microalgae oil. As discussed, both *de novo* synthesis and the transfer of FAs from membrane lipids contribute to the FA composition of TAGs [13].

2.4 Current LC-PUFA Sources and the Potential Benefits of Using Marine Microalgae

Marine oily fish such as sardines, herring, salmon and tuna, together with seafood including crustaceans and mollusks, are the primary sources of

n-3 LC-PUFAs in human diets. When comparing microalgal oils containing LC-PUFAs to those of animal origin, algal oils show several advantages. Firstly, global marine fish stocks are rapidly diminishing as a consequence of overfishing [61], thereby reducing the main current source of EPA and DHA for human consumption. Overfishing can also cause top-down cascade effects on the marine ecosystems, negatively affecting both coastal and pelagic systems [62]. Trawling for fish and invertebrates can have devastating effects on the benthic environment through physical destruction of the seafloor and by removal of benthic communities [63, 64]. Furthermore, fish and fish oils sometimes contain toxic contaminants in levels that restrict their positive properties [65], or in worst case, negatively affect consumers' health [66, 67].

Fish aquaculture is increasing to meet the growing global protein and LC-PUFA demands, resulting in an increased need for fish feed. Traditionally, carnivorous fish are given a diet containing fishmeal and/or oils from wild-caught fish to provide sufficient n-3 LC-PUFAs for development and growth. Some of the feed for farmed fish are now being replaced with more sustainable terrestrial alternatives, such as rapeseed oil and vegetable proteins. This has consequently led to lower concentrations of n-3 LC-PUFAs in farmed fish and, thus, lowered the human intake of these FAs [68, 69]. Lowering levels of n-3 also inevitably increase the n-6 to n-3 ratio, with potentially negative impacts on human health [70]. The FA composition of fish oils is affected by the species composition, season, and the geographic location of the catch, among other factors. Since the FA profile of fish oil is largely influenced by the diet of the fish, complexity (i.e., chain length, number and position of double bonds) increases with species diversity present in the lower trophic levels of the food chain. FA composition of oils derived from microalgae is, on the other hand, relatively simple, and only reflects a single species grown under controlled conditions. As a result, the FA composition is predictable and highly consistent between batches, making extraction of specific FAs relatively simple. Another advantage of substituting fish oils for algal oils in food supplements is that they become available for vegetarians and persons who suffer from seafood or fish allergy [71, 72].

Terrestrial vascular plants can synthesize n-3 and n-6 FAs, although only of C_{18} type (linoleic and α-linolenic acid [18:2 n-6 and 18:3 n-3, respectively]), and to accumulate these in substantial amounts in storage lipids in seeds. As discussed, C_{18} FAs can be metabolized by humans to their longer LC-PUFA counterparts, but the conversion rate is low [73]. To overcome this, there is extensive research to genetically engineer oil plants to produce LC-PUFAs. Genes required for LC-PUFA synthesis in plants are often cloned from microalgae [74, 75]. Although this strategy has proven successful [76, 77], the crops and their products will consequently be labeled as genetically modified (GM), making their introduction into the market for human consumption, not least the European Union [78], challenging. Consumer acceptance for GM plants or plant-derived products is generally low, partly due to limited scientific knowledge about the technique itself [79]. If GM plant oils are instead used as feed in aquaculture, there are no obligations on the producers for GM-labelling the end-product. Therefore, the focus of oil plant LC-PUFA production is partly being shifted from human consumption to an alternative for fish oil in aquafeeds [80, 81]. Microalgal oils offer a ready-to-go solution that does not involve the long diversion through fish nor genetic manipulation. It should, however, be mentioned that gene modification techniques are also applicable and interesting in the context of marine microalgae, as will be discussed in one of the follow sections.

Also animals and fungi have been subjected to genetic manipulation to increase their EPA and DHA status. The company DuPont has metabolically engineered a strain of the single-cell fungus *Yarrowia lipolytica* to synthesize and accumulate EPA under fermentation conditions, with the potential to also be used for the production of DHA and the *n-6* GLA and ARA [82]. Oils derived thereof could be used either as feed or food supplements for humans. Among animals intended for direct human consumption, increased EPA and DHA have been documented in tissue of transgenic common carp (*Cyprinus carpio*) [83], pigs [84], sheep [85], and cattle [86]. Considering that the scientific community advocates a shift towards plant-based diets to cut carbon emissions and improve health [87], these options appear not to have the future on their side.

In addition to algae, several other groups of oleaginous microorganisms are explored for LC-PUFA production. Thraustochytrids, that include the genera *Schizochytrium*, *Aurantiochytrium*, and *Ulkenia*, can grow to high cell densities and accumulate 50-70% of their dry weight as lipids and up to 70% of the total FAs as DHA [88]. Thraustochytrids have been successfully used for commercial DHA production since the 1990s [88]. As for the production of ARA, the filamentous fungus *Mortierella alpina* produces TAGs in high yields with a concentration of ARA in the range of 30-70% of total FAs [89]. *M. alpina* oil is used as a fortifier in most infant formula produced [90].

In summary, there are many existing and potential sources for human LC-PUFA consumption. Among these, oils from non-GM microorganisms are available and can be sustainably produced. To meet the global demand for LC-PUFAs, several sources and biotechnical approaches should be considered.

2.5 Nutraceutical Fatty Acids in Marine Microalgae and Species of Interest

There is a great diversity in the FA composition of different microalgae genera and species. In a study of FA profiles of over 2,000 microalgal strains at stationary phase, 76 different FAs were identified [91]. Despite this great diversity, only about 15 FAs are found in larger quantities, and of these, a handful or so have proven nutraceutical properties. Among the *n*-3 nutraceutical FAs found in microalgae, doubtlessly, EPA and DHA are the ones that have received the most attention, but also their precursors ALA and SDA are interesting from a health perspective. Among the *n*-6 FAs, GLA and ARA have been shown to provide health benefits, especially in infant growth and brain development [92]. The individual role of these FAs for human health and their occurrences in microalgae is discussed in detail below.

The availability of microalgal oil products on the market is rapidly increasing. There are dozens of brand names offering capsules of EPA and/or DHA microalgal oil supplements. Microalgal oils are also added as functional ingredients to foods to increase their nutritional value and include dairy products, beverages, and snacks. The commercially largest product fortified with microalgal oils is infant formula [90]. Infant formulas are usually prepared from a mix of vegetable oils and bovine milk fats and are therefore naturally low in LC-PUFAs if not supplemented. There is evidence that feeding un-supplemented formulas have negative impacts on infant visual acuity, cognitive development and other outcomes [reviewed in 93]. The effects appear to be extra prominent in preterm infants that have not had sufficient time *in utero* to accrete LC-PUFA into fat deposits that could be utilized during the early postnatal period [94]. In the 1990s and early 2000, infant formula started being supplemented with DHA and/or ARA. Today, virtually all infant formulas are fortified with DHA and/or ARA, usually in a ratio of about 1:2 and concentrations close to those found in human breast milk [95]. The current DHA sources for use in formulas are typically fish oils and oils of microalgal origin (e.g., the oil DHASCO® produced from the dinoflagellate *Crypthecodinium cohnii*), while ARA is often derived from fungi [90], particularly the oleaginous fungus *Mortierella alpina* [89]. The infant formula market was valued at

US$45 billion in 2018 and projected to reach US$104 billion by 2026 [96], making it the largest market for microalgal oil supplements.

Microalgae and cyanobacteria provide a plethora of potential species for nutraceutical oil production. Nevertheless, only a limited number of species have so far made their way into commercial production pipelines. As EPA and DHA are the *n*-3 FAs with the largest market value currently, microalgal species producing these FAs have attracted the greatest interest from the industry. Diatoms together with chlorophytes and eustigmatophytes (e.g., *Nannochloropsis* spp.) have been explored for their high *n*-3 LC-PUFA content. *Nannochloropsis* spp. have received particular attention for several reasons: high photosynthetic efficiency, high lipid productivity, a high proportion of EPA of total FAs, a well-established genetic toolbox, and relatively easy to grow in large-scale outdoor systems [reviewed in 97] – all very attractive traits for commercial exploration. Diatom genera commonly explored for *n*-3 FAs include *Chaetoceros, Melosira, Nitzschia, Phaeodactylum*, *Skeletonema*, and *Thalassiosira* [98]. Also, microscopic green algae like the genera *Chlorella*, *Dunaliella* and *Tetraselmis* are of great commercial interest. Other genera include *Isochrysis* (haptophyte), *Storeatula* (cryptophyte), *Porphyridium* (rhodophyte), and several dinoflagellates such as *Paragymnodinium*, *Apocalathium*, *Heterocapsa* among others [98]. The dinoflagellate *Crypthecodinium cohnii* is of particular interest due to its high content of DHA and that it can be grown heterotrophically [99]. Furthermore, significant concentrations of EPA are present in diatom species such as *Chaetoceros gracilis, Chaetoceros calcitrans, Skeletonema costatum,* and *Thalassiosira pseudonana*, and in the prymnesiophyte *Platymonas lutheri*; DHA is found in significant amounts in the prymnesiophytes *Pavlova lutheri* and *Isochrysis* sp. as well as in *Chroomonas salina*, while the rhodophyte *P. cruentum* seems to be a good source of ARA. One of the most advertised diatoms for commercial PUFA/lipid production is *Phaeodactylum tricornutum*: this species can produce 30% TAG (DW) including a large amount of EPA [100].

Marine cyanobacteria generally contain no or very low concentrations of LC-PUFAs such as EPA and DHA [101, 102]. However, certain cyanobacteria species contain considerable amounts of α-linoleic acid (ALA, 18:3 *n*-3) and stearidonic acid (SDA, 18:4 *n*-3), i.e., precursors of EPA and DHA in the human body [101]. Among the cyanobacteria, *Spirulina* (*Arthrospira*) is an outstanding and unique source of the FA γ-linolenic acid (GLA, 18:3 *n*-6). Most *Spirulina* species grow naturally in continental alkaline waters; however, some species (i.e., *Arthrospira subsalsa*) are of marine origin. Furthermore, several attempts have focused on the use of seawater for the cultivation of *Spirulina* in areas with limited access to freshwater [e.g., 103].

2.5.1 α-Linolenic Acid (18:3 *n*-3, Δ9,12,15)

The primary source of dietary ALA is plant oils and includes those derived from flaxseed, rapeseed, walnut, and *Echium*. The rate conversion of ALA to LC-PUFAs is limited in humans, especially the conversion of ALA to DHA [73]. In fact, in a review of intervention studies where ALA from (land) plant oils was used as a supplement, none of the included studies reported an increase in blood erythrocyte or plasma DHA at the study end [104]. It is therefore often advocated that not only ALA but also DHA and EPA should be part of the diet to supply the body with sufficient amounts of *n*-3 LC-PUFAs to meet tissue demands. However, nutritional ALA is partly converted to EPA and there is also evidence that ALA alone can provide enough DHA to meet brain requirements [105].

Algae within the taxon Chlorophyta (green algae) and cyanobacteria often contain comparably large fractions of ALA [101]. Most chlorophytes do not produce FAs beyond 20 carbons (EPA) and few if any cyanobacteria produce FAs longer than 18 carbons. The marine green microalga *Desmodesmus* sp. displayed lipid productivity of 15.9 mg L^{-1} day^{-1} and an ALA content of 24% of total FAs when grown under laboratory conditions [106]. In a bioprospecting project in Portuguese waters, 13 species of microalgae and 14 strains of cyanobacteria were collected and tested for ALA, EPA and DHA production [107]. A strain of *Nannochloropsis* and the cyanobacteria *Nodularia harveyana* were identified for their potential use in commercial ALA production; both producing ALA at 0.6 mg L^{-1} culture per day in controlled indoors growth experiments. Using a similar approach, Thao *et al.* isolated and characterized 50 lipid-producing microalgal strains collected in water sources in Vietnam (freshwater, brackish and marine waters) [108]. Several strains showed acceptable biomass yields and total lipid content and some had ALA >20% of total FAs. Among these strains, at least some appear to be of marine origin (e.g., *Picochlorum oculatum*) and to be good candidates for large-scale oil production. Recently, an oleaginous strain of the green algae *Lobosphaera incisa* was isolated in the Yellow sea in Korea and shown to have high proportions of ALA (23% of total FAs) and ARA (15%) [109]. Freshwater strains of *L. incisa* have previously been investigated for their high ARA content but this is the first report of a strain isolated from a marine environment that could be used for biotechnological applications.

2.5.2 Stearidonic Acid (18:4 *n*-3, Δ6,9,12,15)

The rate-limiting step in converting ALA to EPA in humans is the Δ6 desaturase that catalyzes the production of SDA. SDA is to a greater extent than ALA converted to EPA in humans [110]. Increasing the intake of SDA was shown to effectively raise the omega-3 index, i.e., the fraction of EPA + DHA in red blood cells, a well-established marker that inversely relates to cardiovascular disease [111]. Since SDA is less unsaturated than EPA, it is also less susceptible to oxidation. For this reason, SDA-enriched foods may have longer shelf-life than those containing EPA and be less likely to develop off-flavors. In addition to algae, SDA is naturally found in some seed oils (e.g., hemp, black currant, and *Echium*) and certain fish species. The Monsanto company has developed GM soybeans that produce SDA, which has been granted safety certificate approvals in the US and Canada.

In a recent study, several cryptophyte strains were shown to have growth rates comparable to *Nannochloropsis* species and to produce relatively large quantities of SDA [112]. One strain of *Storeatula major* accumulated 2.4% of DW L^{-1} as SDA with a production rate of 76 μg L^{-1} day^{-1}. While biomass production in the investigated strains was significantly lower than in *Nannochloropsis* species, cryptophytes do not possess heavy cell wall structures that add to the total biomass. The lack of recalcitrant cell walls in cryptophytes eases downstream processing and oil extraction in comparison to other microalgae such as diatoms and dinoflagellates.

The marine haptophyte *Isochrysis galbana* is commercially grown as feed for aquaculture as well as for use as a functional ingredient in foods [e.g., 113]. The lipidome of *I. galbana* biomass is characterized by a large proportion of galactolipids containing SDA and octadecapentaenoic acid (18:5 *n*-3) [114], but also the neutral lipids have a relatively high SDA content [115]. In experimental conditions, *I. galbana* showed maximum biomass and lipid content of 1.25 g DW per liter and 71.1% (w/w), respectively, with SDA being the dominant FA (>50% of total FAs) [115, 116].

2.5.3 Eicosanoid Acid (EPA, 20:5 *n*-3, Δ5,8,11,14,17) and Docosahexaenoic Acid (DHA, 22:6 *n*-3, Δ4,7,10,13,16,19)

Some species of microalgae that seem to be particularly suited for commercial EPA production include *Phaeodactylum tricornutum*, *Isochrysis galbana*, and *Nannochloropsis* spp. [117]. *Nannochloropsis* spp. are well known to accumulate high concentrations of EPA [reviewed in 97], where relative EPA concentrations can be found at around 12% of DW and 30% of total FAs [118]. During nitrogen replete conditions (beneficial for the production of polar lipids), EPA productivity reached 7.7 mg L^{-1} day^{-1} in one study [119]. One of the most advertised diatoms for commercial PUFA/lipid production is *Phaeodactylum tricornutum*: this species can produce 30% TAG (DW) with EPA as the major FA [100]. Furthermore, significant concentrations of EPA are also present in diatom species such as *Chaetoceros gracilis, Chaetoceros calcitrans, Skeletonema costatum* and *Thalassiosira pseudonana*. Lately, eight species of marine raphidophytes were tested for their content of *n*-3 PUFAs [112]. The highest concentration of EPA was 8.5 μg mg DW^{-1} (23.6% of *n*-3 FAs) in *Teleaulax amphioxeia* and 6.1 μg mg DW^{-1} (12.6% of *n*-3 FAs) for *Proteomonas sulcate* [112]. *Monodus subterraneus* is a promising terrestrial/freshwater eustigmatophyte for the production of EPA; however, we mention this species here since PUFA production is enhanced in saline waters [120]. The EPA content in *M. subterraneus* may range from 20 to 37% of total FAs. A group of microalgae of particular interest is the pinguiophytes with five genera of which all have an exceptionally high content of EPA of up to 56% (w/w) [121]. Furthermore, the diatoms *Nitzschia* sp. and *Navicula* sp. are an important source of EPA [122, 123].

Together with ARA, DHA is the most abundant PUFA in the brain and retina [25]. It is therefore not surprising that many of the beneficial health effects of DHA have been linked to visual acuity and various neurological outcomes [124, 125]. DHA is found in significant amounts in the prymnesiophytes *Pavlova* sp. (e.g., *P. lutheri*), *Isochrysis* sp. [117, 126] and the coccolithophore *Emiliania huxleyi* [127] as well as in the cryptophyte *Chroomonas salina* [128 and references therein]. Again, the pinguiophytes are of interest since several genera, in contrast to e.g., *Nannochloropsis* spp., contain both EPA and DHA, together with docosatetraenoic acid (22:4 *n*-3) [121].

2.5.4 Docosapentaenoic Acid (22:5 *n*-3, Δ7,10,13,16,19)

The *n*-3 docosapentaenoic acid (*n*-3 DPA, 22:5 *n*-3) is a new promising source of *n*-3 LC-PUFAs with potential beneficial health properties [129]. *n*-3 DPA is the direct intermediate between EPA and DHA in the *n*-3 biosynthesis pathway from ALA to DHA. This FA is not to be confused with *n*-6 DPA, the final product of the *n*-6 pathway, which is present in several oils extracted from microalgae. Compared to EPA and DHA, fewer studies have focused on *n*-3 DPA and its role in health and disease. Although the fraction of *n*-3 DPA is low in most human tissues, it is the second most abundant *n*-3 FA in brain tissue [25], yet it constitutes less than 1% of total FA. In breast milk, the concentration of *n*-3 DPA often exceeds that of EPA several-fold [130]. Dietary *n*-3 DPA can retroconvert to EPA (through peroxisomal β-oxidation) or be elongated and desaturated to DHA, and thereby increase their blood and tissue levels [131, 132]. Growing evidence also supports that the role of *n*-3 DPA extends beyond being a mere precursor for the production of EPA and DHA; *n*-3 DPA and its oxidized metabolites exert their own unique biological effects. Among the derivates biosynthesized from *n*-3 DPA is a class of specialized pro-resolving mediators with anti-inflammatory properties [133, 134]. The low availability of pure *n*-3 DPA at reasonable prices has limited the use of this FA in clinical intervention studies in humans. Instead, current information on health benefits comes primarily from *in vitro* studies and association studies [135]. Due to the rather recent advantages in our understanding of the biological actions of *n*-3 DPA, the search for algal sources for the specific production of this FA has been limited. To our knowledge, *n*-3 DPA has not yet been purified in commercial-scale from any marine microalgae species. However, this FA may constitute a few percentages of total FAs in, for example, the dinoflagellate *C. cohnii* [136]. Methods for purification of *n*-3 DPA at industrial scale have been proven in strains of the thraustochytrid *Schizochytrium* [137, 138].

2.5.5 γ-Linolenic Acid (18:3 *n*-6, Δ6,9,12)

GLA is a rather rare PUFA (18:3 *n*-6) with some medicinal properties such as lowering low-density lipoprotein in hypercholesterolemic patients [139]; it is 170-fold more effective than LA in this respect [140]. *Spirulina* refers to the commercial name of the biomass produced from cyanobacteria (*Arthrospira platensis* and *Arthrospira maxima*) that is extensively explored in large-scale production, mainly due to its high protein content. However, its nutritional benefits are not limited to this aspect, it also contains a relatively high content of GLA, accounting for up to 20-30% of total FAs. It appears that GLA plays the same role in photosynthetic membranes in *Spirulina* as ALA do in algae and higher plants [141]. Apart from GLA, the main FAs in *Spirulina* strains are palmitic (16:0), LA and oleic acid (18:1 *n*-9). *Spirulina* has a comparably low lipid content, something typical of cyanobacteria, and they represent only 6-13% of its dry weight, half of it being in the form of FAs. Among the lipids, the polar lipids MGDG, DGDG, SQDQ, and PG comprise the bulk [142], while TAGs are a minor component (1-2%). Among algal spices, Miura *et al.* [143] reported that a strain of the marine green microalga *Chlorella* contains about 10% GLA of total FAs, primarily present in chloroplastic membrane lipids.

2.5.6 Arachidonic Acid (20:4 *n*-6, Δ5,8,11,14)

Under most conditions, the body can synthesize sufficient amounts of ARA from dietary LA. Actually, as discussed above, access intake of LA over ALA can increase ARA to levels that potentially can lead to the emergence of or disease progression, e.g., by increased formation of pro-inflammatory mediators [144]. However, ARA is a FA that is abundantly present in most tissues, not least the brain and is essential for normal development and health. Preformed ARA in infant diet is necessary to meet the high metabolic demand during early childhood [92]. Therefore, practically all infant formulas are enriched with ARA, which has been shown to positively contribute to growth, brain development and visual acuity [92, 93]. Moreover, there is an ongoing discussion as to whether ARA, in addition to DHA, should be included in intravenous lipid emulsions given to preterm and very low birth weight infants as nutritional support during the transition to full breast milk feeding [145].

Only a few investigated marine microalgal species have been found to produce ARA in significant quantities. Among these, the red algae (Rhodophyta) *Porphyridium cruentum* can synthesize and accumulate ARA in high concentrations [146, 147]. Apart from this FA, *P. cruentum* is also an important source of EPA [146]. ARA may constitute up to 36% of their total FAs when grown at 25°C, and increase up to 60% at 16°C [148]. Another red microalgal species, *Porphyridium purpureum*, has also been found to be a good accumulator of ARA, with a yield of 160 mg L^{-1} ARA (24% of total FAs) when grown at 30°C and phosphate limiting conditions [149]. *P. purpureum* is found in marine areas (e.g., saltmarshes) but is generally not considered a marine species. Cultivation of *P. purpureum* has also been shown to be successful in scale-up experiments; in fact, conversion of LA to ARA was enhanced when the algae were grown in 250 L photoreactors compared with flask cultures [150].

In a recent interesting study, Du *et al.* cultivated the marine algae *Nannochloropsis oceanica* together with the oleaginous fungus *Mortierella elongate* [151]. Using this co-cultivation strategy, *M. elongata* mycelium could capture *N. oceanica* cells, creating algae-fungi aggregates (bioflocculation). These aggregates are simple to harvest and yielded high levels of TAG and FAs (15 and 22% of total DW, respectively), and ARA (produced by *M. elongate*) and EPA (*N. oceanica*) were present at around 10 molar % each in the total lipids.

Although not a marine species, it is worth mentioning that the green microalga *Lobosphaera incisa* (formerly known as *Parietochloris incisa*) is known as the richest plant source of ARA. Under nitrogen-starvation, TAG biosynthesis and accumulation are promoted, accounting for over 30% of dry weight (more than 95% of total lipids), with ARA reaching as much as 60% of total FAs [152, 153].

2.6 Autotrophic and Heterotrophic Cultivation

Photosynthetic production of microalgae, whether outdoors or indoors, is generally costly since cultures must be maintained at low densities to ensure that sufficient light can penetrate the reactors. As a result, large volumes of water must be handled and processed for algal recovery. However, one way of overcoming these challenges is to grow benthic species in biofilms: one successful example is the Swedish Algae Factory (https://swedishalgaefactory.com/). Heterotrophic fermentation is an alternative method that may yield higher concentrations of the selected species and in turn of the desirable chemicals [154]. A third way to cultivate microalgae in high density, either for the production of biomass or for the extraction of high-value nutraceuticals, is mixotrophy. In mixotrophy, part of the carbon and the energy required for the growth of the microalgae is supplied in the form of dissolved organic substrates; among them, glycerol and glucose are the most used. In contrast to heterotrophy, light is required for mixotrophic cultivation. Some species have been tested both indoor and outdoor with promising results (e.g., *Nannochloropsis, Chlorella, Scenedesmus*), leading to a significant increase in growth rate, cell density, FA production and EPA productivity [155].

Fermentation techniques are used for culturing species from various taxonomic groups of microalgae, including cryptophytes and dinoflagellates. This culturing technique has proven successful for the production of oils with a high content of ARA, EPA, and DHA [123, 156, 157]; for example, the dinoflagellate *C. cohnii* can produce up to 50% of DHA of total FAs [157, 158].

There is an ongoing discussion about heterotrophic versus autotrophic algal cultivation and which method is more cost-efficient. Of course, this depends on the industrial application and if the objective is to produce a high-value product or simply biomass accumulation. For FAs production, it depends on the target FAs, i.e., when considering the production of nutraceutical LC-PUFAs or biodiesel FAs. Heterotrophic growth is potentially easier to control, but not necessarily more cost-efficient, even if artificial light is needed for cultivation in autotrophic cultivation. Blanken *et al.* [159] pointed out that the increased cost using artificial light might be acceptable in the production of high-value products, but should generally be avoided in favor of ambient radiation (sunlight). Furthermore, heterotrophic cultures must be maintained axenically to avoid excessive bacterial growth due to the presence of organic substrates in the medium.

2.7 Cultivation from Laboratory to Industrial Scale

Mass production of microalgae depends mainly on the temperature and the light availability inside the reactor [160, 161]. Intensive research has been carried out since the 1950s to design and improve the growth conditions for maximizing the production of microalgae in culture. With few exceptions, until recently large-scale production of microalgae has been restricted to cultivation in open ponds of a limited number of genera, among them *Dunaliella, Chlorella, Haematococcus,* and *Spirulina*. These microalgae were then marketed mainly as a dry powder or in the form of tablets or pills as food supplements. However, purified β-carotene from *Dunaliella* or astaxanthin from *Haematococcus* can be easily found in a large number of specialized and non-specialized shops. Among open ponds, three types are mainly used: i) raceways, ii) circular and iii) inclined systems (cascades) (Figure 2.4).

It is possible to at least partly control the microalgal cell composition by controlling the environmental conditions of the photobioreactors (PBR) and by multistage or multiphase cultivation. Closed PBRs have clear advantages over open ponds, as they permit better control of certain culture variables (pH, CO_2, temperature, etc.) and the risk of contamination is reduced. Also, the maintenance of selected strains in open cultures for long periods is difficult. However, closed PBRs have some drawbacks that limit their scale-up, among them a high investment, high operational costs, negative energy balance, and difficulty to manage large water volumes, and they are not applicable for the cultivation of all microalgal species [162–165].

PBRs have to accomplish a dual goal: to improve the use of inorganic carbon (CO_2) and to make efficient use of the incident irradiance. Light availability inside the reactor depends on geographical location, time of day, day of the year, as well as on the design of the reactor, position, and biomass concentration. New designs of closed PBR arose since new chemicals from microalgae of high-added value have been introduced to the

Figure 2.4 Examples of cultivation of Spirulina. 100 L bags in indoors (a) and outdoor (b) conditions. Raceway cultivation systems with capacity of 1,350 L (c), 13,500 L (d) and 135,000 L (e).

market. The main types of PBRs can be classified in: i) vertical columns and sleeves; ii) tubular; iii) vertical and inclined flat panels; and iv) flexible film panels (Figure 2.5). Tubular reactors were already designed in the 1950s; however, they were not used for microalgal production until recently due to higher operational costs compared to open systems.

In recent years, the production of selected PUFAs from microalgae has gained attention among researchers and entrepreneurs. As discussed, successful PUFA production from microalgae depends on species and strain selection, design of cultivation system, optimization of culture conditions, and development of methods for FA extraction and purification [166]. Cultivation conditions for improving the yield of selected PUFAs at laboratory scale are discussed in one of the following sections. In general, suboptimal growth conditions promote PUFA accumulation. However, according to Cohen [167] it is preferable to manipulate cell density of the cultures outdoors instead of imposing suboptimal growth conditions (e.g., nitrogen starvation, higher or lower temperature from the optimal, increasing salinity) to achieve maximal LC-PUFA cell content and productivity. In this way, healthy cultures can be ensured, avoiding the risks of massive cell collapse in suboptimal growth conditions. In general, a combination of a high specific growth rate, a high proportion of LC-PUFA (or

Figure 2.5 Cultivation of *N. gaditana* in closed bioreactors, indoors (a and b) and outdoors (c). Photo credit Prof. F. Gabriel Acién (University of Almería, Spain).

the desired FA), and high cell densities are crucial to maximize production at lower costs. High cell density together with a high proportion of LC-PUFA reduces the costs of downstream processing for the extraction and purification of the desired FA.

In outdoor conditions Cohen and Heimer [168] reported an EPA productivity of 2.3 mg L^{-1} day^{-1} in *P. cruentum*; Molina Grima *et al.* [169] found higher EPA productivity in *I. galbana* (8.2 mg L^{-1} day^{-1}) and 50 mg L^{-1} day^{-1} for *P. tricornutum* in tubular bioreactors. Hu *et al.* [170] reported productivity of up to 58.9 mg L^{-1} day^{-1} in a 2.8 cm deep outdoor flat inclined modular bioreactor with *M. subterraneus*. Sukenik [171] cultivated *Nannochloropsis* in large outdoor ponds to be used as feed in the aquaculture industry. Cultures were normally 20 cm deep, mixed by paddlewheels and enriched with CO_2 for pH control. The authors reported a daily biomass production of 7 g DW m^{-2} day^{-1} in winter and 22 g DW m^{-2} day^{-1} in summer [172]. In winter, at a daily temperature of 8-16°C, the highest EPA was obtained [173]. Sukenik *et al.* [174] reported that EPA production was enhanced in nutrient-sufficient conditions, at a daily irradiance below 40 mol quanta m^{-2}, with an areal density of the culture around 65 g m^{-2}.

Hu *et al.* [170] grew *M. subterraneus* in flat plate bioreactors at high cell concentrations. At the optimum cell density for growth (4 g L^{-1}) EPA reached its maximum proportion (32%), while the rest of the FAs were at their lowest. Lower cell densities resulted in a significant decrease in both total FA content and the proportion of EPA. However, at cell concentration above the optimal for growth, EPA proportion started to decline.

Nevertheless, a balance between maximal EPA proportion and maximal areal yield must be found, since in this species highest cell content of EPA coincided with maximal cell mass productivity. In *Porphyridium* sp., Arad and Richmond [175] reported that cell FA dry weight decreased during daylight and that the FA profile is a function of the growth rate. This species contains the PUFAs ARA and EPA. During exponential growth (optimal growth conditions) the main PUFA was EPA. In contrast, under growth-limiting conditions (high cell density, low light, suboptimal pH, high salinity) ARA and LA became dominant. High temperature increased saturated FAs PA and SA, and a reduction of EPA. Diluted cultures (low cell concentration) of *P. cruentum* resulted in an enrichment of EPA and a low concentration of ARA. ARA was high in concentrated cultures in summer, while EPA was higher in winter. Higher yields might be obtained in closed reactors, in which high cell densities are achieved [146]. In addition, it has been described that feeding the cultures with PUFA precursors may result in an increased PUFA production. Okumura *et al.* [in 146] added 18:1, 18:2 and 18:3 to the growth medium of *E. gracilis*, resulting in a significant increase in the production of PUFAs, especially EPA and ARA.

In the case of *Spirulina*, Tanticharoen *et al.* [176] demonstrated that the FA content was higher in outdoors compared to indoors cultures. This increase in total FAs included a 41% increase in GLA. Cheng-Wu *et al.* [177] showed that in outdoor cultivation of *P. incisa*, an irradiance of 250 μmol photons m^{-2} s^{-1} were limiting for algal growth but beneficial for ARA accumulation, whereas irradiance of 2500 μmol photons m^{-2} s^{-1} facilitated rapid growth but with low ARA content in the biomass. Sukenik and Carmeli [178] have shown that the content of TAG in *N. oculata* decreased at the beginning of the light period. Thus, to maximize the proportion of unsaturated FAs and GLA, it is recommended to harvest the algae at the end of the dark period, when the content of neutral lipids and TAG is minimal. Additionally, Molina Grima *et al.* [166] simulated EPA production of *P. tricornutum* taking into account growth rate and irradiance, and concluded that maximal productivity was going to be reached at a dilution rate of 0.04 h^{-1} in spring and fall, with an expected decline in both summer (excess radiation in the surface of the culture) and winter (light limitation).

A new promising and cost-efficient technique for large-scale microalgal growth is foam-bed cultivation (a thin liquid layer at the bottom of the reactor with a large volume of foam exposed to light, above it) that has successfully been tested for *Chlorella* sp. [179]. Alternative bioreactors that have been tested in laboratory scale is, for example, the so-called flat plate bioreactors tested with *Nannochloropsis* sp. [180].

Ruiz *et al.* [181] combined techno-economic models with a market analysis for microalgal cultivation at six locations (The Netherlands, Canary Islands, Turkey, Curacao, Saudi Arabia, Spain). For cultivation in Spain, excluding biorefining products, the cost was estimated to 3.4€ kg^{-1} dry biomass (in 2016). Production of high-value products was shown to be profitable, for example, pigments with a net value of 657 M€ in 15 years [181]. They pointed at different factors for raceways versus closed systems, and that the large water volumes in raceway ponds were one important cost (see Figure 2 in [181]). One way to overcome the large water volumes is to use microalgal (diatom) biofilms (Figure 2.6a), as used for example by Swedish Algae Factory. Here, the cost is reduced by using wastewater from a land-based fish farm in a circular economy set-up (Figure 2.6b).

According to Cohen [167] productivity of EPA (P_{EPA}) (or any other fine chemical with commercial interest) is a function of four parameters: the proportion of EPA ($\%_{EPA}$), the FA content (C_{FA}), the specific growth rate (μ), and the biomass concentration in the cultures at the moment of harvest (X).

Then,

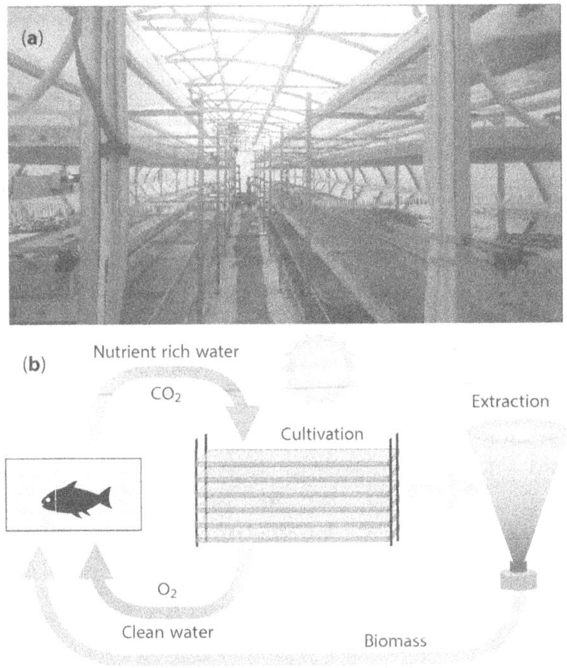

Figure 2.6 Diatoms grown in biofilms (a) in a circular economy set-up (b).

$$P_{EPA} = C_{EPA} \cdot P_B,$$

Where

$$C_{EPA} = \%_{EPA} \cdot C_{FA}, \text{ and } P_B = \mu \cdot X.$$

These parameters must be tested and adjusted to obtain the highest P_{EPA}. For example, in *M. subterraneus* both EPA proportion and FA content were high in dense cultures (high X) [167]. The combination of high X and high $\%_{EPA}$ and C_{FA} yielded EPA productivities as high as 26 mg L^{-1} day^{-1} indoors and 59 mg L^{-1} day^{-1} outdoors.

2.8 Optimizing Growth Condition to Promote Lipid Accumulation and Desired FA Profiles

The chemical composition of microalgae is not an intrinsic constant factor but varies between species, strains, and most importantly, it depends on the environmental factors that control cell growth. The concentration of many cell components may be manipulated by adjusting the culture conditions. In general, carbohydrate and lipid production is stimulated by suboptimal growth conditions (e.g., limiting nutrients, non-optimal irradiance, pH, salinity or temperature, stationary phase of growth) [182]. Under such unfavorable stress conditions, cellular division and growth are arrested and FAs accumulate as components of neutral lipids, primarily TAGs, which are thought to serve as an energy and carbon deposit during the starvation period.

There seems to be a consensus that total FA content is increased under N starvation but at the cost of lower biomass productivity [e.g., 183]. Thus, there is an ongoing debate on whether the highest FA production is achieved at N replete or deplete conditions [e.g., 100]. However, there are differences between different target FAs. For example, the proportion of EPA is generally higher in membrane lipids compared to TAGs. Nitrogen starvation could also result in a redirection of EPA from membrane lipids into TAG molecules [180, 184, 185]. As pointed out by Steinrücken *et al.* [186], due to the complexity in assessing the optimal growth condition for the production of EPA or other high-value FAs, it is important to address additive effects of different (growth) factors and not study one factor irrespective of another.

2.8.1 Temperature Effect

Growth temperature is one of the key environmental factors that modulate microalgal FA quality (degree of unsaturation) and accumulation. In principle, PUFA-producing strains accumulate these chemicals to a greater extent when cultivated at higher temperatures. On the contrary, the degree of unsaturation often increases at lower temperatures. However, these assumptions cannot be generalized to all species, why optimal growth temperature should be empirically tested for each species of interest. Furthermore, this implies that there is often a trade-off between high yield and high PUFA content; and while the proportion of PUFAs may be lower at higher temperatures, total production may still be higher, as exemplified below.

In the unicellular green alga *Dunaliella salina*, an increase in FA unsaturation followed a decrease in the growth temperature from 30 to 12°C [187]. Similarly, in green algae (*Chlorella vulgaris* and *Botryococcus braunii*), increasing growth temperature resulted in a decrease in the content of unsaturated FAs [188]. The haptophyte *P. lutheri* responded by inducing a relative increase of EPA and DHA at lower temperature (15°C) compared to growth at 25°C [189]. In the red microalgae *P. cruentum*, lower growth temperature increased the proportion of MGDG, especially lipid species containing EPA [190]. This species also accumulates ARA that is incorporated into membrane lipids as an acclimation strategy to cope with low temperature [191, 192]. Cohen *et al.* [193] indicated that the proportion of both EPA and ARA in *P. cruentum* followed different trends, and while EPA decreased at high temperature, ARA was reported to increase. According to Iwamoto and Sato [194], EPA content (measured as the percentage of dry weight) of the salt-tolerant freshwater eustigmatophyte *M. subterraneus* was inversely related to temperature. The highest proportion (35%) was found at 20°C; however, higher productivity (mg EPA L^{-1} day^{-1}) was found at 25°C, that is, the optimal temperature for growth. Thus, EPA production is favored at sub-optimal growth temperature; however, higher areal yield is obtained in cultures grown at an optimal temperature, due to higher biomass production. James *et al.* [195] reported that in *Nannochloropsis* sp., total *n*-3 PUFA content increased with growth temperature up to 25°C; above that temperature, a significant decline of total *n*-3 PUFA occurred. Furthermore, Seto *et al.* [196] found that at 25°C, the EPA content of *Nannochloropsis* sp. was reduced compared to when grown at lower temperatures. As a consequence, Sukenik [174] concluded that EPA production declined at a temperature above the optimal for growth, due to low FA synthesis and reduction in growth rate. Also, Van Wagenen *et al.* reported on an inverse relationship between the degree of FA unsaturation and temperature in *Nannochloropsis salina* [197].

In the haptohpyte *Isochrysis galbana*, Zhu *et al.* [198] found an increase in GLA and DHA, with a decrease of LA and monounsaturated and saturated FAs when the temperature was decreased from 30°C to 15°C. Similarly, EPA and DHA were lower in the diatoms *Chaetoceros sp.* and *P. tricornutum*, as well as in the cryptomonads *Rhodomonas* sp. and *Cryptomonas* sp., at higher cultivation temperature [199, 200].

Psychrophilic diatoms can contain very high proportions of PUFAs, reaching >80% of total FAs when grown in subzero temperatures as demonstrated for the ice diatom *Nitzschia lecointei* [201] and the arctic *Porosira glacialis* [202, 203]. Also, the psychrophilic diatom *Attheya longicornis* was shown to produce relatively high amounts of PUFAs at 10°C [204]. When comparing three temperatures (8, 16, and 24°C) in the commercially grown diatom *Odontella aurita*, Pasquet *et al.* [205] observed a significantly higher proportion of both EPA and DHA at 8°C. Cold adapted marine microalgae were also explored in a study by Schulze *et al.* [206], where EPA and DHA concentrations generally were higher in low temperatures, for example, the green algae *Tetraselmis chuii* contained 33% more EPA at 8°C compared to at 15°C. In *Spirulina*, FA content increased when cultivation temperature was elevated in 19 strains tested [207]. Maximal FA content was found in the range 30-35°C; in addition, saturated FAs increased while unsaturated FAs decreased. However, the relative proportion of GLA was higher at 30-35°C.

2.8.2 Irradiance

High cell density has the same effect on the microalgal FA profile as does light limitation, most probably due to the reduction of light availability in dense cultures (i.e., self-shading). The cellular molecular composition of microalgae varies greatly under different cell densities [208]. Many cell components, e.g., chlorophylls, lipids, and proteins, increase with increasing cell density. Maximal concentrations are obtained at the optimal cell density (OCD) for the highest photosynthetic efficiency. Both at low cell density and at cell densities above the OCD, carbohydrates are accumulated in the form of starch grains in the chloroplast.

Cohen [146] found that total FAs of *P. cruentum* increased at low irradiance. Light limitation also induces changes in the FA composition, especially the proportion of C_{20} PUFAs (reduction of C_{20} and increase of 16:0) and the *n*-3/*n*-6 ratio. In high light both EPA and ARA accounted for the major proportion of total FAs in this species (40 and 30%, respectively); when irradiance was reduced, both EPA and ARA proportions dropped to 16 and 22% of the total FAs, respectively, with a parallel increase of 16:0 from 8 to 25% [209].

In general, high irradiance leads to oxidative damage of PUFAs. Molina Grima *et al.* [210] reported that in *P. tricornutum*, very high irradiance and photoinhibiting conditions reduce FA content. An increase in the fraction of EPA was observed in *P. cruentum* following a reduction of irradiance, to values around 4.4% of dry weight at 90 μmol photons m^{-2} s^{-1} [211]. This may be the result of photoadaptation to low irradiance and membrane reorganization since EPA is a major component of the galactolipids of photosynthetic membranes. In addition, in *M. subterraneus* highest EPA productivity (26 mg L^{-1} day^{-1}) was obtained in dense cultures (2-3 g DW L^{-1}) at the end of the exponential phase of growth. In these conditions, at the end of the exponential phase, the growth rate is highly reduced, and a parallel increase of EPA production occurs (together with an increase of the percentage of EPA over total FAs). Thus, maximal EPA cell content and productivity are obtained in semi-continuous cultures maintained at high biomass concentrations. However, in *Nannochloropsis* (in contrast to *Monodum*) growth at saturating light induced an increase of lipids and FAs [212]. Moreover, the degree of unsaturation of the FAs decreased at high irradiance, mainly due to the decrease of EPA [213]. EPA was highest in sub-saturating light conditions, when accumulating galactolipids enriched in EPA [173, 212]. Renaud *et al.* [214] found similar results: reduction of EPA at high irradiance. In *Nannochloropsis*, as in many other microalgal species, the reduction of PUFAs and EPA at high irradiance is probably due to the smaller size of the chloroplast and the light-harvesting antenna. The membranes of the photosynthetic organelles represent up to 75% of the total membranes of the cell [215], of which 90% are in the form of PUFA. As a consequence of increased biomass production, *P. incisa* grown under high light intensity (400 μmol photons m^{-2} s^{-1}) reached higher growth and volumetric contents of total FAs and ARA, in comparison to when grown at medium (200 μmol photons m^{-2} s^{-1}) and low (35 μmol photons m^{-2} s^{-1}) light intensity [216]. In *Spirulina*, FA content increased at low irradiance and high cell density. The concentration of GLA also increased under those conditions. However, the proportion of GLA decreased at high FA content; the increase in FAs was mainly due to an increase of neutral lipids, in which the content of GLA is lower compared to in galactolipids.

The light/dark (L/D) cycle also affects the proportion of PUFAs. Brown *et al.* [217] showed that *T. pseudonana* grown in continuous light increased the percentage of TAG and reduced that of total polar lipids; in addition, 12 h light/12 h dark conditions induced a higher proportion of PUFAs and lower proportions of saturated and monounsaturated FAs. In *Spirulina* the level of desaturation of C_{18} increased under L/D cycles compared to continuous light (CL). During the dark period, mainly saturated FAs in neutral lipids are consumed, thereby increasing the proportion of unsaturated FAs [176]. GLA was also higher in L/D cycles, compared to CL.

2.8.3 Growth Rate

FA profile changes with the growth rate; saturated and monounsaturated FAs decrease and PUFA increases at a higher growth rate. At increasing growth rates there is an increase in protein content and decrease of lipids [169], but the decrease in lipids is mainly of neutral lipids (14:0, 16:0, 16:1), while both glycolipids and phospholipids increased in *P. tricornutum* growing in optimal conditions. Lower cell division implies less need for structural molecules (in our case, glycolipids and phospholipids; [218]), and the lipids are used as carbon storage (neutral lipids). Maximal EPA productivity of 33-50 mg L^{-1} day^{-1} in *P. tricornutum* grown in tubular photobioreactors was achieved at a dilution rate of 0.04 h^{-1} [219]. Low irradiance at the surface of the cultures led to low EPA productivity because of the low growth rate. Very strong irradiance also resulted in low EPA productivity due to small EPA content in photoinhibited cells. Moreover, Veloso *et al.* [220] found that EPA rose from 10.2 to 25.3% and palmitoleic acid (16:1 *n*-7) decreased from 43.3 to 21.0%, at high (1.22 d^{-1}) compared to low growth rates (0.14 d^{-1}). Solovchenko *et al.* [216] found that an increase in the proportion of ARA in *P. incisa* during aging is partially the result of a shift from *n*-3 to *n*-6 PUFA in chloroplastic lipids that are accumulated when the algae entered the stationary phase of growth, which has been reported to occur in many microalgae [153, 193]. In *Spirulina* an increase of polar lipids occurs during active cell division due to the synthesis of cell membranes. Similarly, in the stationary phase, neutral lipids are synthesized as reserve material. Aging implies a decrease in the level of desaturation of C_{18} FAs.

2.8.4 Nitrogen and Phosphorous

It is generally accepted that nitrogen starvation enhances FA production in many species of microalgae [221]. However, as in the case of suboptimal temperature or low irradiance, nitrogen starvation leads to a sharp decline in cell growth and in the proportion of PUFA [222]. Nitrogen limitation associated with excess light ends in an accumulation of non-N reserve compounds such as TAG. Shifrin and Chisholm [221] found a 130-320% increase in oil content in 15 Chlorophyceae species in N-deficient cultures. In general, under conditions of N limitation, the content and proportion of PUFAs are reduced. This correlation between low N and low PUFA content may be because PUFAs are accumulated in the polar lipids, while the TAGs are mainly short-chained saturated and monounsaturated FAs [146, 167]. Under N starvation the synthesis of TAGs takes place at the expense of the polar lipids, which are the main carriers of PUFAs.

Veloso *et al.* [220] found that the FA profile of *P. tricornutum* changed after 10 days of N starvation, and the proportion of EPA decreased from 35 to 10%, while palmitic (16:0) and palmitoleic acids (16:1 *n*-7) increased from 10 to 20%, and from 17 to 25%, respectively. In *P. incisa*, Solovchenko *et al.* [216] found a strong increase in the proportion ARA over total FAs under N starvation. In contrast to eukaryotic algae, no increase of total lipids or TAG was detected in *Spirulina* under limited nitrogen [223]. Most of the FAs were found in the polar fraction. Klyachko-Gurvich [1975 in 224] reported that even after 10 days of N starvation no synthesis of lipids occurred, and no changes in their proportion in the cells were detected.

Recently, Janssen *et al.* [225] observed that during N starvation in the marine eustigmatophyte *Nannochloropsis gaditana*, EPA accumulation in TAG was lower at low biomass-specific radiation, which in turn showed the highest EPA in the polar lipids. Again, demonstrating the complex relationship between different environmental factors and FA production/accumulation.

The cold-adapted green alga *Koliella antarctica* has been proposed as a good candidate for EPA production. When PUFA production was tested in *K. antarctica* under different growth conditions, cells subjected to P starvation showed the highest *n*-3 FA content [226]. For EPA specifically, the control treatment resulted in the highest accumulation (4.9% of total FAs), however, within the same order of magnitude as for P starved cells (3.1% of total FAs) [226].

2.8.5 CO_2

For the psychrophilic diatoms *P. glacialis* and *A. longicornis* different responses to CO_2 aeration (20-25% CO_2) were observed: for *P. glacialis* relative concentration of DHA increased from 3.90 to 5.78% while EPA decreased from 26.59 to 23.66%, but for *A. longicornis* no such responses were observed and the growth rate was hampered [202]. However, in *Chaetoceros muelleri*, the correlation between CO_2 and EPA was positive [227]. Thus, the response of different FAs to CO_2 in diatoms seems species-specific. In the green alga *Dunaliella tertiolecta*, Sydney *et al.* [228] observed an increase in lipid accumulation upon 5% CO_2 enrichment.

2.8.6 Salinity

Iwamoto and Sato [194] found that increasing salinity in the salt-tolerant species *M. subterraneus* from 0 to 30 practical salinity units (PSU) increased lipid content. However, EPA content dropped significantly from 34 to 15%. In *N. oculata* maximal growth occurs at 25-30 PSU [229], while highest amount of FAs was found at 35 PSU; however, at this salinity EPA was at its lowest concentration, and palmitic acid (16:0) was the dominant FA. Thus, optimal salinity for EPA and essential FA production was between 25 and 30 PSU. In *Spirulina*, increased salinity in the range 0.1-1.5 PSU reduced cell growth; however, it did not result in increased FA content.

In conclusion, microalgae cultivated under stress generally accumulate lipids [230], but, due to reduced growth rate, lipid yield in commercial systems might be equally low [231]. Then, for every species or strain a detailed study on specific growth rate under different stressing and non-stressing conditions are necessary to evaluate their capacity for the accumulation of lipids and the proportion of the desired one over the total lipid content, at the time that the maximal potential growth is achieved. The balance between lipid content, the proportion of the desired PUFA and microalgal growth is crucial to maximizing areal yield. Moreover, optimal cell density (that relates to light penetration) for every season must be established to keep the cultures at their optimal concentration for maximal daily productivity, as well as the optimal time of the day for harvesting, due to day/night changes in the amount and proportion of the FAs.

2.9 Genetic Engineering to Promote Lipid Accumulation and Tailoring of Fatty Acid Profiles

Genetic engineering is emerging as a promising tool in manipulating LC-PUFA synthesis in marine microorganisms. The freshwater green microalgae *C. reinhardtii* and the marine diatom *P. tricornutum* have served as the primary models for studies on lipid production and lipid biosynthesis pathways. Today, the number of fully sequenced microalgal genomes that are publically available is in the range of 40-60 [232]. There are several ongoing algae sequencing projects that are expected to dramatically increase this number and serve as valuable resources for future genetic studies [232]. Progress in bioengineering of microalgae has been hampered by difficulties in the delivery of exogenous DNA and targeted gene-editing tools. However, methods for DNA transfer and gene editing in microalgae are under rapid evolution. For example, gene modifications protocols based on TALE and Crisper/Cas9 endonucleases have been successfully adopted for *P. tricornutum* [233–235].

In the context of nutraceutical FA production, two primary targets for genetic engineering are increasing i) total TAG productivity and ii) the fraction of LC-PUFAs. Key enzymes in TAG synthesis are the acyltransferases that catalyze the incorporation of fatty acyls into the glycerol lipid backbone. Particularly, acyl-CoA:DAG acyltransferase (DGAT) is considered as one of the rate-limiting enzymes in TAG synthesis and has therefore been extensively studied [27, 236]. Overexpression of a native DGAT in *P. tricornutum* resulted in a substantial increase in oil production [237]. Neutral lipid content increased by 35% and the proportion of PUFAs increased significantly (76% for EPA) without affecting growth rate. Heterologous co-expression of a yeast DGAT and a plant lipid droplet stabilizing oleosin in the same diatom species resulted in a 3.6-fold increase in TAG content [33]. Characterization of DGATs and their potential to improve TAG formation has also been shown in several other marine microalgae, including *Nannochloropsis oceanica* [238, 239], *Ostreococcus tauri* [240] and *Thalassiosira pseudonana* [241]. Different isoforms of DGAT show a preference for different FA classes (i.e., SFA, MUFA or PUFA). By manipulating the relative expression between DGAT isoforms in *N. oceanica* strains, Xin *et al.* demonstrated that it was possible to influence the degree of TAG unsaturation, thereby producing strains suited for LC-PUFA or biofuel production [242]. Furthermore, overexpressing the native Glycerol-3-phosphate acyltransferase (GPAT), an enzyme that catalyzes the first step in TAG synthesis, in *P. tricornutum* was accompanied by a twofold increase in neutral lipid content as well as a 42% increase in total PUFA [243].

Other targets for increased LC-PUFA production include FA desaturation and elongation enzymes. For example, overexpression of sequences encoding native Δ12 and Δ5 FA desaturases in *N. oceanica* lead to an increase in the final *n*-3 product EPA [244]. In *P. tricornutum*, co-expression of genes encoding a Δ5-elongase and Δ6-desaturase from the algae *Ostreococcus tauri* increased the accumulation of DHA to more than eightfold compared to the wild type strain. In a fast-growing strain of the marine cyanobacterium *Synechococcus* sp., expression of a Δ6 desaturase cloned from *Synechocystis* sp. resulted in successful synthesis of stearidonic acid [245].

An alternative strategy to increase TAG accumulation is to block competing pathways and redirect carbon flux towards lipid biosynthesis. Inactivation of the UDP-glucose pyrophosphorylase gene in *P. tricornutum* resulted in 45 times higher TAG accumulation than in controls

[233]. Thereby, synthesis of the storage polysaccharide chrysolaminarin was inhibited and its precursor UDP-glucose could instead be used for lipid synthesis. Similarly, by knocking down the nitrate reductase-encoding gene in *P. tricornutum*, 43% more carbon was allocated into cellular lipids than in the wild type [246].

Not only the total FA composition should be considered when engineering microalgal lipids, but also the stereoisomeric structure of TAGs should be addressed. Human breast milk TAGs are enriched with palmitic acid (16:0) at the middle (*sn*-2) position of the glycerol backbone while PUFAs are primarily esterified to outer (*sn*-1 and *sn*-3) positions [247]. Vegetable and microalgal oils show a different stereoisomeric composition, where palmitate is located at *sn*-1/3 positions and oleate or linoleate dominate at *sn*-2 [248, 249]. Infant formulas based on vegetable oil blends or bovine milk fat show reduce fat absorption in neonates and possibly hamper calcium uptake compared to human milk fat [90]. *sn*-2 PA enriched formulas are produced by enzyme-catalyzed restructuring of TAGs, a costly process that substantially increases the price of the product. van Erp *et al.* recently demonstrated that it was possible to genetically engineer the TAG biosynthesis pathway in the model plant *Arabidopsis thaliana* to increase the *sn*-2 palmitate content >20-fold, from 3 to 70% [250]. This was achieved by relocating the chloroplast C16:0-specific isoform of the lysophosphatidic acid acyltransferase (LPAT) to the endoplasmic reticulum, in combination with knock-out of PC:DAG choline phosphotransferase activity. These two modifications successfully increased cytosolic glycerolipid biosynthesis that specifically yields *sn*-2 16:0 molecules, and stimulated the flux of newly made DAG into TAG synthesis, respectively. A similar strategy to enrich *sn*-2 palmitate in TAGs could be applied to microalgae and offer a new potential source of human milk fat substitute for infant nutrition.

The discussed examples provide insights into the power of genetic engineering and its potential to enhance LC-PUFA content in microalgae. The growing number of available sequenced genomes along with refined molecular gene-editing techniques will allow adaptation to non-model organisms for industrial-scale production.

2.10 Conclusions

Today, nutritional LC-PUFAs such as EPA and DHA are largely derived from marine fish and seafood, either from wild capture fisheries or from aquaculture farms. These are resource-consuming or limited sources in need of alternatives for long-term sustainable production. Many marine microalgal species produce substantial quantities of LC-PUFAs, and LC-PUFA rich oils produced from microalgae have shown similar positive effects on human health as oils from fish or shellfish (e.g., krill). Microalgal oils have thus emerged as a promising environmentally friendly source of LC-PUFAs. However, microalgae oil production is still faced with some challenges such as implementing results from basic science to industrial-scale production and reducing production costs in a yet sustainable manner.

2.11 Acknowledgements

We thank Mats X. Andersson for critically reading the manuscript. Financial support was provided by Carl Trygger Foundation and J Gustaf Richert Foundation (Wulff).

References

1. Brower, V., Nutraceuticals: Poised for a healthy slice of the healthcare market? *Nat. Biotechnol.* 16(8): p. 728-731, 1998.
2. Zeisel, S.H., Regulation of "Nutraceuticals". *Science.* 285(5435): p. 1853-1855, 1999.
3. Alexander, D.D., Miller, P.E., Van Elswyk, M.E., Kuratko, C.N., and Bylsma, L.C., A Meta-Analysis of Randomized Controlled Trials and Prospective Cohort Studies of Eicosapentaenoic and Docosahexaenoic Long-Chain Omega-3 Fatty Acids and Coronary Heart Disease Risk. *Mayo Clin. Proc.* 92(1): p. 15-29, 2017.
4. Holub, B.J., Docosahexaenoic acid (DHA) and cardiovascular disease risk factors. *Prostaglandins Leukot. Essent. Fatty Acids.* 81(2-3): p. 199-204, 2009.
5. Calder, P.C., Omega-3 fatty acids and inflammatory processes: from molecules to man. *Biochem. Soc. Trans.* 45(5): p. 1105-1115, 2017.
6. Luchtman, D.W. and Song, C., Cognitive enhancement by omega-3 fatty acids from child-hood to old age: findings from animal and clinical studies. *Neuropharmacology.* 64: p. 550-65, 2013.
7. Cardoso, C., Afonso, C., and Bandarra, N.M., Dietary DHA and health: cognitive function ageing. *Nutr. Res. Rev.* 29(2): p. 281-294, 2016.
8. Muskiet, F.A.J., Fokkema, M.R., Schaafsma, A., Boersma, E.R., and Crawford, M.A., Is Docosahexaenoic Acid (DHA) Essential? Lessons from DHA Status

Regulation, Our Ancient Diet, Epidemiology and Randomized Controlled Trials. *J. Nutr.* 134(1): p. 183-186, 2004.

9. Kris-Etherton, P.M., Grieger, J.A., and Etherton, T.D., Dietary reference intakes for DHA and EPA. *Prostaglandins Leukot. Essent. Fatty Acids.* 81(2-3): p. 99-104, 2009.
10. Ervin, R.B., Wright, J.D., Wang, C.Y., and Kennedy-Stephenson, J., Dietary intake of fats and fatty acids for the United States population: 1999-2000. *Adv. Data.* (348): p. 1-6, 2004.
11. Salem, N., Jr. and Eggersdorfer, M., Is the world supply of omega-3 fatty acids adequate for optimal human nutrition? *Curr. Opin. Clin. Nutr. Metab. Care.* 18(2): p. 147-54, 2015.
12. Colombo, S.M., Rodgers, T.F.M., Diamond, M.L., Bazinet, R.P., and Arts, M.T., Projected declines in global DHA availability for human consumption as a result of global warming. *Ambio.* 49: p. 865–880, 2020.
13. Li-Beisson, Y., Thelen, J.J., Fedosejevs, E., and Harwood, J.L., The lipid biochemistry of eukaryotic algae. *Prog. Lipid Res.* 74: p. 31-68, 2019.
14. Blasbalg, T.L., Hibbeln, J.R., Ramsden, C.E., Majchrzak, S.F., and Rawlings, R.R., Changes in consumption of omega-3 and omega-6 fatty acids in the United States during the 20th century. *Am. J. Clin. Nutr.* 93(5): p. 950-62, 2011.
15. Raatz, S.K., Conrad, Z., and Jahns, L., Trends in linoleic acid intake in the United States adult population: NHANES 1999–2014. *Prostaglandins Leukot. Essent. Fatty Acids.* 133: p. 23-28, 2018.
16. Conquer, J.A. and Holub, B.J., Supplementation with an algae source of docosahexaenoic acid increases (n-3) fatty acid status and alters selected risk factors for heart disease in vegetarian subjects. *J. Nutr.* 126(12): p. 3032-9, 1996.
17. Vidgren, H.M., Agren, J.J., Schwab, U., Rissanen, T., Hanninen, O., and Uusitupa, M.I., Incorporation of n-3 fatty acids into plasma lipid fractions, and erythrocyte membranes and platelets during dietary supplementation with fish, fish oil, and docosahexaenoic acid-rich oil among healthy young men. *Lipids.* 32(7): p. 697-705, 1997.
18. Jensen, C.L., Maude, M., Anderson, R.E., and Heird, W.C., Effect of docosahexaenoic acid supplementation of lactating women on the fatty acid composition of breast milk lipids and maternal and infant plasma phospholipids. *Am. J. Clin. Nutr.* 71(1 Suppl): p. 292s-9s, 2000.
19. Otto, S.J., van Houwelingen, A.C., and Hornstra, G., The effect of supplementation with docosahexaenoic and arachidonic acid derived from single cell oils on plasma and erythrocyte fatty acids of pregnant women in the second trimester. *Prostaglandins Leukot. Essent. Fatty Acids.* 63(5): p. 323-8, 2000.
20. Wu, W.H., Lu, S.C., Wang, T.F., Jou, H.J., and Wang, T.A., Effects of docosahexaenoic acid supplementation on blood lipids, estrogen metabolism, and

in vivo oxidative stress in postmenopausal vegetarian women. *Eur. J. Clin. Nutr.* 60(3): p. 386-92, 2006.

21. Bernstein, A.M., Ding, E.L., Willett, W.C., and Rimm, E.B., A meta-analysis shows that docosahexaenoic acid from algal oil reduces serum triglycerides and increases HDL-cholesterol and LDL-cholesterol in persons without coronary heart disease. *J. Nutr.* 142(1): p. 99-104, 2012.
22. Maki, K.C., Yurko-Mauro, K., Dicklin, M.R., Schild, A.L., and Geohas, J.G., A new, microalgal DHA- and EPA-containing oil lowers triacylglycerols in adults with mild-to-moderate hypertriglyceridemia. *Prostaglandins Leukot. Essent. Fatty Acids.* 91(4): p. 141-8, 2014.
23. Craddock, J.C., Neale, E.P., Probst, Y.C., and Peoples, G.E., Algal supplementation of vegetarian eating patterns improves plasma and serum docosahexaenoic acid concentrations and omega-3 indices: a systematic literature review. *J. Hum. Nutr. Diet.* 30(6): p. 693-699, 2017.
24. Ryan, L. and Symington, A.M., Algal-oil supplements are a viable alternative to fish-oil supplements in terms of docosahexaenoic acid (22:6n-3; DHA). *J. Funct. Foods.* 19: p. 852-858, 2015.
25. Lauritzen, L., Hansen, H.S., Jorgensen, M.H., and Michaelsen, K.F., The essentiality of long chain n-3 fatty acids in relation to development and function of the brain and retina. *Prog. Lipid Res.* 40(1-2): p. 1-94, 2001.
26. Kong, F., Romero, I.T., Warakanont, J., and Li-Beisson, Y., Lipid catabolism in microalgae. *The New Phytologist.* 218(4): p. 1340-1348, 2018.
27. Xu, Y., Caldo, K.M.P., Pal-Nath, D., Ozga, J., Lemieux, M.J., Weselake, R.J., and Chen, G., Properties and biotechnological applications of Acyl-CoA:diacylglycerol acyltransferase and phospholipid:diacylglycerol acyltransferase from terrestrial plants and microalgae. *Lipids.* 53(7): p. 663-688, 2018.
28. Du, Z.Y. and Benning, C., Triacylglycerol Accumulation in Photosynthetic Cells in Plants and Algae. *Subcell. Biochem.* 86: p. 179-205, 2016.
29. Guschina, I.A. and Harwood, J.L., Lipids and lipid metabolism in eukaryotic algae. *Prog. Lipid Res.* 45(2): p. 160-186, 2006.
30. Li-Beisson, Y., Shorrosh, B., Beisson, F., Andersson, M.X., Arondel, V., Bates, P.D., . . . Ohlrogge, J., Acyl-lipid metabolism. *Arabidopsis Book.* 8: p. e0133, 2010.
31. Li-Beisson, Y., Beisson, F., and Riekhof, W., Metabolism of acyl-lipids in *Chlamydomonas reinhardtii. Plant J.* 82(3): p. 504-522, 2015.
32. Moellering, E.R., Miller, R., and Benning, C., Molecular genetics of lipid metabolism in the model green alga *Chlamydomonas reinhardtii*, in *Lipids in Photosynthesis: Essential and Regulatory Functions*, H. Wada and N. Murata, Editors., Springer Netherlands: Dordrecht. p. 139-155, 2009.
33. Zulu, N.N., Popko, J., Zienkiewicz, K., Tarazona, P., Herrfurth, C., and Feussner, I., Heterologous co-expression of a yeast diacylglycerol acyltransferase (ScDGA1) and a plant oleosin (AtOLEO3) as an efficient tool for

enhancing triacylglycerol accumulation in the marine diatom *Phaeodactylum tricornutum*. *Biotechnol. Biofuels.* 10: p. 187, 2017.
34. Mori, N., Moriyama, T., Toyoshima, M., and Sato, N., Construction of global acyl lipid metabolic map by comparative genomics and subcellular localization analysis in the red alga *Cyanidioschyzon merolae*. *Front Plant Sci.* 7: p. 958-958, 2016.
35. Leonard, A.E., Pereira, S.L., Sprecher, H., and Huang, Y.-S., Elongation of long-chain fatty acids. *Prog. Lipid Res.* 43(1): p. 36-54, 2004.
36. Park, H.G., Park, W.J., Kothapalli, K.S., and Brenna, J.T., The fatty acid desaturase 2 (FADS2) gene product catalyzes Delta4 desaturation to yield n-3 docosahexaenoic acid and n-6 docosapentaenoic acid in human cells. *FASEB J.* 29(9): p. 3911-9, 2015.
37. Sayanova, O., Mimouni, V., Ulmann, L., Morant-Manceau, A., Pasquet, V., Schoefs, B., and Napier, J.A., Modulation of lipid biosynthesis by stress in diatoms. *Philos. T. Roy. Soc. B.* 372(1728): p. 20160407, 2017.
38. Metz, J.G., Roessler, P., Facciotti, D., Levering, C., Dittrich, F., Lassner, M., . . . Browse, J., Production of polyunsaturated fatty acids by polyketide synthases in both prokaryotes and eukaryotes. *Science.* 293(5528): p. 290-3, 2001.
39. Xie, X., Meesapyodsuk, D., and Qiu, X., Ketoacylsynthase domains of a polyunsaturated fatty acid synthase in *Thraustochytrium* sp. strain ATCC 26185 can effectively function as stand-alone enzymes in *Escherichia coli*. *Appl. Environ. Microbiol.* 83(9): p. e03133-16, 2017.
40. Barclay, W.R., Meager, K.M., and Abril, J.R., Heterotrophic production of long chain omega-3 fatty acids utilizing algae and algae-like microorganisms. *J. Appl. Phycol.* 6(2): p. 123-129, 1994.
41. Leyland, B., Leu, S., and Boussiba, S., Are Thraustochytrids algae? *Fungal Biol.* 121(10): p. 835-840, 2017.
42. Burr, G.O. and Burr, M.M., A new deficiency disease produced by the rigid exclusion of fat from the diet. *J. Biol. Chem.* 82: p. 345–367, 1929.
43. Burr, G.O. and Burr, M.M., On the nature and role of the fatty acids essential in nutrition. *J. Biol. Chem.* 86: p. 587–621, 1930.
44. Spector, A.A. and Kim, H.Y., Emergence of omega-3 fatty acids in biomedical research. *Prostaglandins Leukot. Essent. Fatty Acids.* 140: p. 47-50, 2019.
45. Burr, G.O., Burr, M.M., and Miller, E.S., On the fatty acids essential in nutrition. III. *J. Biol. Chem.* 97(1): p. 1-9, 1932.
46. Jump, D.B., Mammalian fatty acid elongases. *Methods in molecular biology (Clifton, N.J.).* 579: p. 375-389, 2009.
47. Zhang, J.Y., Kothapalli, K.S., and Brenna, J.T., Desaturase and elongase-limiting endogenous long-chain polyunsaturated fatty acid biosynthesis. *Curr. Opin. Clin. Nutr. Metab. Care.* 19(2): p. 103-10, 2016.
48. Chan, J.K., McDonald, B.E., Gerrard, J.M., Bruce, V.M., Weaver, B.J., and Holub, B.J., Effect of dietary alpha-linolenic acid and its ratio to linoleic acid on platelet and plasma fatty acids and thrombogenesis. *Lipids.* 28(9): p. 811-7, 1993.

49. Belury, M.A. and Harris, W.S., Omega-6 fatty acids, inflammation and cardiometabolic health: Overview of supplementary issue. *Prostaglandins Leukot. Essent. Fatty Acids*. 139: p. 1-2, 2018.
50. Innes, J.K. and Calder, P.C., Omega-6 fatty acids and inflammation. *Prostaglandins Leukot. Essent. Fatty Acids*. 132: p. 41-48, 2018.
51. Koletzko, B., Lattka, E., Zeilinger, S., Illig, T., and Steer, C., Genetic variants of the fatty acid desaturase gene cluster predict amounts of red blood cell docosahexaenoic and other polyunsaturated fatty acids in pregnant women: findings from the Avon Longitudinal Study of Parents and Children. *Am. J. Clin. Nutr.* 93(1): p. 211-9, 2011.
52. Lattka, E., Illig, T., Koletzko, B., and Heinrich, J., Genetic variants of the FADS1 FADS2 gene cluster as related to essential fatty acid metabolism. *Curr. Opin. Lipidol.* 21(1): p. 64-9, 2010.
53. Schaeffer, L., Gohlke, H., Muller, M., Heid, I.M., Palmer, L.J., Kompauer, I., . . . Heinrich, J., Common genetic variants of the FADS1 FADS2 gene cluster and their reconstructed haplotypes are associated with the fatty acid composition in phospholipids. *Hum. Mol. Genet.* 15(11): p. 1745-56, 2006.
54. Rzehak, P., Heinrich, J., Klopp, N., Schaeffer, L., Hoff, S., Wolfram, G., . . . Linseisen, J., Evidence for an association between genetic variants of the fatty acid desaturase 1 fatty acid desaturase 2 (FADS1 FADS2) gene cluster and the fatty acid composition of erythrocyte membranes. *Br. J. Nutr.* 101(1): p. 20-6, 2009.
55. Kalisch, B., Dormann, P., and Holzl, G., DGDG and glycolipids in plants and algae. *Subcell. Biochem.* 86: p. 51-83, 2016.
56. Brown, A.E. and Elovson, J., Isolation and characterization of a novel lipid, 1(3),2-diacylglyceryl-(3)-O-4'-(N,N,N-trimethyl)homoserine, from *Ochromonas danica*. *Biochemistry*. 13(17): p. 3476-3482, 1974.
57. Cañavate, J.P., Armada, I., Ríos, J.L., and Hachero-Cruzado, I., Exploring occurrence and molecular diversity of betaine lipids across taxonomy of marine microalgae. *Phytochemistry*. 124: p. 68-78, 2016.
58. Kennedy, E.P., The biological synthesis of phospholipids. *Can. J. Biochem. Physiol.* 34(2): p. 334-48, 1956.
59. Lands, W.E., Metabolism of glycerolipides; a comparison of lecithin and triglyceride synthesis. *J. Biol. Chem.* 231(2): p. 883-8, 1958.
60. Fan, J., Andre, C., and Xu, C., A chloroplast pathway for the de novo biosynthesis of triacylglycerol in *Chlamydomonas reinhardtii*. *FEBS Lett.* 585(12): p. 1985-1991, 2011.
61. Costello, C., Ovando, D., Clavelle, T., Strauss, C.K., Hilborn, R., Melnychuk, M.C., . . . Leland, A., Global fishery prospects under contrasting management regimes. *Proc Natl Acad Sci USA*. 113(18): p. 5125-5129, 2016.
62. Scheffer, M., Carpenter, S., and Young, B.d., Cascading effects of overfishing marine systems. *Trends Ecol. Evol.* 20(11): p. 579-581, 2005.
63. National Research Council, *Effects of Trawling and Dredging on Seafloor Habitat*. Washington, DC: The National Academies Press. 136, 2002.

64. Bueno-Pardo, J., Ramalho, S.P., García-Alegre, A., Morgado, M., Vieira, R.P., Cunha, M.R., and Queiroga, H., Deep-sea crustacean trawling fisheries in Portugal: quantification of effort and assessment of landings per unit effort using a Vessel Monitoring System (VMS). *Sci. Rep.* 7: p. 40795-40795, 2017.
65. Hong, M.Y., Hoh, E., Kang, B., DeHamer, R., Kim, J.Y., and Lumibao, J., Fish oil contaminated with persistent organic pollutants induces colonic aberrant crypt foci formation and reduces antioxidant enzyme gene expression in rats. *J. Nutr.* 147(8): p. 1524-1530, 2017.
66. Bourdon, J.A., Bazinet, T.M., Arnason, T.T., Kimpe, L.E., Blais, J.M., and White, P.A., Polychlorinated biphenyls (PCBs) contamination and aryl hydrocarbon receptor (AhR) agonist activity of Omega-3 polyunsaturated fatty acid supplements: implications for daily intake of dioxins and PCBs. *Food Chem. Toxicol.* 48(11): p. 3093-7, 2010.
67. Mahaffey, K.R., Clickner, R.P., and Jeffries, R.A., Methylmercury and omega-3 fatty acids: Co-occurrence of dietary sources with emphasis on fish and shellfish. *Environ. Res.* 107(1): p. 20-29, 2008.
68. de Roos, B., Sneddon, A.A., Sprague, M., Horgan, G.W., and Brouwer, I.A., The potential impact of compositional changes in farmed fish on its health-giving properties: is it time to reconsider current dietary recommendations? *Public Health Nutr.* 20(11): p. 2042-2049, 2017.
69. Sprague, M., Dick, J.R., and Tocher, D.R., Impact of sustainable feeds on omega-3 long-chain fatty acid levels in farmed Atlantic salmon, 2006-2015. *Sci. Rep.* 6: p. 21892, 2016.
70. Bibus, D. and Lands, B., Balancing proportions of competing omega-3 and omega-6 highly unsaturated fatty acids (HUFA) in tissue lipids. *Prostaglandins Leukot. Essent. Fatty Acids.* 99: p. 19-23, 2015.
71. Howard-Thompson, A., Dutton, A., Hoover, R., and Goodfred, J., Flushing and pruritus secondary to prescription fish oil ingestion in a patient with allergy to fish. *Int. J. Clin. Pharm.* 36(6): p. 1126-9, 2014.
72. Doughman, S.D., Krupanidhi, S., and Sanjeevi, C.B., Omega-3 fatty acids for nutrition and medicine: considering microalgae oil as a vegetarian source of EPA and DHA. *Curr. Diabetes Rev.* 3(3): p. 198-203, 2007.
73. Brenna, J.T., Salem, N., Jr., Sinclair, A.J., and Cunnane, S.C., alpha-Linolenic acid supplementation and conversion to n-3 long-chain polyunsaturated fatty acids in humans. *Prostaglandins Leukot. Essent. Fatty Acids.* 80(2-3): p. 85-91, 2009.
74. Hoffmann, M., Wagner, M., Abbadi, A., Fulda, M., and Feussner, I., Metabolic engineering of omega3-very long chain polyunsaturated fatty acid production by an exclusively acyl-CoA-dependent pathway. *J. Biol. Chem.* 283(33): p. 22352-62, 2008.
75. Domergue, F., Abbadi, A., Zahringer, U., Moreau, H., and Heinz, E., In vivo characterization of the first acyl-CoA Delta6-desaturase from a member of the plant kingdom, the microalga *Ostreococcus tauri*. *Biochem. J.* 389(Pt 2): p. 483-90, 2005.

76. Ruiz-Lopez, N., Usher, S., Sayanova, O.V., Napier, J.A., and Haslam, R.P., Modifying the lipid content and composition of plant seeds: engineering the production of LC-PUFA. *Appl. Microbiol. Biotechnol.* 99(1): p. 143-54, 2015.
77. Andre, C., Buesen, R., Riffle, B., Wandelt, C., Sottosanto, J.B., Marxfeld, H., . . . Lipscomb, E.A., Safety assessment of EPA+DHA canola oil by fatty acid profile comparison to various edible oils and fat-containing foods and a 28-day repeated dose toxicity study in rats. *Food Chem. Toxicol.* 124: p. 168-181, 2019.
78. Halford, N.G., Legislation governing genetically modified and genome-edited crops in Europe: the need for change. *J. Sci. Food Agric.* 99(1): p. 8-12, 2019.
79. Fernbach, P.M., Light, N., Scott, S.E., Inbar, Y., and Rozin, P., Extreme opponents of genetically modified foods know the least but think they know the most. *Nature Human Behaviour.* 3(3): p. 251-256, 2019.
80. Betancor, M.B., Li, K., Sprague, M., Bardal, T., Sayanova, O., Usher, S., . . . Olsen, R.E., An oil containing EPA and DHA from transgenic *Camelina sativa* to replace marine fish oil in feeds for Atlantic salmon (*Salmo salar* L.): Effects on intestinal transcriptome, histology, tissue fatty acid profiles and plasma biochemistry. *PLoS One.* 12(4): p. e0175415, 2017.
81. Ruyter, B., Sissener, N.H., Ostbye, T.K., Simon, C.J., Krasnov, A., Bou, M., . . . Berge, G.M., Omega-3 canola oil effectively replaces fish oil as a new safe dietary source of docosahexaenoic acid (DHA) in feed for juvenile Atlantic salmon. *Br. J. Nutr.*: p. 1-43, 2019.
82. Xie, D., Jackson, E.N., and Zhu, Q., Sustainable source of omega-3 eicosapentaenoic acid from metabolically engineered *Yarrowia lipolytica*: from fundamental research to commercial production. *Appl. Microbiol. Biotechnol.* 99(4): p. 1599-1610, 2015.
83. Zhang, X., Pang, S., Liu, C., Wang, H., Ye, D., Zhu, Z., and Sun, Y., A novel dietary source of EPA and DHA: Metabolic engineering of an important freshwater species-common carp by fat1-transgenesis. *Mar. Biotechnol. (N. Y.).* 21(2): p. 171-185, 2019.
84. Lai, L., Kang, J.X., Li, R., Wang, J., Witt, W.T., Yong, H.Y., . . . Dai, Y., Generation of cloned transgenic pigs rich in omega-3 fatty acids. *Nat. Biotechnol.* 24(4): p. 435-6, 2006.
85. Zhang, P., Liu, P., Dou, H., Chen, L., Chen, L., Lin, L., . . . Ma, R.Z., Handmade cloned transgenic sheep rich in omega-3 Fatty acids. *PLoS One.* 8(2): p. e55941, 2013.
86. Cheng, G., Fu, C., Wang, H., Adoligbe, C., Wei, S., Li, S., . . . Zan, L., Production of transgenic beef cattle rich in n-3 PUFAs by somatic cell nuclear transfer. *Biotechnol. Lett.* 37(8): p. 1565-71, 2015.
87. Clark, M.A., Springmann, M., Hill, J., and Tilman, D., Multiple health and environmental impacts of foods. *Proc Natl Acad Sci USA.* 116(46): p. 23357-23362, 2019.
88. Aasen, I.M., Ertesvåg, H., Heggeset, T.M.B., Liu, B., Brautaset, T., Vadstein, O., and Ellingsen, T.E., Thraustochytrids as production organisms for

docosahexaenoic acid (DHA), squalene, and carotenoids. *Appl. Microbiol. Biotechnol.* 100(10): p. 4309-4321, 2016.

89. Sakuradani, E., Ando, A., Shimizu, S., and Ogawa, J., Metabolic engineering for the production of polyunsaturated fatty acids by oleaginous fungus *Mortierella alpina* 1S-4. *J. Biosci. Bioeng.* 116(4): p. 417-422, 2013.
90. Wei, W., Jin, Q., and Wang, X., Human milk fat substitutes: Past achievements and current trends. *Prog. Lipid Res.* 74: p. 69-86, 2019.
91. Lang, I., Hodac, L., Friedl, T., and Feussner, I., Fatty acid profiles and their distribution patterns in microalgae: a comprehensive analysis of more than 2000 strains from the SAG culture collection. *BMC Plant Biol.* 11(1): p. 124, 2011.
92. Hadley, K.B., Ryan, A.S., Forsyth, S., Gautier, S., and Salem, N., Jr., The essentiality of arachidonic acid in infant development. *Nutrients.* 8(4): p. 216-216, 2016.
93. Carlson, S.E. and Colombo, J., Docosahexaenoic acid and arachidonic acid nutrition in early development. *Adv. Pediatr.* 63(1): p. 453-471, 2016.
94. Lapillonne, A. and Moltu, S.J., Long-chain polyunsaturated fatty acids and clinical outcomes of preterm infants. *Ann. Nutr. Metab.* 69 Suppl 1: p. 35-44, 2016.
95. Brenna, J.T., Varamini, B., Jensen, R.G., Diersen-Schade, D.A., Boettcher, J.A., and Arterburn, L.M., Docosahexaenoic and arachidonic acid concentrations in human breast milk worldwide. *Am. J. Clin. Nutr.* 85(6): p. 1457-64, 2007.
96. Fortune Business Insights, *Infant formula market size, share & industry analysis, by type (Infant milk, Follow-on-Milk, Others), distribution channel (Hypermarkets/Supermarkets, Pharmacy/Medical stores, Specialty stores, Others), and regional forecasts 2019-2026*, 2019.
97. Ma, X.N., Chen, T.P., Yang, B., Liu, J., and Chen, F., Lipid Production from *Nannochloropsis. Mar. Drugs.* 14(4), 2016.
98. Peltomaa, E., Hallfors, H., and Taipale, S.J., Comparison of diatoms and dinoflagellates from different habitats as sources of PUFAs. *Mar. Drugs.* 17(4), 2019.
99. Mendes, A., Reis, A., Vasconcelos, R., Guerra, P., and Lopes da Silva, T., *Crypthecodinium cohnii* with emphasis on DHA production: a review. *J. Appl. Phycol.* 21(2): p. 199-214, 2009.
100. Breuer, G., Lamers, P.P., Martens, D.E., Draaisma, R.B., and Wijffels, R.H., The impact of nitrogen starvation on the dynamics of triacylglycerol accumulation in nine microalgae strains. *Bioresour. Technol.* 124: p. 217-226, 2012.
101. Jonasdottir, S.H., Fatty acid profiles and production in marine phytoplankton. *Mar. Drugs.* 17(3), 2019.
102. Steinhoff, F.S., Karlberg, M., Graeve, M., and Wulff, A., Cyanobacteria in Scandinavian coastal waters — A potential source for biofuels and fatty acids? *Algal Res.* 5: p. 42-51, 2014.

103. Wu, B., Tseng, C.K., and Xiang, W., Large-scale cultivation of *Spirulina* in seawater based culture medium. *Bot. Mar.* 36(2): p. 99–102, 1993.
104. Lane, K., Derbyshire, E., Li, W., and Brennan, C., Bioavailability and potential uses of vegetarian sources of omega-3 fatty acids: a review of the literature. *Crit. Rev. Food Sci. Nutr.* 54(5): p. 572-9, 2014.
105. Domenichiello, A.F., Kitson, A.P., and Bazinet, R.P., Is docosahexaenoic acid synthesis from alpha-linolenic acid sufficient to supply the adult brain? *Prog. Lipid Res.* 59: p. 54-66, 2015.
106. Nagappan, S. and Kumar Verma, S., Co-production of biodiesel and alpha-linolenic acid (omega-3 fatty acid) from microalgae, *Desmodesmus* sp. MCC34. *Energ. Source. Part A.* 40(24): p. 2933-2940, 2018.
107. Guedes, A.C., Amaro, H.M., Barbosa, C.R., Pereira, R.D., and Malcata, F.X., Fatty acid composition of several wild microalgae and cyanobacteria, with a focus on eicosapentaenoic, docosahexaenoic and α-linolenic acids for eventual dietary uses. *Food Res. Int.* 44(9): p. 2721-2729, 2011.
108. Thao, Y.T., Linh, T.D., Si, C.V., Carter, W.T., and Hill, T.R., Isolation and selection of microalgal strains from natural water sources in Viet Nam with potential for edible oil production. *Mar. Drugs*. 15(7), 2017.
109. Lee, S., Lim, S., Jeong, D.G., and Kim, J.H., Characterization of an oleaginous unicellular green microalga, *Lobosphaera incisa* (Reisigl, 1964) Strain K-1, Isolated From a tidal flat in the Yellow Sea, Republic of Korea. *Front. Microbiol.* 9(2159), 2018.
110. Yamazaki, K., Fujikawa, M., Hamazaki, T., Yano, S., and Shono, T., Comparison of the conversion rates of alpha-linolenic acid (18:3(n - 3)) and stearidonic acid (18:4(n - 3)) to longer polyunsaturated fatty acids in rats. *Biochim. Biophys. Acta.* 1123(1): p. 18-26, 1992.
111. Harris, W.S., Lemke, S.L., Hansen, S.N., Goldstein, D.A., DiRienzo, M.A., Su, H., . . . George, C., Stearidonic acid-enriched soybean oil increased the omega-3 index, an emerging cardiovascular risk marker. *Lipids*. 43(9): p. 805-11, 2008.
112. Peltomaa, E., Johnson, M.D., and Taipale, S.J., Marine cryptophytes are great sources of EPA and DHA. *Mar. Drugs*. 16(1), 2017.
113. Gouveia, L., Coutinho, C., Mendonça, E., Batista, A.P., Sousa, I., Bandarra, N.M., and Raymundo, A., Functional biscuits with PUFA-ω3 from *Isochrysis galbana*. *J. Sci. Food Agric.* 88(5): p. 891-896, 2008.
114. de los Reyes, C., Ortega, M.J., Rodríguez-Luna, A., Talero, E., Motilva, V., and Zubía, E., Molecular characterization and anti-inflammatory activity of galactosylglycerides and galactosylceramides from the microalga *Isochrysis galbana*. *J. Agric. Food Chem.* 64(46): p. 8783-8794, 2016.
115. Che, C.A., Kim, S.H., Hong, H.J., Kityo, M.K., Sunwoo, I.Y., Jeong, G.-T., and Kim, S.-K., Optimization of light intensity and photoperiod for *Isochrysis galbana* culture to improve the biomass and lipid production using 14-L photobioreactors with mixed light emitting diodes (LEDs) wavelength under two-phase culture system. *Bioresour. Technol.* 285: p. 121323, 2019.

116. Medina, A.R., Giménez, A.G., Camacho, F.G., Pérez, J.A.S., Grima, E.M., and Gómez, A.C., Concentration and purification of stearidonic, eicosapentaenoic, and docosahexaenoic acids from cod liver oil and the marine microalga *Isochrysis galbana*. *JAOCS*. 72(5): p. 575-583, 1995.
117. Patil, V., Källqvist, T., Olsen, E., Vogt, G., and Gislerød, H.R., Fatty acid composition of 12 microalgae for possible use in aquaculture feed. *Aquac. Int.* 15(1): p. 1-9, 2007.
118. Sharma, K. and Schenk, P.M., Rapid induction of omega-3 fatty acids (EPA) in *Nannochloropsis* sp. by UV-C radiation. *Biotechnol. Bioeng.* 112(6): p. 1243-1249, 2015.
119. Meng, Y., Jiang, J., Wang, H., Cao, X., Xue, S., Yang, Q., and Wang, W., The characteristics of TAG and EPA accumulation in *Nannochloropsis oceanica* IMET1 under different nitrogen supply regimes. *Bioresour. Technol.* 179: p. 483-489, 2015.
120. Khozin-Goldberg, I., Didi-Cohen, S., Shayakhmetova, I., and Cohen, Z., Biosynthesis of eicosapentaenoic acid (EPA) in the freshwater eustigmatophyte *Monodus subterraneus* (Eustigmatophyceae). *J. Phycol.* 38(4): p. 745-756, 2002.
121. Kawachi, M., Inouye, I., Honda, D., O'Kelly, C.J., Bailey, J.C., Bidigare, R.R., and Andersen, R.A., The Pinguiophyceae classis nova, a new class of photosynthetic stramenopiles whose members produce large amounts of omega-3 fatty acids. *Phycol. Res.* 50(1): p. 31-47, 2002.
122. Kyle, D.J., Behrens, P.W., Bingham, S., Arnett, K., and Lieberman, D., Microalgae as a source of EPA-containing oils, in *Biotechnology for the fats and oils industry*, T.H. Applewhite, Editor., American Oil Chemists' Society. p. 117-121, 1989.
123. Apt, K.E. and Behrens, P.W., Commercial developments in microalgal biotechnology. *J. Phycol.* 35(2): p. 215-226, 1999.
124. Lauritzen, L., Brambilla, P., Mazzocchi, A., Harsløf, L.B.S., Ciappolino, V., and Agostoni, C., DHA effects in brain development and function. *Nutrients*. 8(1): p. 6, 2016.
125. Sun, G.Y., Simonyi, A., Fritsche, K.L., Chuang, D.Y., Hannink, M., Gu, Z., . . . Beversdorf, D.Q., Docosahexaenoic acid (DHA): An essential nutrient and a nutraceutical for brain health and diseases. *Prostaglandins Leukot. Essent. Fatty Acids*. 136: p. 3-13, 2018.
126. Tonon, T., Harvey, D., Larson, T.R., and Graham, I.A., Long chain polyunsaturated fatty acid production and partitioning to triacylglycerols in four microalgae. *Phytochemistry*. 61(1): p. 15-24, 2002.
127. Sayanova, O., Haslam, R.P., Calerón, M.V., López, N.R., Worthy, C., Rooks, P., . . . Napier, J.A., Identification and functional characterisation of genes encoding the omega-3 polyunsaturated fatty acid biosynthetic pathway from the coccolithophore *Emiliania huxleyi*. *Phytochemistry*. 72(7): p. 594-600, 2011.

128. Khozin-Goldberg, I., Iskandarov, U., and Cohen, Z., LC-PUFA from photosynthetic microalgae: occurrence, biosynthesis, and prospects in biotechnology. *Appl. Microbiol. Biotechnol.* 91(4): p. 905-15, 2011.
129. Kaur, G., Guo, X.F., and Sinclair, A.J., Short update on docosapentaenoic acid: a bioactive long-chain n-3 fatty acid. *Curr. Opin. Clin. Nutr. Metab. Care.* 19(2): p. 88-91, 2016.
130. Nilsson, A.K., Löfqvist, C., Najm, S., Hellgren, G., Sävman, K., Andersson, M.X., . . . Hellström, A., Long-chain polyunsaturated fatty acids decline rapidly in milk from mothers delivering extremely preterm indicating the need for supplementation. *Acta Paediatr.* 107(6): p. 1020-1027, 2018.
131. Drouin, G., Guillocheau, E., Catheline, D., Baudry, C., Le Ruyet, P., Rioux, V., and Legrand, P., Impact of n-3 docosapentaenoic acid supplementation on fatty acid composition in rat differs depending upon tissues and is influenced by the presence of dairy lipids in the diet. *J. Agric. Food Chem.* 66(38): p. 9976-9988, 2018.
132. Christensen, E., Woldseth, B., Hagve, T.A., Poll-The, B.T., Wanders, R.J., Sprecher, H., . . . Christophersen, B.O., Peroxisomal beta-oxidation of polyunsaturated long chain fatty acids in human fibroblasts. The polyunsaturated and the saturated long chain fatty acids are retroconverted by the same acyl-CoA oxidase. *Scand. J. Clin. Lab. Invest. Suppl.* 215: p. 61-74, 1993.
133. Weylandt, K.H., Docosapentaenoic acid derived metabolites and mediators - The new world of lipid mediator medicine in a nutshell. *Eur. J. Pharmacol.* 785: p. 108-115, 2016.
134. Vik, A., Dalli, J., and Hansen, T.V., Recent advances in the chemistry and biology of anti-inflammatory and specialized pro-resolving mediators biosynthesized from n-3 docosapentaenoic acid. *Bioorg. Med. Chem. Lett.* 27(11): p. 2259-2266, 2017.
135. Drouin, G., Rioux, V., and Legrand, P., The n-3 docosapentaenoic acid (DPA): A new player in the n-3 long chain polyunsaturated fatty acid family. *Biochimie.* 159: p. 36-48, 2019.
136. Mendes, A., Guerra, P., Madeira, V., Ruano, F., Lopes da Silva, T., and Reis, A., Study of docosahexaenoic acid production by the heterotrophic microalga *Crypthecodinium cohnii* CCMP 316 using carob pulp as a promising carbon source. *World J. Microb. Biot.* 23(9): p. 1209-1215, 2007.
137. Yamamura, R. and Shimomura, Y., Industrial high-performance liquid chromatography purification of docosahexaenoic acid ethyl ester and docosapentaenoic acid ethyl ester from single-cell oil. *J. Am. Oil Chem. Soc.* 74(11): p. 1435-1440, 1997.
138. Wang, X., Wang, X., Wang, W., Jin, Q., and Wang, X., Synthesis of docosapentaenoic acid-enriched diacylglycerols by enzymatic glycerolysis of *Schizochytrium* sp. oil. *Bioresour. Technol.* 262: p. 278-283, 2018.
139. Ishikawa, T., Fujiyama, Y., Igarashi, O., Morino, M., Tada, N., Kagami, A., . . . Nakamura, H., Effects of gammalinolenic acid on plasma lipoproteins and apolipoproteins. *Atherosclerosis.* 75(2-3): p. 95-104, 1989.

140. Huang, Y.S., Manku, M.S., and Horrobin, D.F., The effects of dietary cholesterol on blood and liver polyunsaturated fatty acids and on plasma cholesterol in rats fed various types of fatty acid diet. *Lipids*. 19(9): p. 664-672, 1984.
141. Nichols, B.W. and Wood, B.J., The occurrence and biosynthesis of gamma-linolenic acid in a blue-green alga, *Spirulina platensis. Lipids*. 3(1): p. 46-50, 1968.
142. Quoc, K.P. and Dubacq, J.P., Effect of growth temperature on the biosynthesis of eukaryotic lipid molecular species by the cyanobacterium *Spirulina platensis. Biochim. Biophys. Acta*. 1346(3): p. 237-46, 1997.
143. Miura, Y., Sode, K., Nakamura, N., Matsunaga, N., and Matsunaga, T., Production of γ-linolenic acid from the marine green alga *Chlorella* sp. NKG 042401. *FEMS Microbiol. Lett.* 107(2-3): p. 163-167, 1993.
144. Dennis, E.A. and Norris, P.C., Eicosanoid storm in infection and inflammation. *Nat. Rev. Immunol.* 15(8): p. 511-23, 2015.
145. Lapillonne, A., Groh-Wargo, S., Gonzalez, C.H., and Uauy, R., Lipid needs of preterm infants: updated recommendations. *J. Pediatr.* 162(3 Suppl): p. S37-47, 2013.
146. Cohen, Z., *Porphyridium cruentum*, in *Chemicals from microalgae*, Z. Cohen, Editor., Taylor and Francis, 1999.
147. Spolaore, P., Joannis-Cassan, C., Duran, E., and Isambert, A., Commercial applications of microalgae. *J. Biosci. Bioeng.* 101(2): p. 87-96, 2006.
148. Becker, E.W., Microalgae for Aquaculture: Nutritional Aspects, in *Handbook of Microalgal Culture: Applied Phycology and Biotechnology*, A. Richmond and Q. Hu, Editors., Wiley Blackwell, 2013.
149. Su, G., Jiao, K., Li, Z., Guo, X., Chang, J., Ndikubwimana, T., . . . Lin, L., Phosphate limitation promotes unsaturated fatty acids and arachidonic acid biosynthesis by microalgae *Porphyridium purpureum. Bioprocess Biosyst. Eng.* 39(7): p. 1129-1136, 2016.
150. Chang, J., Le, K., Song, X., Jiao, K., Zeng, X., Ling, X., . . . Lin, L., Scale-up cultivation enhanced arachidonic acid accumulation by red microalgae *Porphyridium purpureum. Bioprocess Biosyst. Eng.* 40(12): p. 1763-1773, 2017.
151. Du, Z., Alvaro, J., Hyden, B., Zienkiewicz, K., Benning, N., Zienkiewicz, A., . . . Benning, C., Enhancing oil production and harvest by combining the marine alga *Nannochloropsis oceanica* and the oleaginous fungus *Mortierella elongata. Biotechnol. Biofuels.* 11(1): p. 174, 2018.
152. Merzlyak, M.N., Chivkunova, O.B., Gorelova, O.A., Reshetnikova, I.V., Solovchenko, A.E., Khozin-Goldberg, I., and Cohen, Z., Effect of nitrogen starvation on optical properties, pigments and arachidonic acid content of the unicellular green alga *Parietochloris incisa* (Trebouxiophyceae, Chlorophyta). *J. Phycol.* 43(4): p. 833-843, 2007.
153. Khozin-Goldberg, I., Bigogno, C., Shrestha, P., and Cohen, Z., Nitrogen starvation induces the accumulation of arachidonic acid in the freshwater green alga *Parietochloris incisa* (Trebouxiophyceae). *J. Phycol* 38(5): p. 991-994, 2002.

154. Harel, M. and Place, A.R., Heterotrophic Production of Marine Algae for Aquaculture, in *Handbook of Microalgal Culture*, A. Richmond, Editor., Blackwell publishing. p. 513-524, 2004.
155. Menegol, T., Romero-Villegas, G.I., López-Rodríguez, M., Navarro-López, E., López-Rosales, L., Chisti, Y., . . . Molina-Grima, E., Mixotrophic production of polyunsaturated fatty acids and carotenoids by the microalga *Nannochloropsis gaditana. J. App. Phycol.* 31(5): p. 2823-2832, 2019.
156. Cohen, Z., Norman, H.A., and Heimer, Y.M., Microalgae as a source of ω3 fatty acids, in plants in human nutrition. *World Review of Nutrition and Dietetics.*, A.P. Simopoulos, Editor., Karger: Basel. p. 1-31, 1995.
157. Behrens, P.W. and Kyle, D.J., Microalgae as a source of fatty acids. *J. Food Lipids.* 3(4): p. 259-272, 1996.
158. Barclay, W. and Zeller, S., Nutritional enhancement of n-3 and n-6 fatty acids in rotifers and artemia nauplii by feeding spray-dried *Schizochytrium* sp. *J. World. Aquac. Soc.* 27(3): p. 314-322, 1996.
159. Blanken, W., Cuaresma, M., Wijffels, R.H., and Janssen, M., Cultivation of microalgae on artificial light comes at a cost. *Algal Res.* 2(4): p. 333-340, 2013.
160. Dubinsky, Z., Matsukawa, R., and Karube, I., Photobiological aspects of algal mass culture. *J. Mar. Biotechnol.*(2): p. 61-65, 1995.
161. Hu, Q., Guterman, H., and Richmond, A., A flat inclined modular photobioreactor for outdoor mass cultivation of photoautotrophs. *Biotechnol. Bioeng.* 51(1): p. 51-60, 1996.
162. Zittelli, G.C., Biondi, N., Rodolfi, L., and Tredici, M.R., Photobioreactors for Mass Production of Microalgae, in *Handbook of Microalgal Culture*. Wiley Blackwell. p. 225-266, 2013.
163. Borowitzka, M.A. and Moheimani, N.R., Open Pond Culture Systems, in *Algae for Biofuels and Energy*, M.A. Borowitzka and N.R. Moheimani, Editors., Springer Netherlands: Dordrecht. p. 133-152, 2013.
164. Clarens, A.F., Resurreccion, E.P., White, M.A., and Colosi, L.M., Environmental Life Cycle Comparison of Algae to Other Bioenergy Feedstocks. *Environ. Sci. Technol.* 44(5): p. 1813-1819, 2010.
165. Lardon, L., Hélias, A., Sialve, B., Steyer, J.-P., and Bernard, O., Life-cycle assessment of biodiesel production from microalgae. *Environ. Sci. Technol.* 43(17): p. 6475-6481, 2009.
166. Molina Grima, E., García Camacho, F., and Acién Fernández, F.G., Production of EPA from *Phaeodactylum tricornutum*, in *Chemicals from microalgae*, Z. Cohen, Editor., London: CRC Press, 1999.
167. Cohen, Z., *Monodus subterraneus*, in *Chemicals from microalgae*, Z. Cohen, Editor., Taylor and Francis, 1999.
168. Cohen, Z. and Heimer, Y.M., Production of polyunsaturated fatty acids (EPA, ARA and GLA) by the microalgae *Porphyridium* and *Spirulina*., in *Industrial applications of single cell oil*, D.J. Kyle and C. Ratledge, Editors., American Oil Chemists' Society, 1992.

169. Molina Grima, E., Sánchez Pérez, J.A., García Camacho, F., Fernández Sevilla, J.M., and Acién Fernández, F.G., Effect of growth rate on the eicosapentaenoic acid and docosahexaenoic acid content of *Isochrysis galbana* in chemostat culture. *Appl. Microbiol. Biotechnol.* 41(1): p. 23-27, 1994.
170. Hu, Q., Hu, Z., Cohen, Z., and Richond, A., Enhancement of eicosapentaenoic acid (EPA) and γ-linolenic acid (GLA) production by manipulating algal density of outdoor cultures of *Monodus subterraneus* (Eustigmatophyta) and *Spirulina platensis* (Cyanobacteria). *Eur. J. Phycol.* 32(1): p. 81-86, 1997.
171. Sukenik, A., Carmeli, Y., and Berner, Y., Regulation of fatty acid composition by growth irradiance level in the eustigmatophyte *Nannochloropsis* sp. *J. Phycol.* 25: p. 686-692, 1989.
172. Sukenik, A., Production of eicosapentaenoic acid by the marine eustigmatophyte Nannochloropsis, in *Chemicals from microalgae*, Z. Cohen, Editor., London: CRC Press, 1999.
173. Sukenik, A., Yamaguchi, Y., and Livne, A., Alterations in lipid molecular species of the marine Eustigmatophyte *Nannochloropsis* sp. *J. Phycol.* 29(5): p. 620-626, 1993.
174. Sukenik, A., Levy, R.S., Levy, Y., Falkowski, P.G., and Dubinsky, Z., Optimizing algal biomass production in an outdoor pond: a simulation model. *J. Appl. Phycol.* 3(3): p. 191-201, 1991.
175. Arad, S. and Richmond, A., Industrial production of microalgal cell-mass and secondary products - species of high potential: *Porphyridium* sp. in *Handbook of Microalgal Culture: biotechnology and applied phycology*. Blackwell publishing. p. 289-297, 2004.
176. Tanticharoen, M., Reungjitchachawali, M., Boonag, B., Vonktaveesuk, P., Vonshak, A., and Cohen, Z., Optimization of γ-linolenic acid (GLA) production in *Spirulina platensis. J. Appl. Phycol.* 6(3): p. 295-300, 1994.
177. Cheng-Wu, Z., Cohen, Z., Khozin-Goldberg, I., and Richmond, A., Characterization of growth and arachidonic acid production of *Parietochloris incisa* comb. nov (Trebouxiophyceae, Chlorophyta). *J. Appl. Phycol.*, 14(6): p. 453-460, 2002.
178. Sukenik, A. and Carmeli, Y., Lipid synthesis and fatty acid composition in *Nannochloropsis* sp. (Eustigmatophyceae) grown in a light-dark cycle. *J. Phycol.* 26(3): p. 463-469, 1990.
179. Janoska, A., Lamers, P.P., Hamhuis, A., van Eimeren, Y., Wijffels, R.H., and Janssen, M., A liquid foam-bed photobioreactor for microalgae production. *Chem. Eng. J.*, 313: p. 1206-1214, 2017.
180. Hulatt, C.J., Wijffels, R.H., Bolla, S., and Kiron, V., Production of fatty acids and protein by *Nannochloropsis* in flat-plate photobioreactors. *PLoS One.* 12(1): p. e0170440-e0170440, 2017.
181. Ruiz, J., Olivieri, G., de Vree, J., Bosma, R., Willems, P., Reith, J.H., . . . Barbosa, M.J., Towards industrial products from microalgae. *Energy Environ. Sci.* 9(10): p. 3036-3043, 2016.

182. Hu, Q., Sommerfeld, M., Jarvis, E., Ghirardi, M., Posewitz, M., Seibert, M., and Darzins, A., Microalgal triacylglycerols as feedstocks for biofuel production: perspectives and advances. *Plant J.* 54(4): p. 621-39, 2008.
183. Griffiths, M.J., van Hille, R.P., and Harrison, S.T.L., Lipid productivity, settling potential and fatty acid profile of 11 microalgal species grown under nitrogen replete and limited conditions. *J. Appl. Phycol.*, 24(5): p. 989-1001, 2012.
184. Remmers, I.M., Martens, D.E., Wijffels, R.H., and Lamers, P.P., Dynamics of triacylglycerol and EPA production in *Phaeodactylum tricornutum* under nitrogen starvation at different light intensities. *PLoS One*. 12(4): p. e0175630, 2017.
185. Jia, J., Han, D., Gerken, H.G., Li, Y., Sommerfeld, M., Hu, Q., and Xu, J., Molecular mechanisms for photosynthetic carbon partitioning into storage neutral lipids in *Nannochloropsis oceanica* under nitrogen-depletion conditions. *Algal Res.*, 7: p. 66-77, 2015.
186. Steinrucken, P., Mjos, S.A., Prestegard, S.K., and Erga, S.R., Enhancing EPA content in an arctic diatom: A factorial design study to evaluate interactive effects of growth factors. *Front. Plant Sci.* 9: p. 491, 2018.
187. Lynch, D.V. and Thompson, G.A., Low temperature-induced alterations in the chloroplast and microsomal membranes of *Dunaliella salina. Plant Physiol.* 69(6): p. 1369-75, 1982.
188. Sushchik, N.N., Kalacheva, G.S., Zhila, N.O., Gladyshev, M.I., and Volova, T.G., A temperature dependence of the intra- and extracellular fatty-acid composition of green algae and cyanobacterium. *Russ. J. Plant Phys.*, 50(3): p. 374-380, 2003.
189. Tatsuzawa, H. and Takizawa, E., Changes in lipid and fatty acid composition of *Pavlova lutheri. Phytochemistry*. 40(2): p. 397-400, 1995.
190. Adlerstein, D., Khozin, I., Bigogno, C., and Cohen, Z., Effect on environmental conditions on the molecular species Composition of Galactolipids in the Alga *Porphyridium Cruentum*, in *Physiology, Biochemistry and Molecular Biology of Plant Lipids*, J.P. Williams, M.U. Khan, and N.W. Lem, Editors., Springer Netherlands: Dordrecht. p. 218-220, 1997.
191. Bigogno, C., Khozin-Goldberg, I., and Cohen, Z., Accumulation of arachidonic acid-rich triacylglycerols in the microalga *Parietochloris incisa* (Trebuxiophyceae, Chlorophyta). *Phytochemistry*. 60(2): p. 135-43, 2002.
192. Khozin-Goldberg, I., Yu, H.Z., Adlerstein, D., Didi-Cohen, S., Heimer, Y.M., and Cohen, Z., Triacylglycerols of the red microalga *Porphyridium cruentum* can contribute to the biosynthesis of eukaryotic galactolipids. *Lipids*. 35(8): p. 881-9, 2000.
193. Cohen, Z., Vonshak, A., and Richmond, A., Effect of Environmental Conditions on Fatty Acid Composition of the Red Alga *Porphyridium Cruentum*: correlation to growth rate. *J. Phycol.*, 24(3): p. 328-332, 1988.
194. Iwamoto, H. and Sato, S., EPA production by freshwater unicellular algae. *J. Am. Oil Chem. Soc.*, 63(434), 1986.

195. James, C.M., Al-Hinty, S., and Salman, A.E., Growth and ω3 fatty acid and amino acid composition of microalgae under different temperature regimes. *Aquaculture*. 77(4): p. 337-351, 1989.
196. Seto, A., Kumasaka, K., Hosaka, M., Kojima, E., Kashiwakura, M., and Kato, T., Production of Eicosapentaenoic Acid by a Marine Microalgae and Its Commercial Utilization for Aquaculture, in *Industrial Applications of Single Cell Oils*, D.J. Kyle and C. Ratledge, Editors., AOCS Publishing, 1992.
197. Van Wagenen, J., Miller, T.W., Hobbs, S., Hook, P., Crowe, B., and Huesemann, M., Effects of Light and Temperature on Fatty Acid Production in *Nannochloropsis salina*. *Energies*. 5(3): p. 731-740, 2012.
198. Zhu, C.J., Lee, Y.K., and Chao, T.M., Effects of temperature and growth phase on lipid and biochemical composition of *Isochrysis galbana* TK1. *J. Appl. Phycol.*, 9(5): p. 451-457, 1997.
199. Jiang, H. and Gao, K., Effects of lowering temperature during culture on the production of polyunsaturated fatty acids in the marine diatom *Phaeodactylum tricornutum* (Bacillariophyceae). *J. Phycol.* 40(4): p. 651-654, 2004.
200. Renaud, S.M., Thinh, L.-V., Lambrinidis, G., and Parry, D.L., Effect of temperature on growth, chemical composition and fatty acid composition of tropical Australian microalgae grown in batch cultures. *Aquaculture*. 211(1): p. 195-214, 2002.
201. Torstensson, A., Hedblom, M., Andersson, J., Andersson, M.X., and Wulff, A., Synergism between elevated CO_2 and temperature on the Antarctic sea ice diatom *Nitzschia lecointei*. *Biogeosciences*. 10(10): p. 6391-6401, 2013.
202. Artamonova, E.Y., Vasskog, T., and Eilertsen, H.C., Lipid content and fatty acid composition of *Porosira glacialis* and *Attheya longicornis* in response to carbon dioxide (CO2) aeration. *PLoS One*. 12(5): p. e0177703, 2017.
203. Svenning, J.B., Dalheim, L., Eilertsen, H.C., and Vasskog, T., Temperature dependent growth rate, lipid content and fatty acid composition of the marine cold-water diatom *Porosira glacialis*. *Algal Research*. 37: p. 11-16, 2019.
204. Steinrucken, P., Erga, S.R., Mjos, S.A., Kleivdal, H., and Prestegard, S.K., Bioprospecting North Atlantic microalgae with fast growth and high polyunsaturated fatty acid (PUFA) content for microalgae-based technologies. *Algal Res.*, 26: p. 392-401, 2017.
205. Pasquet, V., Ulmann, L., Mimouni, V., Guihéneuf, F., Jacquette, B., Morant-Manceau, A., and Tremblin, G., Fatty acids profile and temperature in the cultured marine diatom *Odontella aurita*. *J. Appl. Phycol.*, 26(6): p. 2265-2271, 2014.
206. Schulze, P.S.C., Hulatt, C.J., Morales-Sánchez, D., Wijffels, R.H., and Kiron, V., Fatty acids and proteins from marine cold adapted microalgae for biotechnology. *Algal Res.*, 42: p. 101604, 2019.
207. Cohen, Z., Vonshak, A., and Richmond, A., Fatty acid composition of *Spirulina* strains grown under various environmental conditions. *Phytochemistry*. 26(8): p. 2255-2258, 1987.

208. Richmond, A., Biological Principles of Mass Cultivation, in *Handbook of Microalgal Culture*, A. Richmond, Editor., Blackwell publishing. p. 125-177, 2004.
209. Lee, Y.K. and Tan, H.M., Effect of temperature, light intensity and dilution rate on the cellular composition of red alga *Porphyridium cruentum* in light-limited chemostat cultures. *MIRCEN J. Appl. Microbiol. Biotechnol.*, 4(2): p. 231-237, 1988.
210. Grima, E.M., Pérez, J.A.S., Camacho, F.G., Fernández, F.G.A., Sevilla, J.M.F., and Sanz, F.V., Effect of dilution rate on eicosapentaenoic acid productivity of *Phaeodactylum tricornutum* utex 640 in outdoor chemostat culture. *Biotechnol. Lett.*, 16(10): p. 1035-1040, 1994.
211. Cohen, Z., Production potential of eicosapentaenoic acid by *Monodus subterraneus*. *JAOCS.*, 71(9): p. 941-945, 1994.
212. Sukenik, A., Carmeli, Y., and Berner, T., Regulation of fatty acid composition by growth irradiance level in the eustigmatophyte *Nannochloropsis* sp. *J. Phycol.*, 25(4): p. 686-692, 1989.
213. Fábregas, J., Maseda, A., Domínguez, A., and Otero, A., The cell composition of *Nannochloropsis* sp. changes under different irradiances in semicontinuous culture. *World J. Microb. Biot.*, 20(1): p. 31-35, 2004.
214. Renaud, S.M., Parry, D.L., Thinh, L.-V., Kuo, C., Padovan, A., and Sammy, N., Effect of light intensity on the proximate biochemical and fatty acid composition of *Isochrysis* sp. and *Nannochloropsis oculata* for use in tropical aquaculture. *J. Appl. Phycol.*, 3(1): p. 43-53, 1991.
215. Forde, J. and Steer, M.W., The use of quantitative electron microscopy in the study of lipid composition of membranes. *J. Exp Bot.*, 27(6): p. 1137-1141, 1976.
216. Solovchenko, A.E., Khozin-Goldberg, I., Didi-Cohen, S., Cohen, Z., and Merzlyak, M.N., Effects of light intensity and nitrogen starvation on growth, total fatty acids and arachidonic acid in the green microalga *Parietochloris incisa*. *J. Appl. Phycol.*, 20(3): p. 245-251, 2008.
217. Brown, M.R., Dunstan, G.A., Norwood, S.J., and Miller, K.A., Effect of harvest stage and light on the biochemical composition of the diatom *Thalassiosira pseudonana*. *J. Phycol.* 32(1): p. 64-73, 1996.
218. Kates, M. and Volcani, B.E., Lipid components of diatoms. *BBA-Lipid Lipid Met.* 116(2): p. 264-278, 1966.
219. Acién Fernández, F.G., García Camacho, F., Sánchez Pérez, J.A., Fernández Sevilla, J.M., and Molina Grima, E., Modeling of biomass productivity in tubular photobioreactors for microalgal cultures: Effects of dilution rate, tube diameter, and solar irradiance. *Biotechnol. Bioeng.* 58(6): p. 605-616, 1998.
220. Veloso, V., Reis, A., Gouveia, L., Fernandes, H.L., Empis, J.A., and Novais, J.M., Lipid production by *Phaeodactylum tricornutum*. *Bioresour. Technol.* 38(2): p. 115-119, 1991.

221. Shifrin, N.S. and Chisholm, S.W., Phytoplankton lipids: interspecific differences and effects of nitrate, silicate, and light-dark cycles. *J. Phycol.* 17(4): p. 374-384, 1981.
222. Ben-Amotz, A., Tornabene, T.G., and Thomas, W.H., Chemical profile of selected species of microalgae with emphasis on lipids. *J. Phycol.*, 21: p. 72-81, 1985.
223. Piorreck, M., Baasch, K.-H., and Pohl, P., Biomass production, total protein, chlorophylls, lipids and fatty acids of freshwater green and blue-green algae under different nitrogen regimes. *Phytochemistry*. 23(2): p. 207-216, 1984.
224. Cohen, Z., The chemicals of *Spirulina*., in *Spirulina platensis (Arthrospira): Physiology, Cell-biology and Biotechnology*, A. Vonshak, Editor., Taylor and Francis, 1997.
225. Janssen, J.H., Wijffels, R.H., and Barbosa, M.J., Lipid Production in *Nannochloropsis gaditana* during nitrogen starvation. *Biology*. 8(1), 2019.
226. Suzuki, H., Hulatt, C.J., Wijffels, R.H., and Kiron, V., Growth and LC-PUFA production of the cold-adapted microalga *Koliella antarctica* in photobioreactors. *J. Appl. Phycol.* 31(2): p. 981-997, 2019.
227. Wang, X.W., Liang, J.R., Luo, C.S., Chen, C.P., and Gao, Y.H., Biomass, total lipid production, and fatty acid composition of the marine diatom *Chaetoceros muelleri* in response to different CO2 levels. *Bioresour. Technol.*, 161: p. 124-30, 2014.
228. Sydney, E.B., Sturm, W., de Carvalho, J.C., Thomaz-Soccol, V., Larroche, C., Pandey, A., and Soccol, C.R., Potential carbon dioxide fixation by industrially important microalgae. *Bioresour. Technol.*, 101(15): p. 5892-5896, 2010.
229. Renaud, S.M. and Parry, D.L., Microalgae for use in tropical aquaculture II: Effect of salinity on growth, gross chemical composition and fatty acid composition of three species of marine microalgae. *J. Appl. Phycol.*, 6(3): p. 347-356, 1994.
230. Yen, H.-W., Hu, I.C., Chen, C.-Y., Ho, S.-H., Lee, D.-J., and Chang, J.-S., Microalgae-based biorefinery – From biofuels to natural products. *Bioresour. Technol.* 135: p. 166-174, 2013.
231. D'Alessandro, E.B. and Antoniosi Filho, N.R., Concepts and studies on lipid and pigments of microalgae: A review. *Renew. Sust. Energ. Rev.* 58: p. 832-841, 2016.
232. Fu, W., Nelson, D.R., Mystikou, A., Daakour, S., and Salehi-Ashtiani, K., Advances in microalgal research and engineering development. *Curr. Opin. Biotechnol.* 59: p. 157-164, 2019.
233. Daboussi, F., Leduc, S., Marechal, A., Dubois, G., Guyot, V., Perez-Michaut, C., . . . Duchateau, P., Genome engineering empowers the diatom *Phaeodactylum tricornutum* for biotechnology. *Nat. Commun.* 5: p. 3831, 2014.
234. Serif, M., Dubois, G., Finoux, A.-L., Teste, M.-A., Jallet, D., and Daboussi, F., One-step generation of multiple gene knock-outs in the diatom Phaeodactylum tricornutum by DNA-free genome editing. *Nat. Commun.* 9(1): p. 3924, 2018.

235. Nymark, M., Sharma, A.K., Sparstad, T., Bones, A.M., and Winge, P., A CRISPR/Cas9 system adapted for gene editing in marine algae. *Sci. Rep.* 6: p. 24951, 2016.
236. Bhatt-Wessel, B., Jordan, T.W., Miller, J.H., and Peng, L., Role of DGAT enzymes in triacylglycerol metabolism. *Arch. Biochem. Biophys.* 655: p. 1-11, 2018.
237. Niu, Y.F., Zhang, M.H., Li, D.W., Yang, W.D., Liu, J.S., Bai, W.B., and Li, H.Y., Improvement of neutral lipid and polyunsaturated fatty acid biosynthesis by overexpressing a type 2 diacylglycerol acyltransferase in marine diatom *Phaeodactylum tricornutum. Mar. Drugs.* 11(11): p. 4558-69, 2013.
238. Li, D.W., Cen, S.Y., Liu, Y.H., Balamurugan, S., Zheng, X.Y., Alimujiang, A., . . . Li, H.Y., A type 2 diacylglycerol acyltransferase accelerates the triacylglycerol biosynthesis in heterokont oleaginous microalga *Nannochloropsis oceanica. J. Biotechnol.* 229: p. 65-71, 2016.
239. Wei, H., Shi, Y., Ma, X., Pan, Y., Hu, H., Li, Y., . . . Liu, J., A type-I diacylglycerol acyltransferase modulates triacylglycerol biosynthesis and fatty acid composition in the oleaginous microalga, *Nannochloropsis oceanica. Biotechnol. Biofuels.* 10: p. 174, 2017.
240. Wagner, M., Hoppe, K., Czabany, T., Heilmann, M., Daum, G., Feussner, I., and Fulda, M., Identification and characterization of an acyl-CoA:diacylglycerol acyltransferase 2 (DGAT2) gene from the microalga *O. tauri. Plant Physiol. Biochem.* 48(6): p. 407-16, 2010.
241. Manandhar-Shrestha, K. and Hildebrand, M., Characterization and manipulation of a DGAT2 from the diatom *Thalassiosira pseudonana*: Improved TAG accumulation without detriment to growth, and implications for chloroplast TAG accumulation. *Algal Res.*, 12: p. 239-248, 2015.
242. Xin, Y., Lu, Y., Lee, Y.-Y., Wei, L., Jia, J., Wang, Q., . . . Xu, J., Producing designer oils in industrial microalgae by rational modulation of co-evolving type-2 diacylglycerol acyltransferases. *Molecular Plant.* 10(12): p. 1523-1539, 2017.
243. Niu, Y.F., Wang, X., Hu, D.X., Balamurugan, S., Li, D.W., Yang, W.D., . . . Li, H.Y., Molecular characterization of a glycerol-3-phosphate acyltransferase reveals key features essential for triacylglycerol production in *Phaeodactylum tricornutum. Biotechnol. Biofuels.* 9: p. 60, 2016.
244. Poliner, E., Pulman, J.A., Zienkiewicz, K., Childs, K., Benning, C., and Farre, E.M., A toolkit for Nannochloropsis oceanica CCMP1779 enables gene stacking and genetic engineering of the eicosapentaenoic acid pathway for enhanced long-chain polyunsaturated fatty acid production. *Plant Biotechnol. J.*, 16(1): p. 298-309, 2018.
245. Yoshino, T., Kakunaka, N., Liang, Y., Ito, Y., Maeda, Y., Nomaguchi, T., . . . Tanaka, T., Production of ω3 fatty acids in marine cyanobacterium *Synechococcus* sp. strain NKBG 15041c via genetic engineering. *Appl. Microbiol. Biotechnol.*, 101(18): p. 6899-6905, 2017.
246. Levitan, O., Dinamarca, J., Zelzion, E., Lun, D.S., Guerra, L.T., Kim, M.K., . . . Falkowski, P.G., Remodeling of intermediate metabolism in the diatom

Phaeodactylum tricornutum under nitrogen stress. *Proc. Nat. Am. Soc.* 112(2): p. 412-417, 2015.

247. Breckenridge, W.C., Marai, L., and Kuksis, A., Triglyceride structure of human milk fat. *Can. J. Biochem.* 47(8): p. 761-9, 1969.
248. Batista, F.R.M., Lucchesi, K.W., Carareto, N.D.D., Costa, M.C.D., and Meirelles, A.J.A., Properties of microalgae oil from the species *Chlorella protothecoides* and its ethylic biodiesel. *Braz. J. Chem. Eng.*, 35(4): p. 1383-1394, 2018.
249. Danielewicz, M.A., Anderson, L.A., and Franz, A.K., Triacylglycerol profiling of marine microalgae by mass spectrometry. *J. Lipid Res.* 52(11): p. 2101-8, 2011.
250. van Erp, H., Bryant, F.M., Martin-Moreno, J., Michaelson, L.V., Bhutada, G., and Eastmond, P.J., Engineering the stereoisomeric structure of seed oil to mimic human milk fat. *Proc Natl Acad Sci USA*. 116(42): p. 20947-20952, 2019.

3

Production of PUFAs as Dietary and Health Supplements from Oleaginous Microalgae Utilizing Inexpensive Renewable Substrates

Dimitra Karageorgou, Georgios Bakratsas and Petros Katapodis*

Laboratory of Biotechnology, Department of Biological Applications and Technologies, University of Ioannina, Ioannina, Greece

Abstract

Polyunsaturated fatty acids (PUFAs) and especially omega-3 and omega-6 have gained great interest as they are used as health and nutrition supplements for humans. It has been proved that they contribute to the prevention of several diseases like cardiovascular problems and diabetes, the inhibition of cancer tumors and depression, and they exhibit anti-inflammatory and antioxidants properties. In recent years, attention has been focused on the search for new sources of PUFAs, as the human organism isn't able to synthesize them in necessary quantities. Microalgae are a promising candidate for it. They can grow up in a wide range of environments and accumulate specific bioactive compounds in a way dependent on different factors like the genetics of species, their growth phase, the nutrient availability. Their lipid content can reach up to 85% of their biomass under heterotrophic conditions, making them oleaginous microorganisms. The benefits of PUFAs on health, the use of microalgae as lipid source and the factors that affect their productivity are presented in this chapter. Moreover, the search for new alternative, environmentally sustainable carbon sources that can be used as substrates for microalgae growth and be according to biorefinery concept, is a crucial issue and is also analyzed.

Keywords: Microalga, oleaginous, PUFAs, heterotrophic cultures, renewable substrates, omega-3, omega-6

**Corresponding author*: pkatapo@gmail.com

Alok Kumar Patel and Leonidas Matsakas (eds.) Nutraceutical Fatty Acids from Oleaginous Microalgae: A Human Health Perspective, (77–114) © 2020 Scrivener Publishing LLC

3.1 Introduction

Over the last few years, microalgae have received great attention because they are a potential source of important bioactive compounds with commercial uses. They are able to synthesize and produce polyunsaturated fatty acids (PUFAs), pigments, vitamins, saturated and monounsaturated fatty acids, and carbohydrates [1]. Some of them find applications in human health and diet and some others in bioengineering, as they can be used for biodiesel production [2].

One of the most important bioactive compounds that microalgae are able to produce are the PUFAs. PUFAs are fatty acids containing two or more cis or trans double bonds with different length of the aliphatic chain. Two main categories of PUFAs are omega-3 (or n-3) and omega-6 (or n-6). The first category is defined by having the first double bond with three carbons from the methyl terminal and is synthesized by linolenic acid, while the second category by having to the bond six carbons and is synthesized by linoleic acid respectively [3–6]. The most important omega-3 PUFAs are the docosahexaenoic acid (DHA, C22:6), eicosapentaenoic acid (EPA, C20:5) and a-linolenic acid (ALA or C18:3n-3). Omega-6 PUFAs contain g-linolenic acid (GLA, C18:3n-6), linoleic acid (LA) and arachidonic acid (ARA, C20:4n-6) [7–9].

PUFAs show great potential to many applications. One of them is their role in medicine, as they contribute to the prevention of cardiovascular diseases and diabetes, the inhibition of cancer tumors and depression, and they exhibit anti-inflammatory and antioxidants properties [10, 11]. In addition, some PUFAs aren't synthesized by mammals while others like the ω-3 PUFAs (EPA and DHA) are produced by humans in small quantities. As a result, there is a gap between the rate of biosynthesis and the demands of the organisms [5, 7] and the need for new PUFAs producers seems to be necessary. The microorganisms that are able to accumulate more than 20% w/w of their biomass as lipid are referred to as oleaginous and they include microalgae, bacteria, yeast and fungi [12].

So, algae are an appropriate candidate for PUFAs production, as they are able to accumulate lipids as intracellular compounds and these lipids can be characterized as single cell oils (SCO) [7]. Moreover, their lipid content varies from 20 to 50% w/w [10]. Under specific environmental or nutrient stress conditions, like nutrient/nitrogen starvation, high salinity or temperature [13], they can produce up to 85% w/w of lipids [10, 14].

Algae have been estimated to include from 30,000 to 1 million species, while there are scenarios suggesting that there are up to 350 million, most of them still undiscovered [15]. They constitute one of the most diverse organisms on earth, being the result of symbiosis and evolutionary selection [16]. Microalgae consist of unicellular, photosynthetic organisms with fast growing ability [17]. They can grow up in a wide range of environments and accumulate specific bioactive compounds in a way dependent on different factors like the genetics of species, their growth phase and the nutrient availability [2]. Microalgae are able to produce these great amounts of oils [18] in a more efficient way than plants [14] while they don't require fertile land, which is one of the most important problems [10]. They are in principal autotrophic microorganisms, but some species are able to grow through a heterotrophic and mixotrophic metabolic way [13]. So, they can even be cultivated in agro industrial and domestic wastewaters, replacing the expensive carbon substrates and providing bioremediation in parallel with lipid production [12, 19–22]. It has been mentioned that heterotrophic cultivation of microalgae leads to greater biomass production, with higher cell densities and to accumulation of larger lipid quantities if it is compared with autotrophic cultivation [22, 23].

For the above reasons and their health and dietary advantages, there is a high interest for their production of different sources and for their consumption. Microalgae species are one of the best candidates and the use of renewable substrates gives them one more advantage.

The present chapter summarizes the role of PUFAs as dietary and health supplements, the pathways through which they are produced from the microorganism, the different factors that play a crucial role in their production, accumulation and extraction. Finally, the use of renewable substrates are referred to as a way to replace the usual expensive carbon sources as well as to some future perspectives.

3.2 PUFAs as Dietary and Health Supplements

Lipids are one of the necessary bioactive compounds of tissues and membranes. They constitute a group with great heterogeneity and their main compounds are the atoms of carbon, hydrogen and oxygen [24]. Based on the presence and the number of double bonds, they can be characterized as saturated, monounsaturated or polyunsaturated. Lipids are amphiphilic molecules [24] with a hydrophobic tail and a hydrophilic head, a characteristic that makes them part of membranes and gives them the properties that are described above. Moreover, they are present in nature in many forms, having the ability of conversion between them [24].

Two main categories of polyunsaturated fatty acids are the omega-3 and omega-6 FAs. Omega-3 FAs are characterized by the presence of the first double bond, three carbon atoms away from the methyl terminal, while omega-6 FAs have this bond six carbon atoms away [25]. The maintenance of ratio balance between omega-3: omega-6 PUFAs contributes to a healthy metabolism and to the prevention of diseases and disorders. As PUFAs cannot be synthesized by the human organism, they should be acquired through diet. The most usual sources of omega-6 FAs are the vegetables, while for the omega-3 FAs are eggs, meat and fish oil. The most important omega-3 PUFAs are the docosahexaenoic acid (DHA, C22:6), eicosapentaenoic acid (EPA, C20:5) and a-linolenic acid (ALA or C18:3n-3). Omega-6 PUFAs contain g-linolenic acid (GLA, C18:3n-6), linoleic acid (LA) and arachidonic acid (ARA, C20:4n-6) [7–9].

Omega-3 and omega-6 polyunsaturated fatty acids are linked with a variety of health benefits against dietary disorders and diseases. They are necessary compounds for the growth, development and health of many systems of the human organism. However, the most common diets are poor in omega-3 lipids, especially EPA (eicosapentaenoic acid) and DHA (docosahexaenoic acid), while they are rich in omega-6 PUFAs and especially LA (linoleic acid) and ARA (arachidonic acid), something that makes important their enrichments with the first kind of lipids.

The majority of the above lipids cannot be synthesized by mammals. Omega-3 lipids are provided as food supplements from fish, eggs and animal sources, while omega-6 from vegetables. The recommended ratio for ω-3: ω-6 in the human organism should be greater than 10 [3]. In this case, PUFAs have beneficial effects on the human organism. On the other hand, when some lipids of the omega-6 FAs category overcome the omega-3 FAs percentage to a great extent, there is an increase in some pathogenic conditions. As a result, a balance in this ratio seems to be critical for a healthy and balanced life [3].

PUFAs are characterized by the long length of their chain, as well as the number and position of their double bonds, which give them their unique properties [6]. ARA and DHA are compounds of the phospholipid membrane of brain and other tissues. As a result, they have a significant role in blood clotting, in cell signaling, in immune and inflammatory responses [26], while free forms of ARA contribute to the nervous and skeletal system [26]. Moreover, omega-3 and especially EPA and DHA influence cell and tissue responses to external signals. The composition of the membranes changes and that affects their fluidity. Many proteins related with receptors or transports are influenced. As a result, the signaling pathways, gene expressing, and transition factors are affected [27].

It is important to mention that long PUFAs provide protection against cardiovascular diseases and rheumatoid arthritis [28–31]. More specifically, omega-3 PUFAs lead to the decrease of blood pressure, concentration of triglyceride and dyslipidemia. They also affect the cardiac rhythm and the pathways of the arterial walls [6, 27, 28, 31, 32]. Moreover, EPA and DHA decrease the development of atherosclerosis, as they have the ability to decrease the percentage of cholesterol in arteries and to inhibit the present of atrial fibrillation [28, 31]. It is observed that populations that receive the appropriate amounts of omega-3 deal with a minor risk of a stroke episode [6, 28].

The importance of omega-3 FAs supplementation in pregnant and lactating women has been mentioned, as they contribute to the growth of infants and even to long-term neurodevelopment [33]. Omega-3 and omega-6 FAs act at the neuronal process. DHA is the major compound the PUFAs located in nervous system and in membranes of brain and retina [33, 34]. The accumulation and act of DHA at brain starts from the first phases of fetus formation and it is continued in a rate depended way as it grows up [35]. Especially during the second half of pregnancy, accumulation of DHA in cortex and retina is high [36]. In adults, DHA prevents the brain and the central nervous system from damages and nervous losses, something that is related to neurological aging and neurodegeneration diseases like Parkinson's and Alzheimer's [6, 37–39]. Moreover, DHA is the major compound of membranes of the synaptic site [38]. So, it affects the microenvironment of the cells and acts to the neurotransmission. As a result, DHA is necessary over a lifetime, as it contributes to neurogenesis, neurotransmission, neurodegeneration, and protection against oxidation stress and it has the ability to preserve the normal functions of the central nervous system and brain [37, 40].

DHA has also a structural role in the eyes and more specifically in the retina. It is a component of the photoreceptors and it also contributes to the regulation of rhodopsin, to the visual signaling pathways and to the reduction of oxidative stress [27, 35, 37, 41]. PUFAs in coordination with gangliosides, that are sialic acid with glycosphingolipids and exist in all membranes, assist the creation of an enhanced retina with great light adaptation [42].

Moreover, EPA and DHA play an important role in behavior and neurological diseases, inhibiting the risk of suicide and depression [6, 43]. It is proved that low levels of omega-3 fatty acids or/and high ratio of omega-6: omega-3 fatty acids are related to depression [43]. Moreover, deficiency of omega-3 during critical periods of life, may affect the development, the connection, the proper apoptosis of neuronal cells and their pathways, something that is related to psychiatric illnesses. They also affect neurotransmission as it is reported above [44]. EPA in collaboration with the suitable treatment lead to relief of depression symptoms, as they contribute to brain changes [45].

It has been shown that PUFAs have a synergetic effect in cancer therapy, enhancing the efficiency of chemotherapy and/or radiotherapy [46, 47]. DHA and EPA enhance the sensitivity of tumor cells to anticancer drugs leading to their apoptotic death, while they don't affect the non-cancer cells [48, 49]. In addition, DHA seems to have antimetastatic properties, contributing in this way to one of the most important aspects of cancer therapy [50]. Moreover, omega-3 appear to have anti-inflammatory properties [51], which sets a barrier to cancer progression, as patients suffer from metabolic alternations, inflammations and suppression of their immune system [46]. Omega-3 FAs have positive effects against some type of cancers, like lung, skin, stomach, breast and colon cancer. On the other hand, a high percentage of omega-6 and especially ARA, enhance the growth of some cancerous tumors, like colon [52].

Given the great number of effects that PUFAs and especially these of omega-3 FAs present, it is clear that they are able to regulate the human metabolisms and play a role in health maintenance. Moreover, a ratio of omega-3: omega-6 fatty acids is necessary to minimize their harsh effects and keep only the benefits. Last but not least, as most of these lipids aren't synthesized by the human organism, dietary supplements seem to be necessary. Below, we will discuss some new forms of PUFAs sources.

3.3 Microalgae as Source of PUFAs

As noted above, there is a gap between the demand for PUFAs and their sources. In the last few years, more and more sources for these lipids have gained ground on nutrition. These sources are basically organisms, like fungi, bacteria, protozoa and microalgae [25]. The type and quantity of lipids is varying amidst the organisms and even the species. Some species among these categories belong to the oleaginous organisms, as they are able to accumulate up to 70% of lipids in their biomass [53]. The oils that are produced from these oleaginous microorganisms are referred to as single cell oils [54]. So, algae are among the oleaginous microorganisms, as they are able to accumulate single cell oils (SCO) as intracellular compounds [7]. Their lipid content varies from 20 to 50% w/w [10], while under specific conditions, they can reach up to 85% w/w of lipids [10, 14].

Microalgae are a group of microorganisms with great variety and high specialization [55]. They are able to survive under many conditions and in many different habitats [55–58]. Moreover, their metabolic and photosynthetic mechanisms are very complex and appear to have many similarities with those of plants [56, 59]. Microalgae are part of four of the six supergroups of eukaryotic life, which are Archaeplastida, Rhizaria, Chromalveolata, and Excavata [60, 61]. Each of these supergroups is related with special characteristics and is separated into phylums, which can be divided into classes. The Archaeplastida supergroup is important for industry species. Rhodophyta is a phylum of this supergroup and it contains primary marine species with spherical cells including an eccentric nucleus and a large chloroplast. Its most famous members are *Isochrysis galbana* and *Pavlova salina* [62]. Chlorophyta include unicellular and multicellular representatives that contain chlorophyll a and b in a single chloroplast, while they can be found in freshwater, marine and terrestrial habitats. Some well-known members of this phylum are *Chlorella vulgaris*, *Dunaliela salina*, and *Haematococcus pluvialis* [63]. The second basic supergroup is the Chromalveolata, which is split into many phylums and classes and contains a plethora of microalgal species that are used in the food market [60, 61]. It is a wide supergroup, containing multicellular and unicellular species. Moreover, some of these microalgae grow in freshwater ecosystems, others on marine habitants, while some others on terrestrial environments. Some of the most used species are *Nannochloropsis oculata*, *Skeletonema costatum*, *Chaetoceros muelleri*, and *Thalassiosira psudonana* [64]. It is worth mentioning that this group contains species that are able to grow under heterotrophic conditions. For example, Auranthiochytrium, Schizotrichium, Thraustochytrium, and Ulkenia are of commercial interest heterotrophic filamentous organisms [65]. Moreover, *Crypthecodinium cohnii* of dinoplyta phylum, is a heterotrophic microalga that produces DHA. Half percentage of Dinophyta are heterotrophs with a lack of chloroplasts while the others are photosynthetic [66].

Microalgae are able to produce many important high-value biological compounds like proteins, pigments, carbohydrates, lipids [57, 58]. They are among the oleaginous organisms as they are able to produce and accumulate great amounts of lipids [55]. They accumulate them both in the chloroplast and in other parts of the cell in the form of globules [59]. They are able to produce both polar lipids, that are basically the PUFAs and have structural properties and non-polar lipids, that are mainly the triacylglycerols (TAGs) and are storage lipids [2, 10]. In addition, they are able to grow up under autotrophic, heterotrophic and mixotrophic conditions. When they are under the presence of carbon source, they change their metabolic pathways as to consume glucose or any other carbon source and produce lipids and not carbohydrates [55, 59]. The composition of lipids is also related except of the microalgal specie and of the growth phase of every specie of microalga [67]. Table 3.1 represents some examples of microalgal cells and diatoms (Coscinodiscophyceae) under autotrophic conditions and Thraustochytriaceae under heterotrophic conditions that are among the proposed alternative microorganism for PUFAs production. As shown in Table 3.1, the content of the total lipids, the PUFAs and their compounds present differentiation. These differentiations are dependent on the cultural conditions, the classes of microalgae, the species and the strain, and they are discussed below.

The pathways of omega-3 and omega-4 fatty acids appear diverse differences among organisms and species, because of the metabolic differences [10, 82]. But in all organisms, their synthesis requires enzymes, which are called desaturases and have the ability to create double bonds at carbon atoms. Moreover, another category of enzymes is needed that is called elongases and cause the extension of the lipid chain [59, 82, 83]. The human organism doesn't have these enzymes and more especially it has a lack of Δ-12 desaturase or has it in slight amounts. As a result, it is not able to produce linoleic acid (LA: C18:2ω-6) and a-linolenic acid (ALA: C18:3ω-3) [24, 25, 84]. These two molecules are the precursor molecules in the omega-6 and omega-3 FA pathways and the production of arachidonic acid (AA or ARA; C20:4ω-6), eicosapentaenoic acid (EPA; C20:5ω-3) and docosahexaenoic acid (DHA; C22:6ω-3), that make up the important bioactive derivatives [6].

In microalgae, the production of ARA (arachidonic acid), EPA (eicosapentanoic acid), DHA (docosahexaenoic acid) starts from LA and ALA fatty acids. These unsaturated fatty acids are produced from the addition of double bonds to saturated fatty acids. Fatty acids are produced by a primary pathway that uses the acetyl-CoA as source and a complex of multiple synthases enzymes [12, 85]. The lipids are produced by amino acid and carbohydrate metabolism [84]. The conversion of saturated to unsaturated fatty acids is achieved via enzymes that are called desaturases. The responsible enzymes for the production of ALA are the Δ9, Δ12 and Δ15 desaturases [53, 86, 87], while for the production of LA the last desaturase isn't necessary. Starting from stearic acid (C18:0), Δ9 desaturase leads to the production of oleic acid (C18:1), that is further desaturated to linoleic acid (LA; C18:2) by Δ12 desaturase. Finally, Δ15 desaturase produce the C18:3 that constitute the α- linolenic acid (ALA; C20:3) [53, 87]. The Δ12 and Δ15 consist methyl-end desaturases and have the ability to add the double bonds between preexisting double bonds and the methyl-end of the fatty acid [83].

Table 3.1 Fatty acid composition of lipid from different species of microalgae.
Dashes: indicates fatty acid not detected, nd: indicates not determined, results are rounded to one decimal place.

		% Fatty acids							
Class	**Specie**	**18:3 ω-3 (ALA)**	**18:3 ω-6 (GLA)**	**20:4 ω-6 (ARA)**	**20:5 ω-3 (EPA)**	**22:6 ω-3 (DHA)**	**Other PUFA**	**Total PUFA**	**Ref.**
Prymnesiophyceae	*Palova gyran*	–	2.0	3.7	34.2	16.4	1.2	58.2	[68]
	Isocrysis galbana	9.5	6.4	3.4	10.3	33.6	0.8	64.5	[68]
Dinophyceae	*Prorocentrum minimum*	1.3	0.2	13.8	14.0	28.1	3.9	62.3	[68]
Bacillariophyceae	*Phaeodactylum tricomutum*	1.1	0.4	1.0	34.1	1.0	1.4	42.5	[69]
Eustigmatophyceae	*Nannochloropsis oculata*	nd	nd	9.7	21.3	nd	–	36.4	[70]
	Nannochloris atomus	nd	21.7	0.5	3.2	nd	33.6	59.0	[71]
Chlorophyceae	*Chlorella vulgaris*	40.3	4.0	0.4	–	–	–	59.4	[18]
Cyanophyceae	*Spirulina platensis*	nd	23.4	nd	nd	nd	–	42.2	[72]
Chlorodendrophyceae	*Tetraselmis suecica*	0.7	11.1	1.5	4.3	nd	40.5	59.5	[71]
Cryptophyceae	*Oocystis sp.*	30.8	nd	1.9	4.2	–	–	63.9	[73]
	Chroomonas salina	2.6	14.2	0.9	11.9	5.2	33.1	67.9	[71]

(Continued)

Table 3.1 Fatty acid composition of lipid from different species of microalgae.
Dashes: indicates fatty acid not detected, nd: indicates not determined, results are rounded to one decimal place. (*Continued*)

		% Fatty acids							
Class	**Specie**	**18:3 ω-3 (ALA)**	**18:3 ω-6 (GLA)**	**20:4 ω-6 (ARA)**	**20:5 ω-3 (EPA)**	**22:6 ω-3 (DHA)**	**Other PUFA**	**Total PUFA**	**Ref.**
Thraustochytriaceae	*Schizochytrium sp.*	nd	nd	nd	nd	20	nd	60	[74]
	Schizochytrium sp. S31	0.07	0.2	0.6	1.5	29.9	18.1	50.9	[75]
	Schizochytrium sp. LU310	nd	nd	nd	nd	27.5	nd	67.4	[76]
	Schizochytrium limacinum OUC88	0.5	0.3	0.4	0.7	37.5	35.7	75.1	[77]
	Aurantiochytrium sp. TC9	nd	nd	nd	nd	54.3	nd	nd	[78]
	Aurantiochytrium sp. TC20	nd	nd	nd	nd	55.2	nd	nd	[78]
	Aurantiochytrium limacinum	nd	nd	nd	nd	57.3	nd	66.6	[79]
	Aurantiochytrium sp. SD116	nd	nd	0.91	0.81	45.52	nd	53.8	[80]

(*Continued*)

Table 3.1 Fatty acid composition of lipid from different species of microalgae.
Dashes: indicates fatty acid not detected, nd: indicates not determined, results are rounded to one decimal place. (*Continued*)

		% Fatty acids							
Class	**Specie**	**18:3 ω-3 (ALA)**	**18:3 ω-6 (GLA)**	**20:4 ω-6 (ARA)**	**20:5 ω-3 (EPA)**	**22:6 ω-3 (DHA)**	**Other PUFA**	**Total PUFA**	**Ref.**
	Aurantiochytrium limacinum	nd	nd	0.2	21.4	0.9	13.6	36.2	[81]
	Aurantiochytrium mangrovei	nd	nd	0.4	1.3	35.8	11.8	49.3	[81]
	Parietichytrium sp.	nd	nd	5.4	8.2	22.9	18.4	54.9	[81]
Coscinodiscophyceae	*Chaetoceros calcitrans*	nd	0.3	5.7	11.1	0.8	15.7	33.7	[71]
	Chaetoceros gracilis	nd	1.1	6.2	5.7	0.4	11.4	24.8	[71]
	Skeletonema costatum	0.3	0.3	nd	6.0	2.0	17.5	26.1	[71]
	Thalassiosira pseudonana	0.2	0.1	0.3	19.3	3.9	28.1	52.6	[71]

Once produced the initial polyunsaturated fatty acids, which are the linoleic acid (LA; C18:2) and α- linolenic acid (ALA; C18:3), desaturases and elongases start acting to the omega-3 and omega-6 pathways. On omega-6 pathway, as shown in picture 3.1, LA is firstly desaturated by Δ6 desaturase to γ-linolenic acid (GLA; C18:3), which is elongated to form of dihomo-γ-linolenic acid (DGLA; C20:2), with the assistance of an endoplasmic reticulum elongation enzyme system, which involves 2 initiating 3-keto acyl-CoA synthases (ELOVL2 and ELOVL5). After this step, one more double bond is added by Δ5 desaturase and arachidonic acid (ARA; C20:4) is produced.

On the omega-3 pathway, as shown in 3.1, starting from ALA, the act of Δ6 desaturase produces the stearidonic acid (C18:4). It is followed by an endoplasmic reticulum elongation enzyme system, which involves 2 initiating 3-keto acyl-CoA synthases (ELOVL2 and ELOVL5) that produce the eicosatrienoic acid (C20:3). Finally, Δ5 desaturase leads to the production of eicosapentanoic acid (EPA; C20:5). For the production of docosahexaenoic acid (DHA; C22:6), two more steps are necessary. EPA is elongated to docosapentaenoic acid (DPA; C22:5) by the same ER elongation enzyme system as in the production of eicosatrienoic acid step and then a double bond is introduced by D4 desaturase to yield DHA, while peroxisomal β-oxidation removes the 2-carbon acetyl-CoA from the 24:6 fatty acid called tetracosahexaenoic acid [25, 86]. The Δ4, Δ5, Δ6 desaturases are in

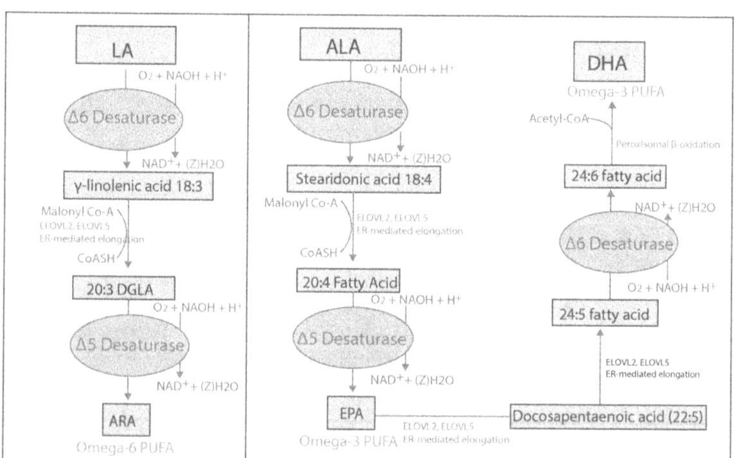

Picture 3.1 On the left: Basic Omega-6 PUFAs metabolic pathway, starting from LA and resulting to ARA with the use of elongases and desaturase. On the right: Basic Omega-3 PUFAs metabolic pathway, starting from ALA and resulting to DHA. They are presented with the intermediate lipids and the enzymes of each step.

front-end category and they add double bonds between other pre-existing and the carboxyl-end [83].

The synthesis and the accumulation of the different kinds of lipids takes place in different compounds of the cell. In most species of microalgae, PUFAs' location is basically the endoplasmic reticulum [82]. This process constitutes an aerobic pathway, as oxygen is necessary [53, 83, 87] and it takes place at the endoplasmic reticulum. Only the first part of the production of stearic acid and the conversation to oleic acid occurs at the chloroplast [87].

Microalgae are able to produce high amounts of PUFAs, under autotrophic, heterotrophic and mixotrophic conditions of cultivation. Environmental and/or nutrient parameters like the starvation, the oxygen supply, temperature etc. lead to changes in metabolic processes of the cell and are able to influence this production, the accumulation and the composition of the lipids [53, 55, 59, 88]. The microalgae have the ability to change their biomass composition after changes in the environment such as the presence of stress conditions [53]. On the other hand, the increase of biomass needs ideal growth conditions. As a result, there is a bottleneck between the biomass and lipid production [89], that may be overcome with the selection of two stages cultivations [55]. Moreover, the selection of the most suitable specie is a major parameter, while the metabolic engineering provides new tools to this direction [55, 82].

3.4 Systems for Microalgal Cultivation

As has been reported above, microalgae are basically photoautotrophic microorganisms. Some of them have the ability to grow under a variety of conditions and in different environments. They are able to be cultivated under autotrophic, heterotrophic and/or mixotrophic conditions, producing high-value products like PUFAs [90]. Some species are also capable of adapting to different conditions, changing their metabolism and morphology if necessary [91]. Microalgae are also known to synthesize great amounts of lipids, as they have similarities with the metabolic system of the plants, but they also have rapid growth rates [92]. This adaption to different environments is related to changes in lipid synthesis, production and accumulation [92].

Under autotrophic growth conditions, the microalgae use atmospheric carbon in the inorganic form of CO_2 and light as carbon and energy source, respectively. At heterotrophic conditions, the microorganisms use an exogenous carbon source. Whereas, at mixotrophic conditions, they use both energy sources and both photosynthetic and respiratory mechanisms at the same time [90, 93, 94], decreasing the rate of the first mechanism and increasing the rate of the second one [95, 96].

Under autotrophic conditions, the microalgal cells cannot reach a high biomass production as the light is a restrictive parameter. As the cell density is increased, the phenomenon of shading appears, and the light isn't able to penetrate to the culture as well [97]. Under heterotrophic and mixotrophic conditions, the microalgae can reach in greater cell densities and bigger lipid production than that of autotrophic conditions [98, 99]. Some microalga species reach 6-7 times greater biomass when they grow mixotrophically compared to autotrophically [100]. Moreover, microalgal cells that grow heterotrophically are considered as an alternative source of DHA [90]. On the other hand, the existence of organic source favors the presence of other microorganisms, creating contamination and competition among them [101]. Moreover, the absence of light leads to the decrease of some compounds whose production is light-dependent. Finally, the demand for carbon source increases the cost of the whole procedure, while the production of CO_2 creates environmental concerns [94, 102]. The most important factor in heterotrophic cultivations is the fact that not all microalgal species can be cultivated under these conditions. The microalgae strain should be equipped with the proper membrane transport systems as to assimilate the carbon source. After that, it should have the suitable enzymes for the carbon metabolism [102].

3.5 Use of Alternative Substrates for Microalgal Growth

As reported above, two of the disadvantages of the heterotrophic cultivations are the increased cost, because of the need of carbon source, and the environmental pollution that some of them cause. Carbohydrates and mainly simple sugars like glucose and sucrose are the main organic source that can be used but constitute expensive substrates [103]. The carbon source represents about 50% of the cost of the culture and at the same time glucose represents almost 80% of the medium cost [104]. Sugars, organic acids, crude glycerol, corn straw, sorghum juice, and waste molasses can be alternative sources, reducing the cost [98, 99, 105–107]. It is also preferable if these sources are renewable. Some of them are not-marketable and/or not-edible products, the organic residuals from agricultural, food and industrial wastes, the lignocellulosic material [105, 108–110]. In any case, the growth media should have the suitable composition [101] and the carbon source has to have the suitable form, as the microalgae may not be capable of metabolizing the initial source, but they could metabolize its derivatives.

Two of these carbon sources are the use of wastes and wastewater. The release of wastewater in the environment can lead to eutrophication [111–113]. The wastewaters contain phosphorus, organic compounds and nitrogen molecules like nitrate, nitrite and ammonia [113]. Microalgae contribute to the removal of pollutants and to wastewater treatment and at the same time they use them as substrate for their growth and the production of high-value compounds [82, 109, 114–116]. Cells are able to remove the nitrogen, ammonium and phosphorus residues and utilize them together with the organic molecules, resulting in the purification of water [112, 117–119]. In some cases, maybe a pretreatment seems to be necessary for the removal of toxic for the algae compounds [103]. Waste can be derived from sources like agricultural and/or food industries. The majority of these wastes contain starch and/or cellulose. Both of these compounds need pretreatment in order to be hydrolyzed to less complex carbohydrates that are suitable for microalgal growth [103]. Cells of *Chlorella sorokiniana* were cultivated heterotrophically at wastewater from agriculture and remove ammonium, nitrates, phosphates and chemical oxygen at rates up to 85%, producing lipids, proteins and carbohydrates in parallel [111]. The same happened with cells of *Chlorella vulgaris* when they grow at artificial wastewater medium, removing up to 97% of ammonium [119]. Cells of the *Botryococcus braunii* were used for the removal of nitrogen and phosphorus from treated sewage from domestic wastewaters, producing lipids. This strain is known for its ability to grow under different environments, producing lipids at a percentage of up to 75% w/w of its biomass [118].

Another important substrate is crude glycerol. Crude glycerol is a by-product of biodiesel production. It was commonly derived from soybean oils. As the sources of biodiesel change, glycerol also derives from animal fats and restaurant oils. In the first case, the source is called white grease, while in the second, yellow grease [120]. Glycerol from these two sources doesn't compete with the food industry, is more ecological and because of the color and the contaminants has a lower price [120]. The procedure for its conversion to purified glycerol that can be used in industry is not affordable [121, 122]. In the case that it comes from grease, the refined procedure has even higher cost [120]. It can be used as carbon source for the heterotrophic cultivation either in the crude or its pure form. Microalgae can transform this carbon source to lipids through fermentation [105]. Some studies on the microalgal *Schizochytrium limacinum* showed that the use of crude glycerol, derived from soybean oils, as substrate led to DHA production [121, 123]. More specifically the lipid content was up to 50% w/w of biomass, with the majority of the fatty acids being palmitic and DHA [123]. Similar study with *Chlorella protothecoides* cells resulted in the production of about 50% w/w of biomass saponifiable lipids under semi-continuous cultures [122]. A similar study on *C. protothecoides* showed glycerol flavors the biomass and lipid production in comparison with glucose [124].

A quite frequently used renewable substrate is cassava. Cassava is a starchy raw material and as a result, a pretreatment is necessary including its hydrolysis to glucose allowing its use by microalgal cells [125]. After the hydrolysis, cassava releases maltose and glucose that are proper substrates for microalgae growth [104]. Cultivation of *C. protothecoides* cells in hydrolyzed cassava resulted in a total lipid production of up to 53% w/w of biomass [126]. Moreover, the saccharification and the cultivation of microalgal cells can be achieved at the same time, reducing the time and the cost of the whole procedure [125]. One more study showed that cassava constitutes better substrate for *C. protothecoides* in comparison to glucose, as it leaded to greater biomass production and enhanced lipid content, among which the C18:2 has the biggest percentage [104].

Another important substrate is sweet sorghum [106, 127]. The advantage of using sweet sorghum is the fact that it doesn't need fertile lands for its cultivation, but it can be grown in poor and semi-land grounds. Its soluble sugars fraction consists of sucrose, glucose and fructose, which are proper carbon sources for microalgae. Sorghum needs pretreatment via hydrolysis to release the carbon compounds. Cells of the microalgal *C. protothecoides* were cultivated in hydrolyzed sweet sorghum, leading to an increase of 36% in lipid production, compared to the use of glucose [127]. On *S. limacinum SR21* cells, the lipid content was about 75% w/w, with the 35% of them being DHA fatty acids [106].

Last, but an important source of carbon, is lignocellulose biomass, which doesn't relate to food [200, 128]. It is a renewable and sustainable resource which consists of cellulose, hemicellulose and lignin and its use as substrate requires pretreatment. The first two compounds can be hydrolyzed to simple sugars, while lignin inhibits this process, blocking the act of cellulases enzymes [129]. As a result, the lignocellulose biomass needs a pretreatment for the separation of the three compounds and the removal of lignin. After that, the hydrolysis of sugars can take place. The main products of the hydrolysis procedure are glucose and xylose, while some other by-products like aromatic and aliphatic acids are released [97]. Lignocellulose biomass usually comes from forestry and/or agricultural residues. The microalgal *Auxenochlorella protothecoides* was cultivated heterotrophically, using as carbon source the hydrolyzed biomass from Norway spruce and birch [130]. The use of the same carbon source on cells of the microalga *Phaeodactylum tricornutum* under mixotrophic conditions, leading to the production of a lipid content of about 38% w/w of biomass, with the 27% of them being PUFAs and mainly EPA and DHA [131].

As shown in Table 3.2, which presents the % lipid content of some microalgal species under heterotrophic or mixotrophic conditions, lipid production and lipid profile present differentiations related to a combination of the used substrate, the presence of light (heterotrophic or mixotrophic mode) and the specie. It is also strongly related with even the strain, as changes in lipid content appeared among different strains of the same species. Moreover, the combination of two different substrates and two stages cultivations could be the next interesting aspect. Already some studies have been achieved, using wastewater with addition of glycerol on *C. vulgaris* cells. In this way, the removal of nutrients from wastes was achieved in a greater level, while in parallel the lipid production was enhanced [132].

As has already been stated, the majority of substrates need a pretreatment in order to be transformed into a proper form that is suitable for hydrolysis form. The pretreatment is used for the separation of the initial compounds of the recalcitrant biomass. There are many pretreatment methods like grinding (physical method), steam/CO_2 explosion (physio-chemical), organosolv process (chemical method) and degradation via fungi (biological) [139]. The aim of the pretreatment process is the deconstruction of biomass and the production of well-defined fractions. Towards this direction, and aiming for the enhancement of the result, new processes have been developed with the combination of techniques, like the hydrid organosolv:steam explosion for lignocellulose biomass [140]. After this, carbohydrates can be hydrolyzed if necessary and used as carbon substrates for heterotrophic and/or mixotrophic cultivation of microalgal cells. The cells are capable of using specific sugars and more preferably monomeric sugars. Acid or enzymatic hydrolysis are the most common methods for the conversion of substrates from complex to suitable forms for culture. In the first one, H_2SO_4 and HCl are the most commonly used reagents. It is a harsh and expensive process, as high temperatures are developed, the reagents can be toxic and dangerous, there is the need of a reactor and the acids must be recovered at the end [141]. Enzymatic hydrolysis is the most usual method, as the conditions are moderate, and it has the advantage of selectivity [142]; the basic limitation of this method is the cost of the required enzyme. Over the past few years, many efforts have been made towards the reuse of enzymes through their immobilization on nanoparticles and as a result the achievement of a more economically affordable process [141]. After this, the carbon compounds can be fermented by the cells, and in some cases, the hydrolysis and the fermentation can be achieved at the same time [125].

Table 3.2 Lipid production (% w/w) of different species of microalgae under heterotrophic and/or mixotrophic condition with the use of different substrates as carbon source.

Strain	Substrate	Heterotrophic/ mixotrophic	% lipids of biomass	Ref.
Schizochytrium limacinum	Crude glycerol	Heterotrophic	>20% DHA 50% lipids/mainly DHA	[121] [123]
Chlorella protothecoides	Glycerol	Heterotrophic	50% (total lipids) 42% (saponifiable)	[122]
	Cassava	Heterotrophic	26.5% (total lipids)	[104]
	Glucose	Heterotrophic	34.5% (total lipids)	[104]
	Glucose	Heterotrophic	57% (total lipids)	[133]
Schizochytrium limacinum SR21	Sweet sorghum	Heterotrophic	75% (total lipids) 35% of them DHA	[106]
Phaeodactylum tricornutum	Norway spruce/birch	Mixotrophic	38% (total lipids) 27% of them PUFAs	[131]
Auxenochlorella protothecoides	Norway spruce	Heterotrophic	63% (total lipids)	[130]
	Norway birch	Heterotrophic	66% (total lipids)	[130]
Chlorella vulgaris	Glucose	Heterotrophic	23% (total lipids)	[99]
		Mixotrophic	21% (total lipids)	[99]

(Continued)

Table 3.2 Lipid production (% w/w) of different species of microalgae under heterotrophic and/or mixotrophic condition with the use of different substrates as carbon source. (*Continued*)

Strain	Substrate	Heterotrophic/ mixotrophic	% lipids of biomass	Ref.
	Glycerol	Heterotrophic	32% (total lipids)	[99]
		Mixotrophic	40% (total lipids)	[98]
	Glucose: glycerol	Mixotrophic	26% (total lipids)	[98]
Nitzschia laevis	Glucose	Heterotrophic	56,3% (total lipids) 10- 11% EPA	[134]
Crypthecodnium cohnii	Glucose	Heterotrophic	36-44% DHA	[135]
Pavlova lutheri	Inositol	Mixotrophic	5-11% SDA 22-29% EPA	[136]
Nannochloropsis sp.	Glucose	Mixotrophic	38,7 % (total lipids) 4,3 of them EPA	[137]
		Heterotrophic	42,7 % (total lipids) 3,1 of them EPA	[137]
	Ethanol	Mixotrophic	34,6 % (total lipids) 4,5 of them EPA	[137]

(*Continued*)

Table 3.2 Lipid production (% w/w) of different species of microalgae under heterotrophic and/or mixotrophic condition with the use of different substrates as carbon source. (*Continued*)

Strain	Substrate	Heterotrophic/ mixotrophic	% lipids of biomass	Ref.
		Heterotrophic	36,9 % (total lipids) 3,8 of them EPA	[137]
Thraustochytrium sp.	Glucose	Heterotrophic	15,2% (total lipids) 52.3% of them DHA	[138]
	Starch	Heterotrophic	15,8% (total lipids) 33,7% of them DHA	[138]
	Maltose	Heterotrophic	6,3% (total lipids) 46,2% of them DHA	[138]
	Sucrose	Heterotrophic	8,8% (total lipids) 28% of them DHA	[138]
	Linseed oil	Heterotrophic	15,6% (total lipids) 33,5% of them DHA	[138]
Schizochytrium sp.	Glucose/corn steep liquor	Heterotrophic	70,3% (total lipids) 37,3% of them DHA	[138]

An important goal of microalgal use for the production of high-value compounds is the minimization of wastes, according to biorefinery concept. In this theory, a successful process leads to if possible zero wastes, via the use of all intermediate wastes and by-products [143]. So, an energy-efficient procedure is achieved. In the case of lipids, all the residues from biomass and by-products should be used for the production of end-products [92]. Microalgae could produce pigments, lipids for biofuel and nutraceutical applications, and carbohydrates simultaneously from the same procedure [143]. The case of wastewater treatment where the cells remove the nutrients and produce lipids is a characteristic paradigm of biorefinery theory [111]. One more example of biorefinery is the case of using sweet sorghum for lipid production as described above. In this case, the produced sucrose after the pretreatment of sorghum could be used for the production of white sugar, while glucose is used for algal growth [106].

Taking everything into consideration, the microalgal cells selected for heterotrophic cultivations should be able to grow without light, metabolize the compounds of carbon source and adjust to changing environments [102]. The lipid biosynthetic pathways are relatively complex, especially when the aim is the production of specific lipids. Moreover, the selection of the suitable, and at the same time cost-effective and environmentally sustainable substrate, constitutes a challenge. If the substrate constitutes waste and after use leads to a valuable product, it is an advantage to the whole procedure and is related with the biorefinery theory. Last but not least, a series of parameters affect the lipid production and they are mentioned below. Finally, the upgrade of a system on industrial scale for the production of EPA, DHA and ARA should solve many demand problems.

3.6 Factors that Affect the Heterotrophic and/or Mixotrophic Cultures

As mentioned above, mixotrophic and heterotrophic cultivations enhance the biomass and lipid production in comparison with autotrophic conditions. There are many factors that affect the microalgae metabolism toward this direction. Some of them are described below.

The substrate and its concentration are major parameters of success, since it has been observed that high concentrations lead to reversal of the desired results. This phenomenon is also related to the stage of culture. A specific carbon concentration may have a negative impact during the first days of culture, but enhance the production at the next cultural phase [99]. For example, cells of *C. protothecoides*, were able to be cultivated heterotrophically with glucose, fructose and sucrose. In the first two substrates, the cell produced total lipids of about 50% w/w, while in sucrose, the biomass and lipid production were significantly lower [127]. Moreover, the mixotrophic growth of the microalga *Neochloris oleoabundans* showed that the concentration of glucose caused a differentiation in biomass and lipid production. In a variety of concentrations ranging from 0 to 30g/L, the best results were observed at 2.5 g/L [96], while at concentrations higher than this, a decrease in production was observed. Moreover, cells of *Schizochytrium limacinum* had better biomass production, having crude glycerol as substrate, comparing with pure glycerol and glucose. At the same study, it was proved that when the concentration of glycerol was greater than 120g/L, inhibition of biomass and lipid production was achieved [121].

Lipid content and as a result PUFAs concentration and composition depends on the growth phase as well [18, 90]. Lipids can be a storage for energy that is needed during cell division and during harsh conditions [90]. During the exponential phase, the main lipids are PUFAs, while during the stationary phase the metabolism changes to the energy storage direction and the cells accumulate TAGs [18].

The nitrogen source and its concentration are also crucial factors. It can be used in many different forms like ammonium, urea, nitrate, nitrate, yeast extract, $NaNO_3$, peptone, tryptone, glycine [100, 121, 144]. All the nitrogen sources are initially transformed to ammonium form that is assimilated to amino acids and used by the microalgal pathways [144]. Every microalgal strain is able to grow better with different nitrogen sources [145]. The nitrogen source and its concentration affect also the biochemical composition of microalgae [144]. Especially in the case of large-scale cultures, a low-cost N source like urea, can decrease the total cost [115]. The cells of *Tetraselmis sp.* preferred organic nitrogen sources for biomass production and ammonium sources for lipid production [144]. In microalgal cells of *Tribonema sp.*, the different nitrogen sources resulted to different biomass production, but the lipid production wasn't affected [100]. Cells of *C. sorokiniana* are able to grow with nitrate and urea, but not with ammonium [145]. Moreover, it is proved that nitrogen deficiency leads to higher content of lipid production and accumulation [146]. Under nitrogen limitation the microalgal growth and protein production are decreased, while the lipid and carbohydrate production are increased [147]. The ways of this result differ among the species of oleaginous microorganisms, but in all cases, the limitation leads to accumulation of acetyl-CoA that is the precursor compound of lipid pathways [148].

The factors reported above are some of the parameters that can affect a heterotrophic and/or mixotrophic cultivation. Some other parameters are the temperature, pH and CO_2 supply. It has been reported that temperature affects the growth of cells and the production of temperature-dependent compounds like lipids. The composition of membrane, its fluidity and its transport system are influenced by temperature changes. As a result, at lower temperature, the cells synthesize and accumulate greater amounts of unsaturated fatty acids [149]. For example, at *S. limacinum* cells, the increase of temperature was leading to decrease of biomass and DHA production. It has been proved that lower temperatures, around 20°C, are favorable for its production [121]. Moreover, in heterotrophic cultures, the pH is decreased as the microalgal cells grow. The biomass and lipid production of *Nannochloropsis salina* cells were affected by pH. The ideal pH rate was pH 8, while pH 9.7 had a drastic effect with a harsh decrease of both biomass and lipid contents [150]. The use of inorganic acid is a solution, but CO_2 is an alternative way for the regulation of pH, raising the acidity [122]. Another use of CO_2 is to increase the carbon content. In the case of the use of wastewater as substrate, the rates of C: N and C: P are usually lower than typical and it leads to inhibition of algae growth. This limitation can be dared with the addition of CO_2 [151]. These parameters increase the cost of the culture, except a low-cost source is available. For example, CO_2 can be derived from boilers or power plants as flue gas [152].

Heterotrophic cultivations of microalgae are easily to be controlled in fermenters and the scale up in bioreactors consist a prospective [101]. The proper combination of cultural parameters and microalgal nature makes the scale up a complex operation [153], especially in the case of mixotrophic cultivation where there is the demand of light and problems with shading and photo inhibition as the light cannot always reach the cells are appeared [154]. Besides the complexity of the procedure, the use of bioreactor gives the opportunity to control the production of biomass and metabolic compounds, check the availability and concentration of nutrients and substrates, regulate and adjust conditions that change during the cells' growth, like pH and temperature, and maintain axenic conditions.

Many different cultivation techniques and bioreactor systems (stirred-tank bioreactors, flat plate reactors, tubular photobioreactors, flat panel airlift bioreactors) are available, with the selection being dependent on the required product. Especially, tubular bioreactors consist of a series of transparent plastic or glass tubes that operate continuously. The limitation of this system is the diameter of the tube in order to have an ideal penetration of light [155, 156]. In contrast, flat-panel bioreactor is a cuboidal construct from transparent glass, plexiglass or polycarbonate. Their two important characteristics are the high ratio of surface area and volume and the open gas disengagement systems. Flat plate bioreactors need less space than coiled tubes because they have narrow U-turns and they also have thinner glass walls. These advantages are important on compactness issue, especially on mixotrophic conditions, where light is an essential factor [155–157]. Another bioreactor category is airlift bioreactors. They are bubble reactors with a draft-tube that promotes gas-liquid mass transfer and mixing. An advantage of airlift bioreactors in contrast to stirred tank reactors is the higher productivity at the cell grown [157, 158]. Finally, a fundamental bioreactor system in industry is stirred-tank bioreactors. The integral part of a stirred tank bioreactor is the agitator, which is responsible for multiple functions, such as aeration, mixing, homogenization as soon as heat and mass transfer. The impeller could be axial or radial considering the flow. A stirred tank bioreactor could be differentiated based on the impeller of bottom clearance, the impeller size, the baffles and their width, the sparger type and position, and the ratio of liquid height to tank diameter [159–161].

The basic key for a successful heterotrophic cultivation in bioreactors is the microalgal strain selection. The cells should have high resistance in mechanical pressure and chemical stress [162]. Some species that are growing with fast rates in heterotrophical bioreactors are *Chlorella, Crypthecodinium*, *Nitzia,* and *Prototheca spp.* which are having specific growth rates above 0.09 h^{-1} [163]. Additionally, the selected strains and bioreactor systems should lead to the production of the desired biomolecules. For instance, *Schizotrichium* sp. and *Thraustochtrium sp.* appear high tolerance in heterotrophically bioreactor cultivation, but they are not favorable for lipid production [164, 165].

Moreover, fed batch cultivations are the most common method for high biomass and product concentrations in short times, as the growth rates can be controlled with the addition of substrate, avoiding the toxicity problems of batch cultures [107, 153]. The addition of carbon source takes place in a pulsed way, making the substrate concentration unable to drop below the desired rate [166–169].

In any case, the effect of all the above parameters is synergistically and strain dependent. As a result, an individual study is necessary for every strain. There are many strains that are capable of producing PUFAs both in fresh water species and marine species and the selection of the right one has crucial significance. Some of the selection criteria could be: the growth rates, the content of the produced lipids and their profile, the environment in which they are going to grow, their ability for adaptation to changes [170]. If the cultural conditions are close to these that cells are used, the acclimatization period is shorter, and the production is enhanced. Physicochemical factors, like nutrient, salinity, temperature, nitrogen, should be also considered in algal selection as they affect their growth [92, 146, 150].

3.7 Conclusions

The benefits of PUFAs and especially omega-3 and omega-6 fatty acids on the human organism have been mentioned in many studies, as they contribute to the prevention and treatment of neurological and cardiovascular diseases; they have anticancer and antioxidant properties. The need for new sources of these lipids has been reported. Single cell microorganisms have been mentioned as a promising source. Among them, microalgae constitute suitable microorganisms for this scope. Some species of them that belong to the oleaginous microorganisms are able to accumulate more than 20% w/w and up to 70% w/w of their biomass as lipids. There are many advantages of the use of microorganisms and especially microalgae for the production of PUFAs. Among the most important of them are the no need for arable land, the short life cycle, their ability to adjust to different environments, their capability of growing under heterotrophic and/or mixotrophic conditions, and their ability for large scale cultures. Moreover, they can make use of a variety of substrates as carbon source, while a combination of parameters affects their biomass and bio compounds production. As a result, microalgae can use renewable and environmentally sustainable substrates for carbon source and others requires nutrients, reducing both the cost of the whole procedure and the environmental footprint. According to the biorefinery concept, microalgae are able to produce many different high-value compounds with parallel and/or chained procedures, leading to zero waste. So, depending on the desired product and the available facilities, the selection of the correct microalga strain and the determination of the suitable cultivation parameters are the main issues. As a consequence, a full study for the optimization of conditions for the biomass and bio compounds production at lab scale is necessary before scaling up to commercial production of PUFAs feedstock.

3.8 Future Perspectives

Single cell oils from oleaginous microorganisms are going to be the substitutes of the traditional sources of PUFAs. Finding new strains and carbon sources and selection of the right one in any case, constitutes a deal and opens a new way for cost reduction of heterotrophic cultures. Moreover, knowledge of the metabolic pathways, the cellular mechanisms and the genes that take part at the lipid production is a crucial chance for the enhancement of their production. Through these techniques, production of PUFAs can be enhanced through the overexpression of genes and enzymes that contribute to their metabolic pathways or via blocking antagonistic pathways. In addition, the search for more accessible alternative substrates that are suitable for microalgae growth and in parallel enhance the production of PUFAs, are environmentally sustainable, economical and affordable and can participate in the biorefinery concept, constitutes a challenge. Metabolic engineering, metabolic models, proteomics and similar 'omics' techniques will provide useful tools towards the lipid production pathways. Finally, after the determination of the above, the design and the destruction of structures for scale up at industrial scale, for the commercial production of PUFAs from microalgae should be necessary.

3.9 Acknowledgements

Dimitra Karageorgou is supported by the Hellenic Foundation for Research and Innovation (HFRI) and the General Secretariat for Research and Technology (GSRT), under the HFRI PhD Fellowship grant (GA. no. 1137).

References

1. Blunt, J.W., Carroll, A.R., Copp, B.R., Davis, R.D., Keyzers, R. A., Prinsep, M.R., Marine natural products. *Nat. Prod. Rep.*, 35, 8, 2018.
2. Schuler, L.M , Schulze, P.C.S., Pereira, H., Barreira, L., Leon, R., Varela J., Trends and strategies to enhance triacylglycerols and high-value compounds in microalgae. *Algal Research*, 25, 263-273, 2017.
3. Shanab, S.M.M., Hafez, R.M., Fouad, A.S., A review on algae and plants as potential source of Arachidonic acid. *Journal of Advanced Research*, 11, 3-13, 2018.
4. Abdo, S.M., Ali, G.H., EL-Baz, F.K., Potential Production of Omega Fatty Acids from Microalgae. *Int. J. Pharm. Sci. Rev. Res.*, 34, 2, 210-215, 2015.
5. Tocher, D.R., Betancor, M.B., Spraque, M., Olsen, R.E., Napier, J.A., Omega-3 Long-Chain Polyunsaturated Fatty Acids, EPA and DHA: Bridging the Gap between Supply and Demand. *Nutrients*, 11, 89, 2019.
6. Deckelbaum, R.J., Torrejon, C., The Omega-3 Fatty Acid Nutritional Landscape: Health Benefits and Sources. *Journal of Nutrition*, 142, 3, 587S-591S, 2012.
7. Beligon, V., Christophe, G., Fontanille P., Larroche, C., Microbial lipids as potential source to food supplements. *Current Opinion in Food Science*, 7, 35-42, 2016.

8. Milledge, J.J, Commercial application of microalgae other than as biofuels: a brief review. *Rev. Environ. Sci. Biotechnol.*, 10, 31-41, 2011.
9. Ursin, V.M., Modification of Plant Lipids for Human Health: Development of Functional Land-Based Omega-3 Fatty Acids. *Journal of Nutrition*, 133, 12, 4271-4274, 2003.
10. Santos-Sanchez, N.F., Valadez-Blanco, R., Hernandez-Carlos, B., Torres-Arino, A., Guadarrama-Mendoza, P.C., Salas-Coronado, R., Lipids rich in ω-3 polyunsaturated fatty acids from microalgae. *Appl. Microbiol. Biotechnol.*, 100, 8667-8684, 2016.
11. Lee, S.A., Whenman, N., Bedford, M.R., Review on docosahexaenoic acid in poultry and swine nutrition: Consequence of enriched animal products on performance and health characteristics. *Animal Nutrition Journal*, 2018.
12. Diwan, B., Parkhey, P., Gupta, P., From agro-industrial wastes to single cell oils: a step towards prospective biorefinery. *Folia Microbiologica*, 63, 547-568, 2018.
13. Markou, G., Nerantzis, E., Microalgae for high-value compounds and biofuels production: A review with focus on cultivation under stress conditions. *Biotechnology Advances*, 31, 8, 1532-1542, 2013.
14. Chisti, Y., Biodiesel from microalgae. *Biotechnology Advances*, 25, 3, 294-306, 2007.
15. Guiry, M.D., How many species of algae are there? *J. Phycol.*, 48, 1057-1063, 2012.
16. Hldebrand, M., Abbriano, R.M., Polle, J.EW., Traller, J.C., Trentacoste, E.M., Smith, S.R., Davis A.K., Metabolic and cellular organization in evolutionarily diverse microalgae as related to biofuels production. *Current Opinion in Chemical Biology*, 17, 506-514, 2013.
17. Swiatkiewicz, S., Arczewska-Wlosek, A., Josefiak, D., Application of microalgae biomass in poultry nutrition. *World's Poultry Science Journal*, 71, 4, 663-672, 2015.
18. Xue, Z., Wan, F., Yu, W., Liu, J., Zhang, Z., X, Ku X., Edible Oil Production from Microalgae: A Review. *Eur. J. Lipid Sci. Technol.*, 120, 6, 2018.
19. Torres, E.M., Hess, D., McNeil, B.T., Guy, T., Quin, J.C., Impact of inorganic contaminants on microalgae productivity and bioremediation potential. *Ecotoxicology and Environmental Safety*, 139, 367-376, 2017.
20. Markou, G., Angelidaki, I., Georgakakis, D., Microalgal carbohydrates: an overview of the factors influencing carbohydrates production, and of main bioconversion technologies for production of biofuels. *Appl. Microbiol. Biotechnol.*, 96, 631-645, 2012.
21. Pittman, J.K., Dean, A.P., Osundeko, O., The potential of sustainable algal biofuel production using wastewater resources. *Bioresource Technology*, 102, 17-25, 2011.
22. Makareviciene, V., Skorupskaite, V., Levisauskas, D., Andruleviciute, V., Kazancev, K., The optimization of biodiesel fuel production from microalgae oil using response surface methodology. *International Journal of Green Energy*, 11, 5, 527-541, 2014.

23. Venkata, M.S., Rohit, M.V., Chiranjeevi, P., Chandra, R., Navaneeth, B., Heterotrophic microalgae cultivation to synergize biodiesel production with waste remediation: Progress and perspectives. *Bioresource Technology*, 184, 169-178, 2015.
24. Wiktorowska-Owczarek, A., Berezinska, M., Nowak, J.Z., PUFAs: Structures, Metabolism and Functions. *Adv. Clin. Exp. Med.*, 24, 6, 931-941, 2015.
25. Abedi, E., Sahari, M.A., Long-chain polyunsaturated fatty acid sources and evaluation of their nutritional and functional properties. *Food Sci. Nutr.*, 2, 5, 443-463, 2014.
26. Tallima, H., Ridi, R.E., Arachidonic acid: Physiological roles and potential health benefits – A review. *Journal of Advanced Research*, 11, 33-41, 2018.
27. Calder, P.C., Yaqoob, P., Omega-3 polyunsaturated fatty acids and human health outcomes. *BioFactors*, 35, 3, 266-272, 2009.
28. Delgado, J.D., Perez-Martinez, P., Lopez-Miranda, J., Perez-Jimenez, F., Long chain omega-3 fatty acids and cardiovascular disease: a systematic review. *British Journal of Nutrition*, 107, S201-S213, 2012.
29. Murray-Taylor, F.M., Ho, Y.Y, Densupsoontorn, N., Chang, C.L., Deckelbaum, R.J., Seo, T., n-3, but not n-6 lipid particle uptake requires cell surface anchoring. *Biochemical and Biophysical Research Communications*, 392, 135-139, 2010.
30. Macchia, A., Grancelli, H., Varini, S., Nul, D., Laffaye, N., Mariani, J., Ferrande, D., Branda, R., Figal, J., Ramos, S., Tognoni, G., Doval, H.C., Omega-3 fatty acids for the prevention of recurrent symptomatic atrial fibrillation: results of the FORWARD (Randomized Trial to Assess Efficacy of PUFA for the Maintenance of Sinus Rhythm in Persistent Atrial Fibrillation) trial. *Journal of the American College of Cardiology*, 61, 4, 463-468, 2013.
31. Sudheendran, S., Chang, C.C., Deckelbaum, R.J., N-3 vs. saturated fatty acids: Effects on the arterial wall. *Prostaglandins, Leukotrienes and Essential Fatty Acids*, 82, 205-209, 2010.
32. Colussi, G., Catena, C., Novello, M., Bertin, N., Sechi, L.A., Impact of omega-3 polyunsaturated fatty acids on vascular function and blood pressure: relevance for cardiovascular outcomes. *Nutrition, Metabolism and Cardiovascular Diseases*, 27, 3, 191-200, 2017.
33. Campoy, C., Escolado, V.M., Anjos, T., Szajewska, H., Uauy, R., Omega 3 fatty acids on child growth, visual acuity and neurodevelopment. *British Journal of Nutrition*, 107, S85-S106, 2012.
34. Mayurasakorn, K., Williams, J.J., Ten, V.S., Deckelbaum, R.J., Docosahexaenoic acid: brain accretion and roles in neuroprotection after brain hypoxia and ischemia. *Curr. Opin. Clin. Nutr. Metab. Care*, 14, 158-167, 2011.
35. Innis, S.M., Omega-3 Fatty acids and neural development to 2 years of age: do we know enough for dietary recommendations? *Journal of Pediatric Gastroenterology and Nutrition*, 48, S16-S24, 2009.
36. Gould, J.F., Smithers, L.G., Makrides, M., The effect of maternal omega-3 (n-3) LCPUFA supplementation during pregnancy on early childhood

cognitive and visual development: a systematic review and meta- analysis of randomized controlled trials. *Am. J. Clin. Nutr.*, 97, 3, 531-544, 2013.
37. Sheila, M.I., Dietary (n-3) Fatty Acids and Brain Development. *J. Nutr.*, 137, 855-859, 2007.
38. Uauy, R., Dangour, A.D., Nutrition in Brain Development and Aging: Role of Essential Fatty Acids. *Nutrition Reviews*, 64, S24-S33, 2006.
39. Valenzuela, R., Sanhueza, J., Valenzuela, A., Docosahexaenoic Acid (DHA), an Important Fatty Acid in Aging and the Protection of Neurodegenerative Diseases. *Journal of Nutritional Therapeutics*, 1, 63-72, 2012.
40. Schuchardt, J.P., Huss, M., Grabo, M.S., Hahn, A, Significance of long-chain polyunsaturated fatty acids (PUFAs) for the development and behaviour of children. *Eur. J. Pediatr.*, 169, 149-164, 2010.
41. Judge, M.P., Harel, O., Lammi-Keefe, C.J., A docosahexaenoic acid–functional food during pregnancy benefits infant visual acuity at four but not six months of age. *Lipids*, 42, 2, 117-122, 2007.
42. Yaqoob, P., Shaikh, S.R., The nutritional and clinical significance of lipid rafts. *Curr. Opin. Clin. Nutr. Metab. Care*, 13, 156-166, 2010.
43. Sublette, M.E., Hibbeln, J.R., Galfavy, H., Oquendo, M.A., Mann, J.J., Omega-3 Polyunsaturated Essential Fatty Acid Status as a Predictor of Future Suicide Risk. *Am. J. Psychiatry*, 163, 6, 1100-1102, 2006.
44. Hibbeln, J.R., Ferguso, T.A., Blasbalg, T.L., Omega-3 fatty acid deficiencies in neurodevelopment, aggression and autonomic dysregulation: Opportunities for intervention. *International Review of Psychiatry*, 18, 2, 107-118, 2006.
45. Puri, B.K., Counsell, S.J., Hamilton, G., Richardson, A.J., Horrobin, D.F., Eicosapentaenoic acid in treatment-resistant depression associated with symptom remission, structural brain changes and reduced neuronal phospholipid turnover. *Int. J. Clin. Pract.*, 55, 8, 560-563, 2001.
46. Mocellin, M.C., Camargo, C.Q., Souza Fabre, M.E., Trindade, E.B.S.M., Fish oil effects on quality of life, body weight and free fat mass change in gastro-intestinal cancer patients undergoing chemotherapy: a triple blind, randomized clinical trial. *Journal of Functional Foods*, 31, 113-122, 2017.
47. Silva, J.A.P., Fabre, M.E.S., Waitzberg, D.L., Omega-3 supplements for patients in chemotherapy and/or radiotherapy: A systematic review. *Clinical Nutrition*, 34, 4, 359-366, 2015.
48. Hajjaji, N., Bougnoux, P., Selective sensitization of tumors to chemotherapy by marine-derived lipids: A review. *Cancer Treatment Reviews*, 39, 5, 473-488, 2013.
49. D' Eliseo, D., Velotti, F., Omega-3 Fatty Acids and Cancer Cell Cytotoxicity: Implications for Multi-Targeted Cancer Therapy. *J. Clin. Med.*, 5, 15, 2016.
50. Merendino, N., Constantini, L., Manzi, L., Molinari, R., D' Eliseo, D., Velloti, F., Dietary ω-3 Polyunsaturated Fatty Acid DHA: A Potential Adjuvant in the Treatment of Cancer. *Biomed Res. Int.*, 22, 310186, 2013.
51. Mocellin, M.C., Camargo, C.Q., Nunes, E.A., Fiates, G.M.R., B.S.M.T., A systematic review and meta-analysis of the n-3 polyunsaturated fatty acids

effects on inflammatory markers in colorectal cancer. *Clinical Nutrition*, 35, 2, 359-369, 2016.
52. Sakai, M., Kakutani, S., Horikawa, C., Tokuda, H., Kawashima, H., Shibata, H., Okubo, H., Sasaki, S., Arachidonic acid and cancer risk: a systematic review of observational studies. *BMC Cancer*, 12, 606, 2012.
53. Sun, X.M., Geng, L.J., Ren, L.J., Ji, X.J., Hao, N., Chen, K.Q., Huang, H., Influence of oxygen on the biosynthesis of polyunsaturated fatty acids in microalgae. *Bioresource Technology*, 250, 868-876, 2018.
54. Yan-ling, M., Microbial oils and its research advance. *Chin J Bioprocess Eng*, 4, 4, 7–11, 2006.
55. Lari, Z., Moradi-kheibari, N., Ahmadzadeh, H., Abrishamchi, P., Moheimani, N.R., Murry, M.A., Bioprocess engineering of microalgae to optimize lipid production through nutrient management. *J. Appl. Phycol.*, 28, 6, 3235-3250, 2016.
56. Guschina, I.A., Harwood, J.L., Lipids and lipid metabolism in eukaryotic algae. *Progress in Lipid Research*, 45, 160-186, 2006.
57. Borowitzka, M.A., High-value products from microalgae—their development and commercialization. *J. Appl. Phycol.*, 25, 3, 743-756, 2013.
58. Harwood, J.L., Guschina, I.A., The versatility of algae and their lipid metabolism. *Biochimie*, 91, 6, 679-684, 2009.
59. Minhas, A.K., Hodgson, P., Barrow, C.J., Adholeya, A., A Review on the Assessment of Stress Conditions for Simultaneous Production of Microalgal Lipids and Carotenoids. *Front. Microbiol.*, 7, 546, 2016.
60. Adl, S.M., Simpson, A.G.B., Farmer, M.A., Andersen, R.A.,Anderson, O.R., Barta, J.R., Bowser, S.S., Brugerolle, G.U.Y., Fensome, R.A., Fredericq, S., James, T.Y., Karpov, S., Kugrens, P., Krug, J., Lane, C.E., Lewis, L.A., Lodge, J., Lynn, D.H., Mann, D.G., McCourt, R.M., Mendoza, L.,Moestrup, Ø., Mozley-Standridge, S.E., Nerad, T.A.,Shearer, C.A., Smirnov, A.V., Spiegel, F.W., Taylor, M.F.J.R. The new higher level classification of eukaryotes with emphasis on the taxonomy of protists. *J. Eukaryot. Microbiol.* 52, 399-451, 2005.
61. Keeling, P.J., Burger, G., Durnford, D.G., Lang, B.F.,Lee, R.W., Pearlman, R.E., Roger, A.J., Gray, M.W. The tree of eukaryotes. *Trends Ecol. Evol.* 20, 670-676, 2005.
62. Gantt, E., Conti, S.F. The ultrastructure of Porphyridium cruentum. *J. Cell Biol.* 26, 365-381, 1965.
63. Leliaert, F., Smith, D.R., Moreau, H., Herron, M.D., Verbruggen, H., Delwiche, C.F., de Clerk, O. Phylogeny and molecular evolution of the green algae. *Crit. Rev. Plant Sci.* 31, 1-46, 2012.
64. Sarno, D., Kooistra, W.H.C.F. Diversity in the genus Skeletonema (Bacillariophyceae). II. An assessment of the taxonomy of S. costatum-like species with the description of four new species. *J. Phycol.* 41, 151-176, 2005.
65. Tsui, C.K.M., Marshall, W., Yokoyama, R., Honda, D., Lippmeier, J.C., Craven, K.D., Peterson, P.D., Berbee, M.L., Labyrinthulomycetes phylogeny

and its implications for the evolutionary loss of chloroplasts and gain of ecto-plasmic gliding. *Mol. Phylogen. Evol.* 50, 129-140, 2009.

66. Prabowo, D.A., Hiraishi, O., Suda, S. Diverisity of crypthecodinium spp. (Dinophyceae) from Okinawa prefecture. *Jpn. J. Mar. Sci. Technol.* 21, 181-191, 2013.
67. Shuang, L., Jilin, X., Jiao, C., Juanjuan, C., Chengxu, Z., Xiaojun, Y., The major lipid changes of some important diet microalgae during the entire growth phase. *Aquaculture*, 428-429, 104-110, 2014.
68. Suh, S.S., Kim, S.J., Hwang, J., Park, M., Lee, T.K., Kil, E.J., Lee, S., Fatty acid methyl ester profile and nutritive values of 20 marine microalgae in Korea. *Asian Pacific Journal of Tropical Medicine*, 8, 3, 191-196, 2015.
69. Matos, A.P., Feller, R., Moecke, E.H.S., Oliveira, J.V., Junior, A.F., Derner, R.B., Sant'Anna, E.S., Chemical characterization of six microalgae with potential utility for food application. *Journal of American Oil Chemists' Society*, 93, 3, 963-972, 2016.
70. Wei, L., Huang, X., Long-Duration effect of multi-factor stresses on the cellular biochemistry, oil-yielding performance and morphology of *Nannochlorospis oculate*. *PLoS ONE*, 12, 3, 2017.
71. Volkman, J.K., Jeffrey, S.W., Nichols, P.D., Rogers, G.I., & Garland, C.D., Fatty acid and lipid composition of 10 species of microalgae used in mariculture. *Journal of Experimental Marine Biology and Ecology*, 128(3), 219-240, 1989.
72. Ramadan, M.F., Asker, M.M.S., Zeinab, K.I., Functional Biactive Compounds and Biological Activities of *Spirulina platensis* Lipids. *Czech J. Food Sci.*, 26, 3, 211-222, 2008.
73. Patil, V., Kallqvist, T., Olsen, E., Vogt, G., Gislerod, H.R., Fatty acid composition of 12 microalgae for possible use in aquaculture feed. *Aquacult. Int.*, 15, 1-9, 2007.
74. Zhang, L., Zhao, H., Lai, Y., Wu, J., Chen, H., Improving docosahexaenoic acid productivity of *Schizochytrium* sp. by a two-stage AEMR/shake mixed culture mode. *Bioresource Technology*, 142, 719-722, 2013.
75. Song, X., Zhang, X., Guo, N., Zhu, L., & Kuang, C. Assessment of marine thraustochytrid Schizochytrium limacinum OUC88 for mariculture by enriched feeds. *Fisheries Science*, 73(3), 565-573, 2007.
76. Ling, X., Guo, J., Liu, X., Zhang, X., Wang, N., Lu, Y., Ng, I-S., Impact of carbon and nitrogen feeding strategy on high production of biomass and docosahexaenoic acid (DHA) by *Schizochytrium* sp. LU310. *Bioresource Technology*, 184, 139-147, 2015.
77. Sato, T., Ishihara, K., Shimizu, T., Aoya, J., & Yoshida, M. Laboratory Scale Culture of Early-Stage Kuruma Shrimp Marsupenaeus japonicus Larvae Fed on Thraustochytrids Aurantiochytrium and Parietichytrium. *Journal of Shellfish Research*, 37(3), 571-580, 2018.
78. Chang, K.J.L., Paul, H., Nichols, P.D., Koutoulis, A., Blackburn, S.I., Australian thraustochytrids: potential production of dietary long-chain omega-3 oils using crude glycerol. *Journal of Functional Foods*, 19, B, 810-820, 2015.

79. Huang, T.Y., Lu, W.C., Chu, I.M., A fermentation strategy for producing docosahexaenoic acid in *Aurantiochytrium limacinum* SR21 and increasing C22:6 proportions in total fatty acid. *Bioresource Technology*, 123, 8-14, 2012.
80. Sahin, D., Tas, E., & Altindag, U.H. Enhancement of docosahexaenoic acid (DHA) production from Schizochytrium sp. S31 using different growth medium conditions. *AMB Express*, 8(1), 7, 2018.
81. Riesenberg, D., Guthke, R., High-cell-density cultivation of microorganisms. *Appl Microbiol Biotechnol* 51(4),422–430, 1999.
82. Gimpel, J.A., Henriquez, V., Mayfield, S.P., In Metabolic Engineering of Eukaryotic Microalgae: Potential and Challenges Come with Great Diversity. *Front. Microbiol.*, 6, 1376, 2015.
83. Gong, Y., Wian, X., Jian, M., Hu, C., Hu, H., Huang, F., Metabolic engineering of microorganisms to produce omega-3 very long-chain polyunsaturated fatty acids. *Progress in Lipid Research*, 56, 19-35, 2014.
84. Ganesan, B., Brothersen, C., McMahon, D.J., Fortification of Foods with Omega-3 Polyunsaturated Fatty Acids. *Critical Reviews in Food Science and Nutrition*, 54, 98-114, 2014.
85. Castro, L.F., Tocher, D.R., Monroig, O., Long-chain polyunsaturated fatty acid biosynthesis in chordates: Insights into the evolution of Fads and Elovl gene repertoire. *Progress in Lipid Research*, 62, 25-40, 2016.
86. Bell, M., Tocher, D., Biosynthesis of Polyunsaturated Fatty Acids in Aquatic Ecosystems: General Pathways and New Directions, In: *Lipids in Aquatic Ecosystems. Biomedical and Life Sciences,* Brett M.T., Kainz M. & Arts MT (eds.), pp. 211-236, Dordrecht, Heidelberg, London, New York: Springer.
87. Martins, D.A., Custodio, L., Barreira, L., Pereira, H., Ben-Hamadou, R., Varela, J., Abu-Salah, K.M., Alternative Sources of n-3 Long-Chain Polyunsaturated Fatty Acids in Marine Microalgae. *Mar. Drugs*, 11, 2259-2281, 2013.
88. Fernades, T., Fernades, I., Andrade, C.A.P., Cordeiro, N., Changes in fatty acid biosynthesis in marine microalgae as a response to medium nutrient availability. *Algal Research*, 18, 314-320, 2016.
89. Muhlroth, A., Li , K., Rokke, G., Winge, P., Olsen, Y., Hohmann-Marriot, M.F., Vadstein, O., Bones, A.M., Pathways of Lipid Metabolism in Marine Algae, Co-Expression Network, Bottlenecks and Candidate Genes for Enhanced Production of EPA and DHA in Species of Chromista. *Mar. Drugs*, 11, 4662-4697, 2013.
90. Adarme-Vega, T.C., Lim, D.K.Y., Timmins, M., Vermen, F., Li, Y., Schenk, P.M., Microalgal biofactories: a promising approach towards sustainable omega-3 fatty acid production. *Microbial Cell Factories*, 11, 96, 2012.
91. Leonardi, P.I., Popovich, C.A., Damiani, C., Feedstocks for second-generation biodiesel: microalgae's biology and oil composition, in: *Economic Effects of Biofuel Production*, Dr. Marco Aurelio Dos Santos Bernardes (Ed.), ISBN: 978-953-307-178-7, InTech, 2011.

92. Rawat, I., Ranjith, R.K., Mutanda, T., Bux, F., Biodiesel from microalgae: A critical evaluation from laboratory to large scale production. *Applied Energy*, 103, 444-467, 2013.
93. Chen, H.H., Jiang, J.G., Lipid accumulation mechanisms in auto- and heterotrophic microalgae. *J. Agric. Food Chem.*, 65, 8099-8110, 2017.
94. Perez-Garcia, O., Escalente, F.M.E., de-Bashan, L.E., Bashan, Y., Heterotrophic cultures of microalgae: Metabolism and potential products. *Water Research*, 45, 11-36, 2011.
95. Liu, X., Duan, S., Li, A., Xu, N., Cai, Z., Hu, Z., Effects of organic carbon sources on growth, photosynthesis, and respiration of *Phaeodactylum tricornutum*. *J. Appl. Phycol.*, 21, 239-246, 2009.
96. Giovanardi, M., Baldisserotto, C., Ferroni, L., Longoni, P., Cella, R., Pancaldi, S., Growth and lipid synthesis promotion in mixotrophic *Neochloris oleoabundans* (Chlorophyta) cultivated with glucose. *Protoplasma*, 251, 1, 115-125, 2014.
97. Liang, Y., Producing liquid transportation fuels from heterotrophic microalgae. *Applied Energy*, 104, 860-868, 2013.
98. Heredia-Arroyo, T., Wei, W., Ruan, R., Hu, B., Mixotrophic cultivation of Chlorella vulgaris and its potential application for the oil accumulation from non-sugar materials. *Biomass and Bioenergy*, 35, 2245-2253, 2011.
99. Liang, Y., Sakany, N., Cui, Y., Biomass and lipid productivities of Chlorella vulgaris under autotrophic, heterotrophic and mixotrophic growth conditions. *Biotechnol. Lett.*, 31, 1043-1049, 2009.
100. Wang, H., Zhou, W., Shao, H., Liu, T., A comparative analysis of biomass and lipid content in five *Tribonema* sp. strains at autotrophic, heterotrophic and mixotrophic cultivation. *Algal Research*, 24, 284-289, 2017.
101. Bumbak, F., Cook, S, Zachleder, V., Hauser, S., Kovar, K., Best practices in heterotrophic high-cell-density microalgal processes: achievements, potential and possible limitations. *Appl. Microbiol. Biotechnol.*, 91, 31-46, 2011.
102. Morales-Sanchez, D., Tinoco-Valencia, R., Kyndt J., Martinez, A., Heterotrophic growth of Neochloris oleoabundans using glucose as a carbon source. *Biotechnology for Biofuels*, 6, 100, 2013.
103. Kapdan, I.K., Kargi, F., Bio-hydrogen production from waste materials. *Enzyme and Microbial Technology*, 38, 569-582, 2006.
104. Wei, A., Zhang, X., Wei, D., Cheng, G., Wu , O., Yang, S.T., Effects of cassava starch hydrolysate on cell growth and lipid accumulation of the heterotrophic microalgae *Chlorella protothecoides*. *J. Ind. Microbiol. Biotechnol.*, 36, 1383-1389, 2009.
105. Liang, M.H., Jiang, J.G., Advancing oleaginous microorganisms to produce lipid via metabolic engineering technology. *Progress in Lipid Research*, 52, 395-408, 2013.
106. Liang, Y., Sarkany, N., Cui, Y., Yesuf, J., Trushenski, J., Blackburn, J.W., Use of sweet sorghum juice for lipid production by *Schizochytrium limacinum* SR21. *Bioresource Technology*, 101, 3623-3627, 2010.

107. Yan, D., Lu, Y., Chen, Y.F., Wu, O., Waste molasses alone displaces glucose-based medium for microalgal fermentation towards cost-saving biodiesel production. *Bioresource Technology*, 102, 6487-6493, 2011.
108. Whalen, J., Xu, C.C., Shen, F., Kumar, A., Eklund, M., Yan, J., Sustainable biofuel production from forestry, agricultural and waste biomass feedstocks. *Applied Energy*, 198, 281-283, 2017.
109. Lizzul, A.M., Hellier, P., Purton, S., Baganz, F., Ladommatos, N., Campos, L., Combined remediation and lipid production using Chlorella sorokiniana grown on wastewater and exhaust gases. *Bioresource Technology*, 151, 12-18, 2014.
110. Matsakas, L., Giannakou, M., Voros, D., Effect of synthetic and natural media on lipid production from *Fusarium oxysporum*. *Electronic Journal of Biotechnology*, 30, 95-102, 2017.
111. Guldhe, A., Ansari, F.A., Singh, P., Bux, F., Heterotrophic cultivation of microalgae using aquaculture wastewater: A biorefinery concept for biomass production and nutrient remediation. *Ecological Engineering*, 99, 47-53, 2017.
112. Ruiz-Marin, A., Medoza-Espinoza, L.G., Stephenson, T., Growth and nutrient removal in free and immobilized green algae in batch and semi-continuous cultures treating real wastewater. *Bioresource Technology*, 101, 58-64, 2010.
113. Nasir, N.M., Bakar, N.S.A., Lananan, F., Hamid, S.H.A., Lam, S.S., Juson, A., Treatment of African catfish, *Clarias gariepinus* wastewater utilizing phytoremediation of microalgae, *Chlorella* sp. with *Aspergillus niger* bio-harvesting. *Bioresource Technology*, 190, 492-498, 2015.
114. Wang, L., Min, M., Li, Y., Chen, P., Chen, Y., Liu, Y., Wang, Y., Ruan, R., Cultivation of Green Algae *Chlorella* sp. in Different Wastewaters from Municipal Wastewater Treatment Plant. *Appl. Biochem. Biotechnol.*, 162, 1174-1186, 2010.
115. Ramanna, L., Gulde, A., Rawat, I., Bux, F., The optimization of biomass and lipid yields of *Chlorella sorokiniana* when using wastewater supplemented with different nitrogen sources. *Bioresource Technology*, 168, 127-135, 2014.
116. Subramaniam, R., Dufresche, S., Zappi, M., Bajpai, R., Microbial lipids from renewable resources: production and characterization. *J. Ind. Microbiol. Biotechnol.*, 37, 1271-1287, 2010.
117. Chan, Y., Wu, Z., Bian, L., Feng, D., Leung, D,Y.C., Cultivation of *Spirulina platensis* for biomass production and nutrient removal from synthetic human urine. *Applied Energy*, 102, 427-431, 2013.
118. Orpez , R., Martinez, M.E, Hodaifa, G., Youfsi, F.E., Jbari, N., Sanchez, S., Growth of the microalga *Botryococcus braunii* in secondarily treated sewage. *Desalination*, 246, 625-630, 2009.
119. Feng, Y., Li, C., Zhang, D., Lipid production of *Chlorella vulgaris* cultured in artificial wastewater medium. *Bioresource Technology*, 102, 101-105, 2011.
120. Liang, Y., Sarkany, N., Cui, Y., Blackburn, J.W., Batch stage study of lipid production from crude glycerol derived from yellow grease or animal fats through microalgal fermentation. *Bioresource Technology*, 101, 6745-6750, 2010.

121. Chi, Z., Pyle, D., Wen, Z., Frear, C., Chen, S., A laboratory study of producing docosahexaenoic acid from biodiesel-waste glycerol by microalgal fermentation. *Process Biochemistry*, 42, 1537-1545, 2007.
122. Garcia, M.C.C., Sanchez, M.D.M., Miron, A.S., Camacho, F.G., Grima, E.M., A process for biodiesel production involving the heterotrophic fermentation of Chlorella protothecoides with glycerol as the carbon source. *Applied Energy*, 103, 341-349, 2013.
123. Pyle, D.J., Garcia, R.A., Wen, Z., Producing Docosahexaenoic Acid (DHA)-Rich Algae from Biodiesel-Derived Crude Glycerol: Effects of Impurities on DHA Production and Algal Biomass Composition. *J. Agric. Food Chem.*, 56, 3933-3939, 2008.
124. Grady, O.J., Morgan, J.A., Heterotrophic growth and lipid production of *Chlorella protothecoides* on glycerol. *Bioprocess Biosyst. Eng.*, 34, 121-125, 2011.
125. Lu, Y., Ding, Y., Wu, Q., Simultaneous saccharification of cassava starch and fermentation of algae for biodiesel production. *J. Appl. Phycol.*, 23, 115-121, 2011.
126. Lu, Y., Zhai, Y., Liu, M., Wu, Q., Biodiesel production from algal oil using cassava (*Manihot esculenta* Crantz) as feedstock. *J. Appl. Phycol.*, 22, 573-578, 2010.
127. Gao, C., Zhai, Y., Ding, Y., Wu, Q., Application of sweet sorghum for biodiesel production by heterotrophic microalga *Chlorella protothecoides*. *Applied Energy*, 87, 756-761, 2010.
128. Patel, A., Matsakas, L., Hrůzová, K., Rova, U., Christakopoulos, P., Biosynthesis of Nutraceutical Fatty Acids by the Oleaginous Marine Microalgae *Phaeodactylum tricornutum* Utilizing Hydrolysates from Organosolv-Pretreated Birch and Spruce Biomass. *Mar. Drugs*, 17, 119, 2019.
129. Khare, S.K., Pandey, A., Larroche, C., Current perspectives in enzymatic saccharification of lignocellulosic biomass. *Biochemical Engineering Journal*, 102, 38-44, 2015.
130. Patel, A., Matsakas, L., Hrůzová, K., Rova, U., Christakopoulos, P., Heterotrophic cultivation of *Auxenochlorella protothecoides* using forest biomass as a feedstock for sustainable biodiesel production. *Biotechnol. Biofuels*, 11, 169, 2018.
131. Patel, A., Matsakas, L., Hrůzová, K., Rova, U., Christakopoulos, P., Biosynthesis of Nutraceutical Fatty Acids by the Oleaginous Marine Microalgae *Phaeodactylum tricornutum* Utilizing Hydrolysates from Organosolv-Pretreated Birch and Spruce Biomass. *Mar. Drugs*, 17, 119, 2019.
132. Ma, X., Zheng, H., Addy, M., Anderson, E., Liu, Y., Chen, P., Ruan, R., Cultivation of *Chlorella vulgaris* in wastewater with waste glycerol: strategies for improving nutrients removal and enhancing lipid production. *Bioresource Technology*, 207, 252-261, 2016.
133. Xiong, W., Li, X., Xiang, J., Wu, Q., High-density fermentation of microalga Chlorella protothecoides in bioreactor for microbio-diesel production. *Appl. Microbiol. Biotechnol.*, 78, 29–36, 2008.

134. Tan, C.K., Johns, M.R., Screening of diatoms for heterotrophic eicosapentaenoic acid production. *Journal of Applied Phycology*, 8(1), 59-64 1996.
135. de Swaaf, M.E., de Rijk, T.C., Eggink, G., & Sijtsma, L., Optimisation of docosahexaenoic acid production in batch cultivations by Crypthecodinium cohnii. In *Progress in Industrial Microbiology*, 35, 185-192, 1999.
136. Guihéneuf, F., Mimouni, V., Ulmann, L., & Tremblin, G., Combined effects of irradiance level and carbon source on fatty acid and lipid class composition in the microalga Pavlova lutheri commonly used in mariculture. *Journal of Experimental Marine Biology and Ecology*, 369(2), 136-143, 2009.
137. Fang, X., Wei, C., Zhao-Ling, C., & Fan, O., Effects of organic carbon sources on cell growth and eicosapentaenoic acid content of Nannochloropsis sp. *Journal of Applied Phycology*, 16(6), 499-503, 2004.
138. Singh, A., Wilson, S., & Ward, O. P., Docosahexaenoic acid (DHA) production by Thraustochytrium sp. ATCC 20892. *World Journal of Microbiology and Biotechnology*, 12(1), 76-81, 1996.
139. Rani, V., Mohanram, S., Tiwari, R., Nain, L., Arora, A., Beta-Glucosidase: Key Enzyme in Determining Efficiency of Cellulase and Biomass Hydrolysis. *J. Bioproces. Biotech.*, 5, 1, 2014.
140. Matsakas, L., Nitsos, C., Righavendram, V., Yakimenko, O., Persson, G., Olsson, E., Rova, U., Olsson, L., Christakopoulos, P., A novel hydrid organosolv: steam explosion method for the efficient fractionation and pretreatment of birch biomass. *Biotechnol. Biofuels*, 11, 160, 2018.
141. Sun, Y., Cheng, Y., Hydrolysis of lignocellulosic materials for ethanol production: a review. *Bioresource Technology*, 83, 1-11, 2002.
142. Ingle, A.P., Rathod, J., Pandit, R., Silva, S.S., Rai, M., Comparative evaluation of free and immobilized cellulase for enzymatic hydrolysis of lignocellulosic biomass for sustainable bioethanol production. *Cellulose*, 24, 2, 2017.
143. Rawat, I., Kumar, R.R., Mutanda, T., Bux, F., Dual role of microalgae: Phycoremediation of domestic wastewater and biomass production for sustainable biofuels production. *Applied Energy*, 88, 3411-3424, 2011.
144. Kim, G., Mujtaba, G., Lee, K., Effects of nitrogen sources on cell growth and biochemical composition of marine chlorophyte *Tetraselmis* sp. for lipid production. *Algae*, 31, 3, 257-266, 2016.
145. Wan, M.X., Wang, R.M., Xia, J.L., Julian, N.R., Nie, Z.Y., Naoko, K., George, A.O., Michael, J.B., Physiological Evaluation of a New Chlorella sorokiniana Isolate for its Biomass Production and Lipid Accumulation in Photoautotrophic and Heterotrophic Cultures. *Biotechnology and Bioengineering*, 109, 8, 1958-1964, 2012.
146. Kim, G., Mujtaba, G., Rizwan, M., Lee, K., Environmental stress strategies for stimulating lipid production from microalgae for biodiesel. *Appl. Chem. Eng.*, 25, 6, 553–558, 2014.
147. Ho, S-H., Ye, X., Hasunuma, T., Chang, J-S., Kondo, A., Perspectives on engineering strategies for improving biofuel production from microalgae: a critical review. *Biotechnol. Adv.*, 32, 8, 1448–1459, 2014.

148. Garay, L.A., Boundy-Mills, K.B., German, J.B., Accumulation of High-Value Lipids in Single-Cell Microorganisms: A Mechanistic Approach and Future Perspectives. *J. Agric. Food Chem.*, 62, 13, 2709-2727, 2014.
149. Jiang, Y., Chen, F., Effects of Temperature and Temperature Shift on Docosahexaenoic Acid Production by the Marine Microalga *Crypthecodinium cohnii*. *Journal of the American Oil Chemists' Society*, 77, 6, 613-617, 2000.
150. Bartley, M.L., Boeing, W.J., Daniel, D., Dungan, B.N., Schaub, T., Optimization of environmental parameters for *Nannochloropsis salina* growth and lipid content using the response surface method and invading organisms. *Journal of Applied Phycology*, 28, 1, 15-24, 2016.
151. Woertz, I., Feffer, A., Lundquist, T., Nelson, Y., Algae Grown on Dairy and Municipal Wastewater for Simultaneous Nutrient Removal and Lipid Production for Biofuel Feedstock. *Journal of Environmental Engineering*, 135, 1115-1122, 2009.
152. Medeiros, D.L., Sales, E.A., Kiperstok, A., Energy production from micro-algae biomass: carbon footprint and energy balance. *Journal of Cleaner Production*, 96, 493-500, 2015.
153. Doucha, J., Lívansky, K., Production of high-density Chlorella culture grown in fermenters. *J. Appl. Phycol.* 24, 1, 35-43, 2011.
154. Oncel, S., Sabankay, M., Microalgal biohydrogen production considering light energy and mixing time as the two key features for scale-up. *Bioresource Technology*, 121, 228–234, 2012.
155. Duan, Y., & Shi, F. Bioreactor design for algal growth as a sustainable energy source. In *Reactor and Process Design in Sustainable Energy Technology*, 27-60, 2014.
156. Xu, L., Weathers, P.J., Xiong, X.R., & Liu, C.Z. Microalgal bioreactors: challenges and opportunities. *Engineering in Life Sciences*, 9(3), 178-189, 2009.
157. Kumar, K., Dasgupta, C.N., Nayak, B., Lindblad, P., & Das, D., Development of suitable photobioreactors for CO2 sequestration addressing global warming using green algae and cyanobacteria. *Bioresource Technology*, 102(8), 4945-4953, 2011.
158. Chiang, S.H., Shi, F., & Gu, X., A new development in flotation process. *Journal of the Chinese Institute of Chemical Engineers*, 34(1), 7-15, 2003.
159. Pulz, O., & Scheibenbogen, K, Photobioreactors: design and performance with respect to light energy input. In *Bioprocess and Algae Reactor Technology, Apoptosis*, 123-152, 1998.
160. Sevilla, J.F., Grima, E.M., Camacho, F.G., Fernández, F.A., & Pérez, J.S., Photolimitation and photoinhibition as factors determining optimal dilution rate to produce eicosapentaenoic acid from cultures of the microalga Isochrysis galbana. *Applied Microbiology and Biotechnology*, 50(2), 199-205, 1998.
161. Gao, M., Song, X., Feng, Y., Li, W., & Cui, Q., Isolation and characterization of Aurantiochytrium species: high docosahexaenoic acid (DHA) production by the newly isolated microalga, Aurantiochytrium sp. SD116. *Journal of Oleo Science*, 62(3), 143-151, 2013.

162. Bumbak, F., Cook, S., Zachleder, V., Hauser, S., & Kovar, K., Best practices in heterotrophic high-cell-density microalgal processes: achievements, potential and possible limitations. *Applied Microbiology and Biotechnology*, 91(1), 31, 2011.
163. Shi, X.M., Wu, Z.Y., Chen, F., Kinetic modeling of lutein production by heterotrophic Chlorella at various pH and temperatures. *Mol. Nutr. Food. Res.* 50(8), 763–768, 2006.
164. Jain, R., Raghukumar, S., Sambaiah, K., Kumon, Y., Nakahara, T., Docosahexaenoic acid accumulation in thraustochytrids: search for the rationale. *Mar. Biol.* 151(5), 1657–1664, 2007.
165. Jakobsen, A.N., Aasen, I.M., Josefsen, K.D., Strom, A.R., Accumulation of docosahexaenoic acid-rich lipid in thraustochytrid Aurantiochytrium sp. strain T66: effects of N and P starvation and O–2 limitation. *Appl. Microbiol. Biotechnol.* 80(2), 297–306, 2008.
166. Sansawa, H., Endo, H., Production of intracellular phytochemicals in Chlorella under heterotrophic conditions. *J. Biosci. Bioeng.* 98 (6), 437–444, 2004.
167. Sun, N., Wang, Y., Li, Y.T., Huang, J.C., Chen, F., Sugar-based growth, astaxanthin accumulation and carotenogenic transcription of heterotrophic Chlorella zofingiensis (Chlorophyta). *Process Biochem.* 43(11),1288–1292, 2008.
168. Ogbonna, J.C., Tomiyama, S., Tanaka, H., Heterotrophic cultivation of Euglena gracilis Z for efficient production of alphatocopherol. *J. Appl. Phycol.* 10(1), 67–74, 1998.
169. Schmidt, R.A., Wiebe, M.G., Eriksen, N.T., Heterotrophic high celldensity fed-batch cultures of the phycocyanin-producing red alga Galdieria sulphuraria. *Biotechnol. Bioeng.* 90(1), 77–84, 2005.
170. Brennan, L., Owende, P., Biofuels from microalgae—a review of technologies for production, processing, and extractions of biofuels and co-products. *Renew. Sustain. Energy Rev.*, 14, 2, 557-577, 2010.

4

Lipid and Poly-Unsaturated Fatty Acid Production by Oleaginous Microorganisms Cultivated on Hydrophobic Substrates

Markella Tzirita[1], Bríd Quilty[2,3] and Seraphim Papanikolaou[1*]

[1] Laboratory of Food Microbiology and Biotechnology, Department of Food Science and Technology, Agricultural University of Athens, Athens, Greece
[2] School of Biotechnology, Dublin City University, Dublin 9, Ireland
[3] National Institute for Cellular Biotechnology (NICB), Dublin City University, Dublin 9, Ireland

Abstract

Microorganisms that accumulate more than 20% (w/w) of their biomass as lipid are known as oleaginous microorganisms. The lipids produced by these microorganisms are referred to as single-cell oil (SCO). SCOs are cleaner and less expensive to produce than oils obtained from plant and animal sources. They can have similar composition to valuable fats such as cacao-lipid or they may contain polyunsaturated fatty acids (PUFAs) such as dihomo-γ-linolenic acid, eicosapentaenoic acid, docosahexaenoic acid and arachidonic acid often not present in the plant or animal kingdoms. High-value lipids produced by microorganisms are important in the food and healthcare industry. Their production is cost effective while the price of production of lipids and fats of the plant and animal kingdom can vary considerably. SCO production can also have a beneficial effect on the environment. Hydrophobic substrates, including waste streams such as wastewater containing fats, oils and greases (FOGs) can be used thereby preventing any negative environmental impact from the waste stream. The ability of oleaginous microorganisms to change the composition of waste fats and their properties into value-added end products with no need for chemical catalysts, which are difficult to remove, is a further advantage and of particular interest.

Keywords: biodegradation, oleaginous, poly-unsaturated fatty acids, single-cell oil value-added lipids, waste fats, lipids

**Corresponding author*: spapanik@aua.gr

Alok Kumar Patel and Leonidas Matsakas (eds.) Nutraceutical Fatty Acids from Oleaginous Microalgae: A Human Health Perspective, (115–144) © 2020 Scrivener Publishing LLC

4.1 Lipid Production (Single Cell Oil)

There are two metabolic pathways involved in microbial lipid production or lipid biosynthesis. The *ex novo* pathway uses hydrophobic substrates and the *de novo* pathway which is used when microorganisms are cultivated on hydrophilic substrates [1, 2]. The *de novo* biosynthesis of lipids is a secondary metabolic process. It occurs when one or more of the nutrients, such as N, P, K, Mg, S or Fe, required for cell proliferation, are limiting, while carbon continues to be available. Nitrogen (N) is usually the limiting factor for lipid accumulation. Under these conditions and following complete nitrogen consumption, the existing cells continue to assimilate the available carbon source and convert it into intracellular lipids. Single-cell oil (SCOs) are mainly composed of triacylglycerols (TAGs) which account for 90% of storage lipids. Of these TAGs, 44% contain unsaturated fatty acids, followed by free fatty acids, monoacylglycerols, diacylglycerols and steryl esters, as well as low levels of sterols and polar lipids (glyco-, sphingo-, phospholipids) [3–6]. TAGs are formed through the metabolic pathway Kennedy, which uses 3-phospho-glycerol and acetyl-CoA as substrate (Figure 4.1). According to the literature, the main fatty acids formed are myristic acid (C14:0), palmitic acid (C16:0), palmitoleic acid (C16:1), stearic acid (C18:0), oleic acid (C18:1), linoleic acid (C18:2) and γ-linolenic acid (C18:3) [2]. Enhancement of SCO accumulation is possible after optimizing the process parameters as well as the culture conditions regarding

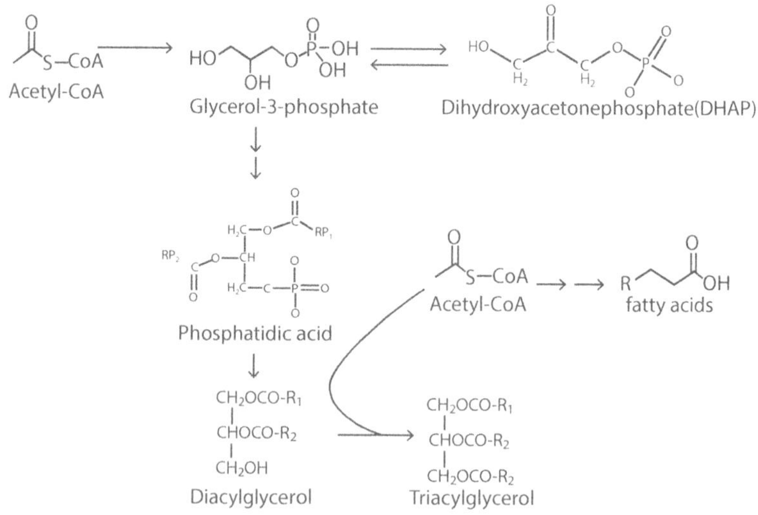

Figure 4.1 Biosynthesis of intracellular lipids [1].

C/N ratio, nitrogen source and oxygen demand using novel process modulations such as solid-state/semi-solid-state fermentation (SSF) [7]. A wide variety of hydrophilic substrates have been investigated for *de novo* SCO production from oleaginous yeasts and fungi and include sugar-based substrates both simple (glucose, fructose, xylose, lactose, sucrose) and complex sugars (starch, pectin), whey, glucose-enriched wastes, molasses, ethanol, glycerol, low-molecular weight organic acids such as citric acid and acetic acid and other low-cost renewable substrates including starch hydrolysate and tomato waste hydrolysate [8].

The *ex novo* process occurs in the presence of hydrophobic substrates and is a growth- associated process not requiring nitrogen-limiting conditions [9–12]. Following hydrolysis of the hydrophobic substrate, the free fatty acids enter the cell and are either completely degraded to acetyl-CoA which enters the Krebs cycle or they accumulate in the cells. Therefore, production of cellular lipids occurs through the "biomodification" of the employed fatty substrates (i.e., the creation of cocoa-butter or other exotic fat substitutes or the synthesis of SCOs containing FAs that are "bioactive", like γ-linolenic acid, eicosapentaenoic acid etc., through elongation or dehydrogenation reactions of the *ex novo* pathway incorporated inside the cell's FAs) [1]. The ability of the oleaginous microorganisms to modify the structure of fats and change their properties is called bioconversion. The most advantageous is that the bioconversion occurs without needing chemical catalysts, which are difficult to remove [1, 13–15]. In general, hydrolysis, esterification, interesterification and transesterification are the most important biological processes in the industrial sector [16–19]. Moreover, the approach to bio-convert long-chain fatty acids to more useful fatty acids has been investigated instead of a complete oxidation process. Cheap and low-cost fatty substrates have been successfully used for the production of value-added products, such as lipid production from palm oil and stearin [20]. Moreover, lipid production from industrial lipids has been reported [21]. The produced SCOs from those substrates have a similar composition to equivalent valuable plant and animal fats such as cacao-lipid, cocoa butter, sal fat and shea fat [12]. Additional, animal fats, which are commonly rich in saturated fatty acids, have been also reported to be bio-converted to lipids rich in unsaturated fatty acids [22]. Produced SCO may contain polyunsaturated fatty acids (PUFAs) that in many cases are not present in the plant or animal kingdoms. These PUFAs such as dihomo-γ-linolenic acid, eicosapentaenoic acid, docosahexaenoic acid and arachidonic acid are valuable because they can be used for medical and dietary purposes. In general, high-value lipids produced by microorganisms are important in the food and healthcare industry; they are cost effective to produce while the price of various naturally occurring lipids and fats of the plant and animal kingdom can vary considerably [12].

4.2 Lipid Biodegradation and Synthesis

Biodegradation of lipids is difficult due to their low bioavailability and the hydrophobic properties of the fatty acids they contain [23]. Biodegradation by microorganisms is generally a growth-associated process, in which the microbial population takes up the carbon source of the substrate for its growth needs. More specifically, the energy required for the biosynthetic reactions is released and the by-products of the reactions are converted to cell constituents, with a consequently increase of the microbial population in number and biomass [24].

Specific enzymes, called lipases, catalyze lipid hydrolysis, such as the enzyme lipase (triacylglycerol acylhydrolase, EC 3.1.1.3) which catalyzes the hydrolysis of TAGs to diacylglycerols, monoacylglycerols, fatty acids and glycerol under suitable conditions [25, 26]. The microbes will degrade the fat and consume the produced fatty acids for synthesis of cellular material and growth, but they may also transform them, changing the intracellular fatty acid concentration and producing new fatty acids. Figure 4.2 illustrates the main principles of an oleaginous heterotrophic microorganism cultivated on lipids. Aerobic degradation of lipids involves the following three steps;

1. Emulsification of the substrate (ideally producing biosurfactants): Metabolic processes for optimising the contact between the microbial cell and the organic pollutants which require biosurfactant production and emulsification (Figure 4.3),

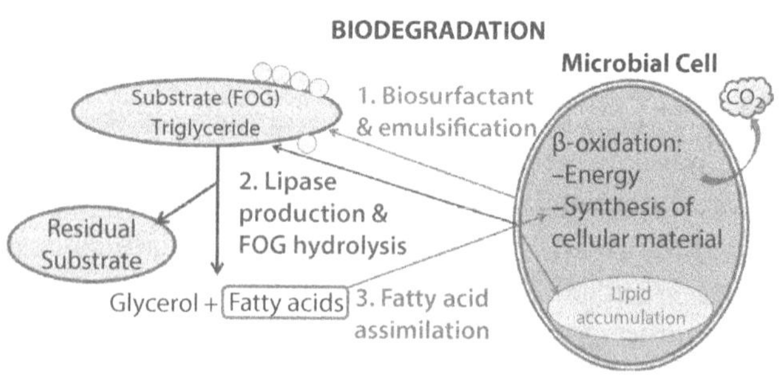

Figure 4.2 Main principles of aerobic biodegradation of lipids [31, 32].

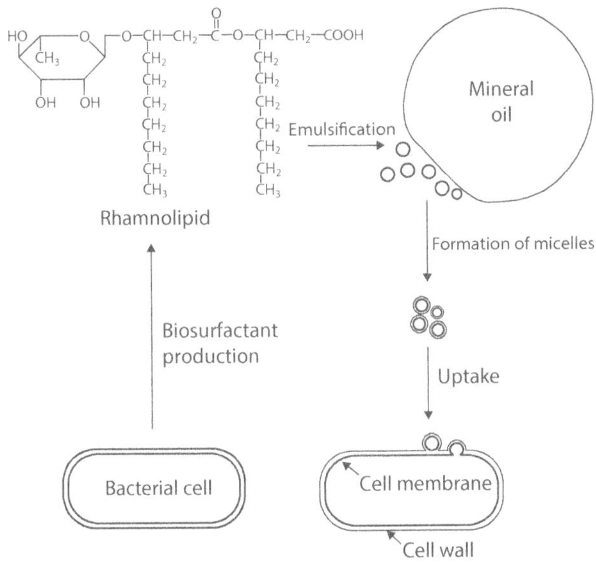

Figure 4.3 Biosurfactant production and lipid emulsification [27].

2. Production of the enzymes to hydrolyze the substrate: Hydrolysis of the fatty substrate to glycerol and fatty acids by extracellular enzymes (lipases) (Figure 4.4). Glycerol, following lipid hydrolysis or when supplied as a substrate, is taken up by the cells and enters the glycolytic pathway (Figure 4.5).

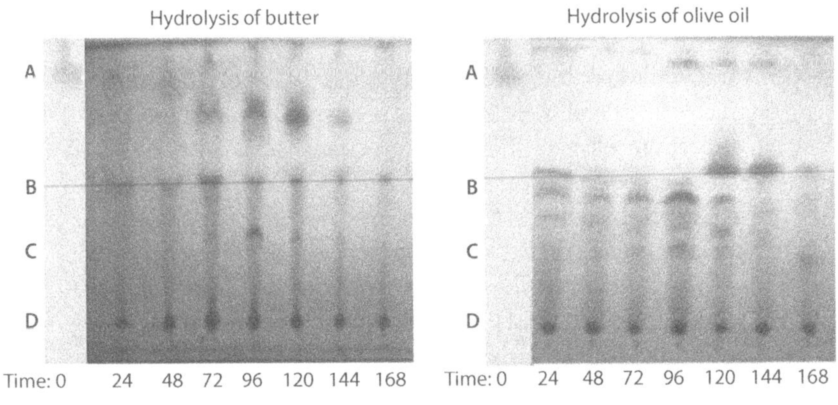

Figure 4.4 Hydrolysis of fat substrates (waste butter and waste olive oil) using TLC analysis (A – triacylglycerols; B – free fatty acids; C – diacylglycerols; D – monoacylglycerols) [33].

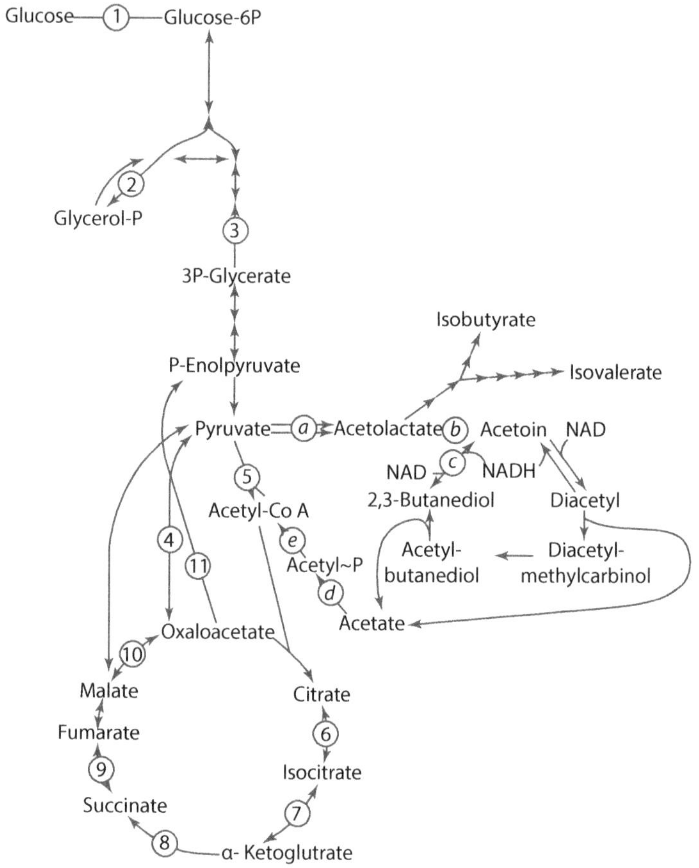

Figure 4.5 Glycerol degradation through biochemical pathways of glycolysis and TCA [34].

3. Assimilation of the released fatty acids: Transportation into the microbial cell followed by the act of oxygenases and peroxidases to activate oxygen. Conversion of fatty acids to acetyl-CoA which enters the tricarboxylic acid (TCA) cycle takes place in degradation pathways (β-oxidation) [27–29]. In case of an odd-chain fatty acid, then the last fragment is propionyl-CoA, which is converted to acetyl-CoA through a variety of possible pathways. Combination of β-oxidation of fatty acids, the TCA cycle and respiratory chain, provides more energy per carbon atom than any other energy source (Figure 4.6) [30].

The enzymatic capability of its microorganism expressed by the incorporation of the fat substrate into the microbial cell, followed by modifications

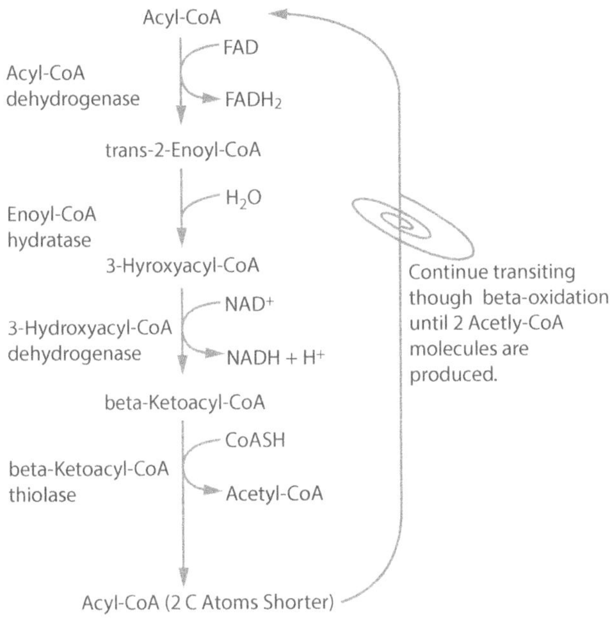

Figure 4.6 Beta oxidation process (http://flipper.diff.org/).

of the intracellular fatty acids. The produced oil is called SCO [3, 17, 35–37]. Further intracellular metabolism involves three distinct phases (Figure 4.2), which are, in general, associated with oleaginous heterotrophic microorganisms cultivated on substrates with a high C:N ratio [38, 39]:

1. The balanced growth phase: Biosynthesis of cell biomass from the central precursor metabolites, e.g., acetyl-CoA, succinate, pyruvate, occurs. All nutrients in the growth environment are in sufficient amounts to sustain growth.
2. The lipid production phase: Cell proliferation ceases due to the exhaustion of some essential nutrients such as nitrogen and magnesium and the carbon source, found in the growth medium in excess, is converted into storage lipid. Moreover, the microbes will hydrolyze the fat and assimilate the produced fatty acids for growth and/or will transform them changing the intracellular fatty acid concentration and producing new fatty acids. Commonly, citric acid (in low concentrations) is excreted in the growth medium in the oleaginous phase.

3. The phase of reserve lipid degradation: Once the carbon source of the substrate is totally consumed, the cells will use the storage lipid as an intracellular carbon source, entering that way to the lipid degradation phase.

4.3 Hydrophobic Substrates

A variety of hydrophobic substrates produced during industrial processes have been used for SCO production both as substrate and co-substrate (in combination with a hydrophilic substrate). Examples of fatty material used as substrates for lipid accumulation are vegetable oils, fatty alkyl esters (methyl, ethyl, butyl or vinyl esters of fatty acids), pure free fatty acids, waste cooking oil, fish oils and industrial fats composed of free fatty acids of animal or vegetable origin [2, 3, 33, 37, 40–42]. SCO production using wastewater rich in fats, oils and greases also leads to remediation of the wastewater with a consequent positive impact on the environment.

4.3.1 Waste Fats, Oils and Grease (FOG)

FOG deposits are becoming a global challenge for sanitation development schemes [43–47]. The presence of FOGs in wastewater can lead to the production of foul odors, the blockage of sewer lines and consequently floods, as well as interference with the proper operation of sewage treatment operation, causing foaming and the growth of filamentous bacteria [47, 48]. FOG is generated in high amounts from food facilities, including the restaurant trade, fast-food outlets, hotels, canteens and domestic sources, the dairy industry and food processing plants and leads to potential blockages and wastewater management problems [49]. FOG deposits can be formed also as a result of a saponification reaction between fatty acids and calcium ions released from biologically induced concrete corrosion [44, 50, 51]. The majority of blockages in sewers are caused by FOG and so solutions to the problems caused by FOG are very topical. The impact on the UK sewer system of FOG has grown over the last 20 years. According to British Water, water companies are responding to over 370,000 sewer blockages every year, 80% of which are caused by excessive build-up of FOG. In extreme situations enormous "fatbergs" develop in sewers. Fatbergs are large masses of waste material bound together by FOGs. The most notorious fatberg was one discovered in Whitechapel, London, in September 2017. It was 250m long and was estimated to weigh

130 tonnes. These blockages are costing the country (government, businesses and customers) more than £88 million a year, excluding additional costs, such as when sewers collapse and the effect this has on people's lives and businesses. Situations of extreme rainfall, combined with FOG blockages and sewer failure, will lead to severe flooding, environmental pollution and property damage. The removal of FOG from wastewater is critically important to ensure that wastewater is disposed of efficiently and economically, avoiding blockages of sewers and problems in municipal wastewater treatment plants [43–45, 47]. In Ireland, all food producing facilities must comply with the Water Pollution Act 2007 which limits FOG discharge to less than 100 mg/L. In order to reduce the levels of FOG, physical and biological treatments are used. For the physical treatment, FOG is intercepted at source using grease traps, specially designed units that capture the fats, oils and greases allowing the water element of the waste to enter the drainage system [47]. Biological treatment involves the use of microorganisms usually used in the form of bioaugmentation products [49, 52].

4.3.2 Olive-Mill Wastewater (OMW)

OMW, the effluent from the production of olive oil, is considered one of the most challenging agro-industrial wastes to treat. This important residue of the olive oil industry is difficult to treat largely because of its high content in phenolic compounds [53, 54]. The strong odor and dark color associated with this wastewater as well as the relatively high organic load has a direct negative impact on the environment if left untreated. OMW with high concentrations of organic matter and phenolic compounds will result in a reduction of available oxygen to organisms in the aquatic environment if released without prior treatment. Lack of oxygen is disadvantageous to the environment and the soil porosity, upsetting the balance of ecosystems, having as a result contaminated and polluted aquifers and environments [53, 55, 56]. Consequent problems are the production of odors and the appearance of eutrophication. The organic matter in OMW consists of sugars, phenols, tannins, polyphenols, aromatic molecules, ash and in some cases lipids and nitrogen [54]. The phenolic compounds are mainly considered responsible for the toxicity of the OMW constituting in several cases the limiting step of their large-scale management [57–60]. In some cases where the traditional press extraction process is applied, OMW containing very high concentrations of sugars (mostly glucose, in concentrations ≥65 g/L) are produced [53]. Recently, a new approach to the valorization of these wastewaters has developed involving the use of the waste stream as substrate in various fermentation processes. Metabolites of high added-value can be produced during these fermentation processes with simultaneous partial detoxification, decolorization and phenol removal, of the residue [54–56, 61–64]. Species of the yeast *Yarrowia lipolytica* are ideal candidates for remediation and valorization of OMW as well as of other recalcitrant wastes and residues, due to their ability to grow vigorously on a variety of substrates and to tolerate hostile growth media [64, 65]. Many studies have reported fermentations with OMW blended with hydrophilic substrates, such as glycerol, for enhanced production of metabolic compounds [56, 64]. Sarris *et al.* [56] reported significant increased amounts of produced lipids after the addition of OMW to a glycerol-based medium. The continuous development of biofuels, biodiesel and bioethanol, leads to the production of crude glycerol as a by-product [66]. The synthesis of 10.0 kg of biodiesel derived from trans-esterification of various oils, generates c. 1.0 kg of glycerol (purity ≈90.0% w/w) [67, 68]. In recent years, biodiesel production has a global increased development. In 2016, more than 30.8×10^6 tons were produced (7.5% more than in 2015) and it is estimated that more than 3.0×10^6 tons of glycerol as an important industrial side-product were produced from biodiesel manufacture [69]. Glycerol is possible to be converted to a range of value-added composites, included SCO, by eukaryotic microorganisms (SCOs) [70–73]. Biodiesel-derived glycerol when used as a carbon source in industrial microbiology has many advantages over conventional substrates such as commercial sugars. One such advantage is the significantly low cost of this renewable material. The production of biodiesel currently generates highly purified (e.g., purity of ≈85–90% w/w, sometimes >90% w/w) industrial glycerol as a by-product. At the beginning of 2016, the cost of this substrate in the Asian market was as low as *ca* 0.16 US $/kg, and falling. The substrate also has excellent nutritional properties, containing inorganic components such as potassium, calcium, sulfur and magnesium favorable for microbial growth [8]. In some cases, biodiesel production generates low-quality glycerol, the so-called "pitch" glycerol. The disposal of this waste stream is usually by incineration, adding to climate relevant emissions (CO_2 and N_2O) and to the cost of the entire biodiesel production chain. The use of this by-product in microbial fermentations can mitigate any negative environmental impact.

4.4 Oleaginous Microorganisms

Microorganisms which can accumulate over 20% (w/w) of their biomass as lipid are known from the early 1940s and have been termed "oleaginous". The lipids obtained from these microorganisms are considered as single-celled oil (SCO), and it is cleaner and less expensive to produce than oils from agricultural and animal sources. These microorganisms are of particular interest because of their capacity to convert cheap raw materials (such as fatty substrates, carbohydrates and alcohols) into value-added end products such as fats and oils with a similar composition to cacao-lipid and other valuable plant derived fats [2, 33, 37, 40–42]. According to Papanikolaou and Aggelis [2], a sum of the accumulated saturated fatty acids (%) higher than 50% corresponds to a quality similar to that of plant-derived cocoa-butter or other exotic fats. The intracellular synthesis of lipids is limited to a relatively small number of microorganisms, including specific genera and species of microalgae, yeasts, filamentous fungi as well as bacteria [74–76]. Oleaginous microorganisms are significant alternative sources of oils having considerable biotechnological interest and low production cost by using low-cost renewable materials, such as agro-industrial by-products as substrates [4–6, 39, 77–81]. SCOs are in most cases rich in PUFAs, which are currently used in animal and human nutrition [5, 82]. Furthermore, hen grown on certain wastes such as vegetable oil, animal fat, cooking oil or waste grease, oleaginous microorganisms are considered potential candidates for the production of non-conventional oil such as "2nd generation" biodiesel, a more environmentally friendly alternative to fossil fuels [2].

In terms of SCOs production, oleaginous yeast have many advantages over bacteria, molds and microalgae due to their high growth rate and lipid yield. The advantages of using yeast as lipid producers include 1) the production of lipids similar to vegetable oils and fats, 2) the good growth on cheap agro-industrial and food industrial wastes, 3) the fast rate of the lipid production in bulk in large capacity reactors than the usual time-consuming agricultural practices and 4) the non-toxicity of the yeasts and their products to humans [40]. The most well-known lipid-producing genera are: *Candida*, *Cryptococcus*, *Rhodotorula*, *Trichosporon*, and *Yarrowia* [12], which can accumulate 90% w/w of the total storage lipids as TAGs and sometimes higher than 80% w/w of this total lipids include unsaturated fatty acids (such as linoleic and oleic acids). A restricted number of yeast strains have been recorded to be capable of growing on fats and at the same time accumulate significant lipid quantities. The oleaginous yeast *Y. lipolytica* is one of the most extensively studied "unconventional" yeast. *Y. lipolytica* yeasts can produce a range of value-added compounds during growth on several types of agro-industrial co-products and residues of either fatty or hydrophilic chemical structure [1, 83–86]. Investigations of *Y. lipolytica* grown on animal derived fats and glycerol revealed the production of SCO rich in saturated fatty acids, such as stearic acid, with potential as cocoa butter substitute, due to equivalent synthesis and value [10, 21, 87, 88]. Very appealing have been identified the possible commercial of cocoa-butter-like oil [10, 11, 21]. *Y. lipolytica* NCIM 3450 grown on waste-cooking oil successfully produced SCO, with intracellular lipid content of 0.45 g/g and a lipid production of 2.45 g/L [89]. Glucose (65 g/L) was used in combination with OMW using strain *Y. lipolytica* ACA-DC 50109 [63]. SCO production was 2 g/L with a conversion yield of 0.2 g/g. Animal fat, a by-product of the meat industry, was also used for the production of SCO by the *Y. lipolytica* strain ACA-DC 50109 and intracellular lipid content accumulated to 0.44–0.54 g/g [11]. Oleaginous yeast, *Cryptococcus curvatus*, also showed great potential to bioconvert a waste substrate. When grown on 20 g/L waste cooking oil very high lipid content of 70.13 ± 1.65% w/w was achieved with a lipid concentration of 13.06 ± 0.92 g/L. When 40 g/L waste cooking-oil was used as substrate, the yeast produced 20.34 g/L lipids with a lipid content of 57.05% w/w [90].

The production of microbial PUFAs, with high nutritional value such as eicosapentaenoic and arachidonic acids, has also been investigated [91, 92]. In general, oleaginous yeasts and moulds can produce lipids rich in PUFAs of medical and nutritional interest such as γ-linolenic acid (Δ6,9,12 C18:3), dihomo-γ-linolenic acid, arachidonic acid, docosahexaenoic acid, eicosapentanoic acid [2]. *Mucor circinelloides* was used to convert linoleic acid from sunflower oil to γ- linolenic acid, an acid of particular interest in both pharmaceutical and cosmetic industries [93]. *M. circinelloides* is a widely studied microorganism for the production of intracellular lipids. Szczęsna-Antczak *et al.* [94] reported production of 54% lipid in DCW (dry cell weight; w/w) of *M. circinelloides* cultivated on olive oil. The addition of glucose as a co-substrate increased the lipid yield to 65% (w/w). lipid in DCW in the presence of glucose as a co-substrate on the hydrophobic medium. The authors also reported high lipid production by *Mucor racemosus.* When the organism was grown on olive oil, 46.7% (w/w) lipid was produced which increased to 70.8% (w/w) lipid following the addition of glucose to the medium. High lipid production values ranging from 40-56.6% (w/W) lipid has been also reported by Szczęsna-Antczak *et al.* [95] for a variety of hydrophobic media [95]. Oleaginous Zygomycetes including *Amylomyces rouxii, M. isabellina* and *M. circinelloides* have been shown to be excellent producers of SCO and also to be capable of synthesizing lipolytic enzymes in significant quantities [96]. A yield of 14% (w/w) lipid has been reported for the bacterium *Rhodococcus opacus* PD630 cultivated on dairy wastewater [97, 98], while the oleaginous bacterial strain *Gordonia* sp. DG produced 40-52% (w/w) lipid when on various oils. However, only 13% (w/w) lipid content was achieved when the organism was cultivated on olive oil [98, 99]. Tzirita *et al.* [33] reported significantly high lipid values in terms of dry cell weight when a bacterial consortium (mixture of *Bacillus* spp. and a *Pseudomonas putida* strain CP1) was grown on waste-cooking oil and waste-cooking butter. Maximum values achieved for growth on the waste cooking butter were up to *ca* 63% w/w (with 1.38 g/L lipid production) with up to *ca* 42% w/w (with 1 g/L lipid production) for growth of the consortium on waste-cooking olive oil. This is the first known report of such high lipid yield values by bacteria of these genera cultivated on hydrophobic substrates.

The production of intracellular lipid production by microorganisms growing on hydrophobic substrates or blends of hydrophobic and hydrophilic substances, is summarised in Table 4.1. The table illustrates the range of values for lipid production that can be achieved.

In many cases, the accumulated lipid composition is almost the same with that of the substrate, as previously reported [16, 33, 90], whereas, in a few cases there are remarkable differences of the fatty acid composition of the intra-cellular free-fatty acids present in comparison with the substrate TAGs (Table 4.2). Table 4.2 describes the fatty acid composition (%, w/w) of a range of substrate fats (S) and the fatty acid composition (%, w/w) of the cellular lipids (C) produced during growth of various oleaginous microorganisms on these substrates.

4.5 Conclusions

SCO is produced by oleaginous microorganisms and compares very favorably to oils produced by members of the plant and animal kingdoms. The SCO can have a similar composition to equivalent valuable plant fats such as cacao-lipid and it can contain unique PUFAs which are high value-added lipids and important in the food and healthcare industry. The microbial process also has many advantages over other oil producing methods. The microorganisms used are non-pathogenic, no chemical catalysts are needed and cheap, readily available substrates may be used. The possibility of converting waste streams to valuable product means the process can be sustainable and environmentally friendly.

Table 4.1 Intracellular lipid production by microorganisms growing on hydrophobic substrates (or blends of hydrophobic and hydrophilic substances).

Carbon source	Microorganism	DCW (g/L)	Lipid in DCW (%, w/w)	Reference
	Zygomycetes			
Sunflower oil	*M. circinelloides*[a]	5.4	36.8	[9]
Sunflower oil and glucose	*M. mucedo*[a]	19.4	62.0	[100]
Sunflower oil and glucose	*M. plumbeus*[a]	19.8	60.0	[100]
Sunflower oil and glucose	*C. echinulata*[b]	18.3	58.0	[100]
Sunflower oil and glucose	*C. elegans*[b]	18.0	53.0	[100]
Sunflower oil and glucose	*T. elegans*[c]	18.6	56.0	[100]
Sunflower oil and glucose	*R. microscopus*[d]	18.2	43.0	[100]
Sunflower oil and glucose	*R. stolonifer*[d]	18.9	63.0	[100]
Edible-oil-containing waste (sunflower oil ~80% w/w)	*C. echinulata*[b]	24.9	24.0	[101]
Olive oil	*M. circineloides*[a]	24.1	54.0	[102]
Olive oil and glucose	*M. circineloides*[a]	39.9	65.0	[102]
Olive oil	*M. racemosus*[a]	23.1	46.7	[102]
Olive oil and glucose	*M. racemosus*[a]	32.2	70.8	[102]
Fried soybean oil	*Rhizopus* sp.	10.1	28.4	[103]

(Continued)

Table 4.1 Intracellular lipid production by microorganisms growing on hydrophobic substrates (or blends of hydrophobic and hydrophilic substances). (*Continued*)

Carbon source	Microorganism	DCW (g/L)	Lipid in DCW (%, w/w)	Reference
Canola oil	*Rhizopus* sp.	10.0	40.6	[103]
Palm oil	*Rhizopus* sp.	11.9	54.3	[103]
Sesame oil	*M. circinelloides*[a]	12.9	40.0	[103]
Canola oil	*M. circinelloides*[a]	9.6	36.6	[103]
Olive oil and glycerol	*M. circinelloides*[a]	33.8	47.3	[104]
Olive oil	*M. circinelloides*[a]	31.4	47.8	[104]
Technical fatty acids	*M. circinelloides*[a]	49.4	56.7	[104]
Fatty acid ethyl esters	*M. circinelloides*[a]	41.2	49.8	[104]
Lecithin from rape oil processing	*M. circinelloides*[a]	39.6	56.6	[104]
Sunflower oil and glucose	*Mortierella isabellina* CCF-14	9.4	50.9	[100]
Sunflower oil and glucose	strain CCF-1088	10.1	56.6	[100]
Sunflower oil and glucose	strain CCF-1098	10.3	55.2	[100]
	Bacteria			
Waste-cooking oil	*Bacillus* spp. with *Pseudomonas putida* CP1	2.4	42	[33]

(*Continued*)

Table 4.1 Intracellular lipid production by microorganisms growing on hydrophobic substrates (or blends of hydrophobic and hydrophilic substances). (*Continued*)

Carbon source	Microorganism	DCW (g/L)	Lipid in DCW (%, w/w)	Reference
Waste-cooking butter	*Bacillus* spp. with *Pseudomonas putida* CP1	2.2	63	[33]
Olive oil	*Rhodococcus opacus PD630*	0.21	19	[99]
Sesame oil		0.45	67	[99]
Castor oil		0.38	58	[99]
Cotton oil		0.32	38	[99]
Peanut oil		0.23	52	[99]
Sunflower oil		1.06	44	[99]
Olive oil	*Gordonia sp. DG*	0.56	13	[99]
Sesame oil		1.21	50	[99]
Castor oil		0.41	49	[99]
Cotton oil		0.48	50	[99]
Peanut oil		0.35	40	[99]
Sunflower oil		1.18	52	[99]
	Yeasts			
Waste cooking oil	*Yarrowia lipolytica* LFMB 20	4.4	24	[33]
Waste cooking butter	LFMB 20	4.3	20	[33]
Waste cooking oil	*Cryptococcus curvatus*	13	70	[90]

[a]: *Mucor*; [b]: *Cunninghamella*; [c]: *Thamnidium*; [d]: *Rhizopus*.

Table 4.2 Fatty acid composition (%, w/w) of substrate fat (S) and cellular lipids (C) during growth of oleaginous microorganisms on fatty substrates.

Substrate		C16:0	C18:0	C18:1	C18:2	C18:3	C20:5	C22:6	Strain	Ref.
Methylstearate	S	–	100	–	–	–	–	–	*Torupopsis* sp.	[105]
	C	4	40	48	5	1	–	–		
Vinylesters	S	35	65	–	–	–	–	–	*Torupopsis* sp.	[105]
	C	26	37	22	8	T.	–	–		
Methylesters	S	30	70	–	–	–	–	–	*Candida tropicalis*	[105]
	C	35	17	32	6	2	–	–		
Methylesters	S	30	70	–	–	–	–	–	*Trichosporon* sp.	[105]
	C	39	18	32	5	1	–	–		
Pure stearic acid	S	–	100	–	–	–	–	–	*Rhodospodidium toruloides*	[106]
	C	13	49	26	4	T.	–	–		
Corn oil	S	12	2	22	62	1	–	–	*Candida lipolytica*	[107]
	C	11	2	36	41	T.	–	–		
Palm oil	S	30	2	45	11	2	–	–	*Candida lipolytica*	[107]
	C	27	8	47	10	T.	–	–		

(Continued)

Table 4.2 Fatty acid composition (%, w/w) of substrate fat (S) and cellular lipids (C) during growth of oleaginous microorganisms on fatty substrates. (*Continued*)

Substrate		C16:0	C18:0	C18:1	C18:2	C18:3	C20:5	C22:6	Strain	Ref.
Olive oil	S	13	3	69	11	T.	–	–	*Candida lipolytica*	[107]
	C	12	3	70	13	T.	–	–		
Palmitate soaps	S	52	5	31	7	T.	–	–	*Candida lipolytica*	[108]
	C	34	3	38	17	2	–	–		
Refined soybean oil	S	10	4	22	56	8	–	–	*Aspergillus flavus*	[16]
	C[a]	9	2	9	66	12	–	–		
Soybean oil	S	10	4	22	56	8	–	–	*Candida lipolytica*	[16]
	C[a]	7	7	21	60	9	–	–		
EPO[b]	S	7	1	10	71	10	–	–	*Langermania gigantea*	[109]
	C	7	1	9	74	7	–	–		
EPO	S	7	2	11	69	7	–	–	*Rhodotorula* sp.	[109]
	C	30	5	36	25	T.	–	–		
EPO	S	7	2	11	69	7	–	–	*Candida lipolytica*	[109]
	C	8	1	13	76	T.	–	–		

(*Continued*)

Table 4.2 Fatty acid composition (%, w/w) of substrate fat (S) and cellular lipids (C) during growth of oleaginous microorganisms on fatty substrates. (*Continued*)

Substrate		C16:0	C18:0	C18:1	C18:2	C18:3	C20:5	C22:6	Strain	Ref.
Crude fish oil	S	16	6	22	3	1	9	13	*Geotrichum* sp.	[110]
	C	20	7	21	5	2	7	16		
Crude fish oil	S	16	6	22	3	1	9	13	*Candida guillermondii*	[111]
	C	20	7	29	6	2	4	10		
Refined sardine oil	S	17	5	15	2	–	22	13	*Geotrichum* sp.	[112]
	C	7	3	13	4	–	26	21		
Tuna head oil	S	18	4	26	1	–	7	27	*Geotrichum* sp.	[112]
	C	11	2	31	5		8	37		
Tuna head oil	S	18	4	26	1	–	7	27	*Candida guilliermondii*	[112]
	C	22	2	38	4		2	14		
Stearin/ HORO[c] 50/50	S	15	26	38	10	3	–	–	*Yarrowia lipolytica*	[10]
	C	14	50	24	4	–	–	–		

(*Continued*)

Table 4.2 Fatty acid composition (%, w/w) of substrate fat (S) and cellular lipids (C) during growth of oleaginous microorganisms on fatty substrates. (*Continued*)

Substrate		C16:0	C18:0	C18:1	C18:2	C18:3	C20:5	C22:6	Strain	Ref.
Stearin	S	25	52	2	T.	–	–	–	*Yarrowia lipolytica*	[11]
	C	16	80	1	–	–	–	–		
Stearin/HORO 70/30	S	20	38	22	5	1	–	–	*Yarrowia lipolytica*	[113]
	C	21	68	8	1	–	–	–		
Stearin/HORO 40/60	S	13	22	44	9	5	–	–	*Yarrowia lipolytica*	[113]
	C	16	57	23	4	T.	–	–		
Olive oil	S	11	5	79	4	T.	–	–	*Yarrowia lipolytica*	[114]
	C[a]	3	1	83	8	T.	–	–		
Olive oil	S	11	5	79	4	T.	–	–	*Yarrowia lipolytica*	[114]
	C[d]	30	24	35	5	T.	–	–		
Waste cooking olive oil	S	16	2	72	7	T.	–	–	*Aspergillus* sp.	[15]
	C	9	19	65	6	T.	–	–		

(*Continued*)

Table 4.2 Fatty acid composition (%, w/w) of substrate fat (S) and cellular lipids (C) during growth of oleaginous microorganisms on fatty substrates. (*Continued*)

Substrate		C16:0	C18:0	C18:1	C18:2	C18:3	C20:5	C22:6	Strain	Ref.
Waste cooking olive oil	S	16	2	72	7	T.	–	–	*Aspergillus niger*	[15]
	C	14	3	78	5	T.	–	–		
Waste cooking olive oil	S	11.5	2.8	73.7	11.0	–	–	–	*Y. ipolytica LEFB20*	[33]
	C	10.0	1.9	72.7	12.6	–	–	–		[33]
Waste cooking butter	S	34.4	4.6	37.4	21.1	–	–	–	*Y. ipolytica LEFB20*	[33]
	C	42.1	4.9	38.5	12.1	–	–	–		[33]
Waste cooking olive oil	S	12.0	2.6	77.6	7.0	–	–	–	Mixture of *Bacillus spp.* and *Pseudomonas putida* CP1	[33]
	C	T.	T.	54.5	45.5	–	–	–		[33]
Waste cooking butter	S	45.5	16.5	33.7	1.9	–	–	–		[33]
	C	30.7	55.0	14.2	0.0	–	–	–		[33]

T. <0.6%, *w/w*, [a]: Representation of the TAG fraction of total cellular lipids produced, [b]: EPO is evening primrose oil, [c]: Stearin: Industrial derivative of tallow composed of fully saturated free fatty acids; HORO: Fully hydrolyzed oleic rapeseed oil, [d]: Representation of the free-fatty acids fraction of total cellular lipids.

References

1. Papanikolaou, S. and Aggelis, G., *Yarrowia lipolytica*: A model microorganism used for the production of tailor-made lipids. *Eur. J. Lipid Sci. Technol.*, 112(6), 639-654, 2010.
2. Papanikolaou, S. and Aggelis, G., Lipids of oleaginous yeasts. Part I and Part II. *Eur. J. Lipid. Sci. Technol.*, 113 (8), 1031 - 1073, 2011.
3. Donot, F., Fontana, A., Baccou, J.C., Strub, C. and Schorr-Galindo, S., Single cell oils (SCOs) from oleaginous yeasts and moulds: production and genetics. *Biomass Bioenerg.*, 68, 135-150, 2014.
4. Bellou, S., Baeshen, M.N., Elazzazy, A.M., Aggeli, D., Sayegh, F., Aggelis, G., Microalgal lipids biochemistry and biotechnological perspectives. *Biotechnol. Adv.*, 32(8),1476–1493, 2014.
5. Bellou, S., Triantaphyllidou, I.E., Aggeli, D., Elazzazy, A.M., Baeshen, M.N., Aggelis, G., Microbial oils as food additives: recent approaches for improving microbial oil production and its polyunsaturated fatty acid content. *Curr. Opin. Biotechnol.*, 37, 24–35, 2016.
6. Athenaki, M., Gardeli, C., Diamantopoulou, P., Tchakouteu, S.S., Sarris, D., Philippoussis, A., Papanikolaou, S., Lipids from yeasts and fungi: physiology, production and analytical considerations. *J. Appl. Microbiol.*, 124, 336–367, 2018.
7. Kosa, M., Ragauskas, A.J., Lipids from heterotrophic microbes: advances in metabolism research FA. *Trends Biotechnol.*, 29, 53–61, 2011.
8. Huang, C., Chen, X., Xiong, L., Ma, L., Chen, Y., Single cell oil production from low-cost sub-strates: the possibility and potential of its industrialization. *Biotechnol. Adv.*, 31, 129–139, 2013.
9. Aggelis, G., Komaitis, M., Papanikolaou, S., Papadopoulos, G., A mathematical model for the study of lipid accumulation in oleaginous microorganisms. I. Lipid accumulation during growth of *Mucor circinelloides* CBS 172-27 on a vegetable oil. *Gracas y Aceites*, 46, 169–173, 1995.
10. Papanikolaou, S., Chevalot, I., Komaitis, M., Aggelis, G. and Marc, I., Kinetic profile of the cellular lipid composition in an oleaginous *Yarrowia lipolytica* capable of producing a cocoa-butter substitute from industrial fats. *Ant. Van Leeu.*, 80(3), 215-224, 2001.
11. Papanikolaou, S., Chevalot, I., Komaitis, M., Marc, I. and Aggelis, G., Single cell oil production by *Yarrowia lipolytica* growing on an industrial derivative of animal fat in batch cultures. *Applied Microbiology and Biotechnology*, 58(3), 308-312, 2002.
12. Carsanba, E., Papanikolaou, S. and Erten, H., Production of oils and fats by oleaginous microorganisms with an emphasis given to the potential of the nonconventional yeast *Yarrowia lipolytica*. *Critical Reviews in Biotechnology*, 38(8), 1230-1243, 2018.

13. Bednarski, W., Adamczak, M., Kowalewska-Piontas, J. and Zadernowski, R., Biotechnological methods for the up-grading and modification of animal waste fats. *Acta Biotechnologica,* 14(4), 387-393, 1994.
14. Aggelis, G. and Sourdis, J., Prediction of lipid accumulation-degradation in oleaginous micro-organisms growing on vegetable oils. *Antonie Van Leeuwenhoek,* 72(2), 159-165, 1997.
15. Papanikolaou, S., Dimou, A., Fakas, S., Diamantopoulou, P., Philippoussis, A., Galiotou-Panayotou, M. and Aggelis, G., Biotechnological conversion of waste cooking olive oil into lipid-rich biomass using *Aspergillus* and *Penicillium* strains. *Journal of Applied Microbiology,* 110(5), 1138-1150, 2011.
16. Koritala, S., Hesseltine, C., Pryde, E. and Mounts, T., Biochemical modification of fats by microorganisms: A preliminary survey. *Journal of the American Oil Chemists' Society,* 64(4), 509-513, 1987.
17. Ratledge, C. and Wynn, J.P., The biochemistry and molecular biology of lipid accumulation in oleaginous microorganisms. *Advances in Applied Microbiology,* 51, 1-51, 2002.
18. Gupta, R., Gupta, N. and Rathi, P., Bacterial lipases: an overview of production, purification and biochemical properties. *Applied Microbiology and Biotechnology,* 64(6), 763-781, 2004.
19. Kontkanen, H., Rokka, S., Kemppinen, A., Miettinen, H., Hellström, J., Kruus, K., Marnila, P., Alatossava, T. and Korhonen, H., Enzymatic and physical modification of milk fat: A review. *International Dairy Journal,* 21(1), 3-13, 2011.
20. Xu, X., Fomuso, L.B. and Akoh, C.C., Modification of menhaden oil by enzymatic acidolysis to produce structured lipids: optimization by response surface design in a packed bed reactor. *Journal of the American Oil Chemists' Society,* 77(2), 171-176, 2000.
21. Papanikolaou, S., Muniglia, L., Chevalot, I., Aggelis, G. and Marc, I.. Accumulation of a cocoa-butter-like lipid by *Yarrowia lipolytica* cultivated on agro-industrial residues. *Current Microbiology,* 46(2), 124-130, 2003.
22. Hou, C.T., Biotechnology for fats and oils: new oxygenated fatty acids. *New Biotechnology,* 26(1–2), 2-10, 2009.
23. Cammarota, M. and Freire, D., A review on hydrolytic enzymes in the treatment of wastewater with high oil and grease content. *Bioresource Technology.* 97 (17), 2195-2210, 2006.
24. Martin, A.M., *Biological degradation of wastes.* Kluwer Academic Pub, 1991.
25. Sharma, R., Chisti, Y., Banerjee, U.C., Production, purification, characterization and applications of lipases. *Biotechnol. Adv.*, 19 (8), 627-662, 2001.
26. Snellman, E.A., Colwell, R.R., Acinetobacter lipases: molecular biology, biochemical properties and biotechnological potential. J. *Ind. Microbiol. Biotechnol.*, 31 (9), 391-400, 2004.
27. Fritsche, W. and Hofrichter, M., Aerobic degradation by microorganisms. *Biotechnology Set,* 144-167, 2006.

28. Matsumiya, Y., Wakita, D. and Kimura, A., Isolation and characterization of a lipid-degrading bacterium and its application to lipid-containing wastewater treatment. *J. Biosci. Bioengineer.*, 103 (4), 325-330, 2007.
29. Chipasa, K.B. and Mdrzycka, K., Characterization of the fate of lipids in activated sludge. *Journal of Environmental Sciences*, *20*(5), 536-542, 2008.
30. Al-Khatib, M.A.F.R., Alam, M.Z. and Shabana, H., Isolation of bacterial strain for biodegradation of fats, oil and grease. *Malaysian Journal of Analytical Sciences* (MJAS), 19(1), 138-143, 2015.
31. Tzirita, M., A characterisation of bioaugmentation products for the treatment of waste fats, oils and grease (FOG) (Doctoral dissertation, Dublin City University), 2012.
32. Tzirita, M., Papanikolaou, S. and Quilty, B., Degradation of fat using bioaugmentation: Optimization of a bioaugmentation product. LAP LAMPERT Academic Publishing, pp. 69, ISBN: 978-3-659-96479-9, 2018.
33. Tzirita, M., Papanikolaou, S., Chatzifragkou, A., Quilty, B., Waste fat biodegradation and biomodification by *Yarrowia lipolytica* and a bacterial consortium composed of *Bacillus* spp. and *Pseudomonas putida*. *Engineering in Life Sciences*, 18, 932-942, https://doi.org/10.1002/elsc.201800067, 2018.
34. Speck, E. and Freese, E., Control of metabolite secretion in *Bacillus subtilis*. *Journal of General Microbiology*, 78(2), 261, 1973.
35. Aggelis, G. and Komaitis, M., Enhancement of single cell oil production by *Yarrowia lipolytica* growing in the presence of *Teucrium polium* L. aqueous extract. *Biotechnology Letters*, 21(9), 747-749, 1999.
36. Papanikolaou, S., Muniglia, L., Chevalot, I., Aggelis, G. and Marc, I., *Yarrowia lipolytica* as a potential producer of citric acid from raw glycerol. *Journal of Applied Microbiology*, 92(4), 737-744, 2002.
37. Papanikolaou, S., Chevalot, I., Galiotou-Panayotou, M., Komaitis, M., Marc, I. and Aggelis, G., Industrial derivative of tallow: a promising renewable substrate for microbial lipid, single-cell protein and lipase production by *Yarrowia lipolytica*. *Electronic Journal of Biotechnology*, 10(3), 425-435, 2007.
38. Dourou, M., Mizerakis, P., Papanikolaou, S. and Aggelis, G., Storage lipid and polysaccharide metabolism in *Yarrowia lipolytica* and *Umbelopsis isabellina*. *Appl Microbiol Biotechnol.*, 101, 7213–7226, 2017.
39. Dourou, M., Aggeli, D., Papanikolaou, S. and Aggelis, G., Critical steps in carbon metabolism affecting lipid accumulation and their regulation in oleaginous microorganisms. *Appl. Microbiol. Biotechnol.*, 102, 2509–2523, 2018.
40. Papanikolaou, S. and Aggelis, G., Lipid production by *Yarrowia lipolytica* growing on industrial glycerol in a single-stage continuous culture. *Bioresource Technology*, 82(1), 43-49, 2002.
41. Papanikolaou, S., Oleaginous yeasts: Biochemical events related with lipid synthesis and potential biotechnological applications. *Fermentat. Tech.*, 1 (1), e103, 2011.
42. Tzirita, M., Papanikolaou, S., Quilty, B., Enhanced fat degradation following the addition of a *Pseudomonas* species to a bioaugmentation product used

in grease traps. *Journal of Environmental Sciences*. https://doi.org/10.1016/j.jes.2018.07.008, 2018.

43. Chipasa, K.B., Mędrzycka, K., Behavior of lipids in biological wastewater treatment processes. *J. Ind. Microbiol. Biotechnol.*, 33 (8), 635-645, 2006.
44. He, X., de los Reyes, F.L., Leming, M.L., Dean, L.O., Lappi, S.E., Ducoste, J.J., Mechanisms of Fat, Oil and Grease (FOG) deposit formation in sewer lines. *Water Res.*, 47 (13), 4451-4459, 2013.
45. Williams, J.B., Clarkson, C., Mant, C., Drinkwater, A., May, E., Fat, oil and grease deposits in sewers: Characterisation of deposits and formation mechanisms. *Water Res.*, 46 (19), 6319-6328, 2012.
46. Frank, R.R, Cipullo, S., Garcia, J., Davies, S., Wagland, S.T., Villa, R., et al., Compositional and physicochemical changes in waste materials and biogas production across 7 landfill sites in UK. *Waste Manage.*, 63, 11-17, 2017.
47. He, X., So, M. J., de los Reyes, F.L., Shifts in microbial communities in bioaugmented grease interceptors removing fat, oil, and grease (FOG). *Appl. Microbiol. Biotechnol.*, 100 (16), 7025–7035, 2016.
48. Srinivasan, A., Viraraghavan, T., Biological processes for the removal of oil from wastewater - a review. *Fresenius Environmental Bulletin*, 16(12A), 1532 -1543, 2007.
49. Brooksbank, A.M., Latchford, J.W., Mudge, S.M., Degradation and modification of fats, oils and grease by commercial microbial supplements. *World J. Microbiol. Biotechnol.*, 23 (7), 977-985, 2007.
50. He, X., Iasmin, M., Dean, L.O., Lappi, S.E., Ducoste, J.J., De Los Reyes III, F.L., Evidence for fat, oil, and grease (FOG) deposit formation mechanisms in sewer lines. *Environmental Science & Technology* 45(10), 4385-4391, 2011.
51. Tang, H.L., Xie, Y.F., Chen, Y., Use of Bio-Amp, a commercial bio-additive for the treatment of grease trap wastewater containing fat, oil and grease. *Bioresour. Technol.*, 124, 52-58, 2012.
52. Tzirita, M., Quilty, B., Papanikolaou, S. A study of the suitability of three commercial bioaugmentation products for use in grease traps. *Biomass Convers. Bior.*, Received: 2 August 2019 /Revised: 25 October 2019 /Accepted: 31 October 2019, https://doi.org/10.1007/s13399-019-00549-4
53. Mantzavinos, D., Kalogerakis, N., Treatment of olive mill effluents, Part I. Organic matter degradation by chemical and biological processes-an overview. *Environ. Int.*, 31, 289–295, 2005.
54. Crognale, S., D'Annibale, A., Federici, F., Fenice, M., Quaratino, D., Petruccioli, M., Olive oil mill wastewater valorization by fungi. *J. Chem. Technol. Biotechnol.*, 81, 1547–1555, 2006.
55. Sarris, D., Stoforos, N.G., Mallouchos, A., Kookos, I.K., Koutinas, A.A., Aggelis, G., Papanikolaou, S., Production of added-value metabolites by *Yarrowia lipolytica* growing in olive mill wastewater-based media under aseptic and non-aseptic conditions. *Eng. Life Sci.*, 17, 695–709, 2017.
56. Sarris, D., Rapti, A., Papafotis, N., Koutinas, A.A., Papanikolaou, S., Production of Added-Value Chemical Compounds through Bioconversions

of Olive-Mill Wastewaters Blended with Crude Glycerol by a *Yarrowia lipolytica* Strain. *Molecules*, 24, 222, 2019.
57. Hamdi, M., Toxicity and biodegradability of olive mill wastewaters in batch anaerobic digestion. *Appl. Biochem. Biotechnol.*, 37, 155–163, 1992.
58. Zervakis, G., Balis, C., Bioremediation of olive mill wastes water through the production of fungal biomass. In *Proceedings of the Second International Conference on Mushrooms Biology and Mushrooms Products; June 9-12 1996, University Park, Pennsylvania,* Daniel Joseph Royse Ed.; Pen State University Press, pp. 311–323, 1996.
59. Tsioulpas, A., Dimou, D., Iconomou, D., Aggelis, G., Phenolic removal in olive oil mill wastewater by strains of *Pleurotus* spp. in respect to their phenol oxidase (laccase) activity. *Bioresour. Technol.*, 84, 251–257, 2002.
60. Aggelis, G., Iconomou, D., Christou, M., Bokas, D., Kotzailias, S., Christou, G., Tsagou, V., Papanikolaou, S., Phenolic removal in a model olive oil mill wastewater using *Pleurotus ostreatus* in bioreactor cultures and biological evaluation of the process. *Water Res.*, 37, 3897–3904, 2003.
61. Crognale, S., Federici, F., Petruccioli, M., β-Glucan production by *Botryosphaeria rhodina* on undiluted olive-mill waste waters. *Biotechnol. Lett.*, 25, 2013–2015, 2003.
62. D'Annibale, A., Sermani, G.G., Federici, F., Petruccioli, M., Olive-mill wastewaters: A promising substrate for microbial lipase production. *Bioresour. Technol.*, 97, 1828–1833, 2006.
63. Papanikolaou, S., Galiotou-Panayotou, M., Fakas, S., Komaitis, M., Aggelis, G., Citric acid production by *Yarrowia lipolytica* cultivated on olive-mill wastewater-based media. *Bioresour. Technol.*, 99, 2419–2428, 2008.
64. Tzirita, M., Kremmyda, M., Sarris, D., Koutinas, A.A. and Papanikolaou, S., Effect of Salt Addition upon the Production of Metabolic Compounds by *Yarrowia lipolytica* Cultivated on Biodiesel-Derived Glycerol Diluted with Olive-Mill Wastewaters. *Energies*, 12(19), 3649, 2019.
65. Aly, A.A., Hasan, Y.N., Al-Farraj, A.S., Olive mill wastewater treatment using a simple zeolite-based low-cost method. *J. Environ. Manag.*, 145, 341–348, 2014.
66. Sarris, D. and Papanikolaou, S., Biotechnological production of ethanol: biochemistry, processes and technologies. *Engineering in Life Sciences, 16*(4), 307-329, 2016.
67. Rivaldi, J.D., Sarrouh, B.F., da Silva, S.S., Development of biotechnological processes using glycerol from biodiesel production. In *Current Research Topics in Applied Microbiology and Microbial Biotechnology*; Mendez-Vilas, A., Ed.; World Scientific Publishing Co. Formatex Research Center: Madrid, Spain, pp. 429–433, 2009.
68. Wen, Z., Pyle, D.J., Athalye, S.K., Glycerol waste from biodiesel manufacturing. In *Microbial Conversions of Raw Glycerol;* Aggelis, G., Ed.; Nova Science Publishers Inc.: New York, NY, USA, pp. 1–7, 2009.

69. Monteiro, M.R., Kugelmeier, C.L., Pinheiro, R.S., Batalha, M.O., da Silva César, A., Glycerol from biodiesel production: Technological paths for sustainability. *Ren. Sustain. Energy Rev.*, 88, 109–122, 2018.
70. Bellou, S., Moustogianni, A., Makri, A., Aggelis, G., Lipids containing polyunsaturated fatty acids synthesized by Zygomycetes grown on glycerol. *Appl. Biochem. Biotechnol.*, 166, 146–158, 2012.
71. Dedyukhina, E.G., Chistyakova, T.I., Mironov, A.A., Kamzolova, S.V., Morgunov, I.G., Vainshtein, M.B., Arachidonic acid synthesis from biodiesel-derived waste by *Mortierella alpina*. *Eur. J. Lipid Sci. Technol.*, 116, 429–437, 2014.
72. Fontanille, P., Kumar, V., Christophe, G., Nouaille, R., Larroche, C., Bioconversion of volatile fatty acids into lipids by the oleaginous yeast *Yarrowia lipolytica*. *Bioresour. Technol.*, 114, 443–449, 2012.
73. Tchakouteu, S.S., Kalantzi, O., Gardeli, C., Koutinas, A.A., Aggelis, G., Papanikolaou, S., Lipid production by yeasts growing on biodiesel-derived crude glycerol: Strain selection and impact of substrate concentration on the fermentation efficiency. *J. Appl. Microbiol.*, 118, 911–927, 2015.
74. Ratledge, C., Biotechnology of oils and fats. In: Ratledge C, Wilkinson S (eds) *Microbial Lipids*. Academic, London, pp 567–668, 1989.
75. Ratledge, C., Fatty acid biosynthesis in microorganisms being used for Single Cell oil Production. *Biochemie*, 86, 807-815, 2004.
76. Cohen, Z., Ratledge, C., Single Cell oils, American Oil Chemists' Society, Champaign, IL, USA, 2005.
77. Beopoulos, A., Cescut, J., Haddouche, R., Uribelarrea, J.L., Molina-Jouve, C., Nicaud, J.M., *Yarrowia lipolytica* as a model for bio-oil production. *Prog. Lipid Res.*, 48, 375–387, 2009.
78. Bommareddy, R.R., Sabra, W., Maheshwari, G., Zeng, A.P., Metabolic network analysis and experimental study of lipid production in *Rhodosporidium toruloides* grown on single and mixed substrates. *Microb. Cell Factories*, 14, 36, 2015.
79. Arous, F., Mechichi, T., Nasri, M., Aggelis, G., Fatty acid biosynthesis during the life cycle of *Debaryomyces etchellsii*. *Microbiology - SGM*, 162, 1080–1090, 2016.
80. Arous, F., Azabou, S., Triantaphyllidou, I-E., Aggelis, G., Jaouani, A., Nasri, M., Mechichi, T., Newly isolated yeasts from *Tunisian microhabitats*: lipid accumulation and fatty acid composition. *Eng. Life. Sci.*, 17, 226–236, 2017.
81. Lazar, Z., Liu, N., Stephanopoulos, G., Holistic approaches in lipid production by *Yarrowia lipolytica*. *Trends Biotechnol.*, 36, 1157 – 1170, 2018.
82. Čertík, M., Adamechová, Z., Guothová, L., Simultaneous enrichment of cereals with polyunsaturated fatty acids and pigments by fungal solid-state fermentations. *J. Biotechnol.*, 168, 130–134, 2013.
83. Sarris, D., Rapti, A., Papafotis, N., Koutinas, A.A., Papanikolaou, S., Production of Added-Value Chemical Compounds through Bioconversions

of Olive-Mill Wastewaters Blended with Crude Glycerol by a *Yarrowia lipolytica* Strain. *Molecules*, 24, 222, 2019.
84. Papanikolaou, S., Aggelis, G., Biotechnological valorization of biodiesel derived glycerol waste through production of single cell oil and citric acid by *Yarrowia lipolytica*. *Lipid Technol.*, 21, 83–87, 2009.
85. Morgunov, I.G., Kamzolova, S.V., Lunina, J.N., The citric acid production from raw glycerol by *Yarrowia lipolytica* yeast and its regulation. *Appl. Microbiol. Biotechnol.*, 97, 7387–7397, 2013.
86. Poli, J.S., da Silva, M.A.N., Siqueira, E.P., Pasa, V.M.D., Rosa, C.A., Valente, P., Microbial lipid produced by *Yarrowia lipolytica* QU21 using industrial waste: A potential feedstock for biodiesel production. *Bioresour. Technol.*, 161, 320–326, 2014.
87. Xiong, D., Zhang, H., Xie, Y., Tang, N., Berenjian, A., Song, Y., Conversion of mutton fat to cocoa butter equivalent by increasing the unsaturated fatty acids at the sn-2 position of triacylglycerol through fermentation by *Yarrowia lipolytica*. *Am. J. Biochem. Biotechnol.*, 11, 57–65, 2015.
88. Zhao, L., Li, B., Xiong, D., Zhang, H., Song, Y., Yang, S., Cocoa-butter equivalent production from *Yarrowia lipolytica* by optimization of fermentation technology. *Am. J. Biochem. Biotechnol.*, 12, 196–205, 2016.
89. Katre, G., Joshi, C., Khot, M., Zinjarde, S., RaviKumar, A., Evaluation of single cell oil (SCO) from a tropical marine yeast *Yarrowia lipolytica* NCIM 3589 as a potential feedstock for biodiesel. *AMB Express*, 2, 36, 2012.
90. Patel, A. and Matsakas, L., A comparative study on de novo and ex novo lipid fermentation by oleaginous yeast using glucose and sonicated waste cooking oil. *Ultrasonics sonochemistry, 52*, 364-374, 2019.
91. Zeng, Y., Ji, X.J., Chang, S.M., Nie, Z.K. and Huang, H., Improving arachidonic acid accumulation in *Mortierella alpina* through B-group vitamin addition. *Bioprocess and Biosystems Engineering*, 1-6, 2011.
92. Liang, Y., Zhao, X., Strait, M. and Wen, Z., Use of dry-milling derived thin stillage for producing eicosapentaenoic acid (EPA) by the fungus *Pythium irregulare*. *Bioresource Technology*, 111(0), 404-409, 2012.
93. Aggelis, G., Balatsouras, G., Comaitis, M., Anagnostopoulou, G., Dimitroulias, G., Pina, M. and Graille, J., Gamma linolenic acid production by bioconversion of the linoleic acid of some oils. Revue Francaise des Corps Gras (France), 1991.
94. Szczęsna-Antczak, M., Antczak, T., Piotrowicz-Wasiak, M., Rzyska, M., Binkowska, N., Bielecki, S., Relationships between lipases and lipids in mycelia of two *Mucor* strains. *Enzyme Microb. Technol.*, 39, 1214–1222, 2006.
95. Szczęsna-Antczak, M., Struszczyk-Świta, K., Rzyska, M., Szeląg, J., Stańczyk, Ł., Antczak, T., Oil accumulation and in situ trans/esterification by lipolytic fungal biomass. *Bioresour. Technol.*, 265, 110–118, 2018.
96. Papanikolaou, S. and Aggelis, G., Sources of microbial oils with emphasis to *Mortierella* (*Umbelopsis*) *isabellina* fungus. *World J. Microbiol. Biotechnol.*, 35(4), 63, 2019.

97. Kumar S, Gupta N, Pakshirajan K. Simultaneous lipid production and dairy wastewater treatment using *Rhodococcus opacus* in a batch bioreactor for potential biodiesel application. *J Environ Chem Eng.*, 3, 1630–6. https://doi.org/10.1016/j.jece.2015.05.030, 2015.
98. Patel, A., Antonopoulou, I., Enman, J., Rova, U., Christakopoulos, P. and Matsakas, L., Lipids detection and quantification in oleaginous microorganisms: an overview of the current state of the art. *BMC Chemical Engineering, 1*(1), 13, 2019.
99. Gouda MK, Omar SH, Aouad LM., Single cell oil production by *Gordonia* sp. DG using agro-industrial wastes. *World J Microbiol Biotechnol.*, 24, 1703–11, 2008.
100. Čertik, M., Balteszova, L., Sajbidor, J., Lipid formation and γ-linolenic acid production by Mucorales fungi grown on sunflower oil. *Lett. Appl. Microbiol.*, 25, 101–105, 1997.
101. Roux-Van der Merwe, MP, Badenhorst, J., Britz, T.J., Fungal treatment of an edible-oil-containing industrial effluent. *World J. Microbiol. Biotechnol.*, 21, 947–953, 2005.
102. Szczęsna-Antczak, M., Antczak, T., Piotrowicz-Wasiak, M., Rzyska, M., Binkowska N, Bielecki, S., Relationships between lipases and lipids in mycelia of two *Mucor* strains. *Enzyme Microb. Technol.*, 39, 1214–1222, 2006.
103. Tauk-Tornisielo, S.M., Arasato, L.S., de Almeida, A.F., Govone, J.S., Malagutti, E.N., Lipid formation and γ-linolenic acid production by *Mucor circinelloides* and *Rhizopus* sp., grown on vegetable oil. *Braz. J. Microbiol.*, 40, 342–345, 2009.
104. Szczęsna-Antczak, M., Struszczyk-Świta, K., Rzyska, M., Szeląg, J., Stańczyk, Ł., Antczak, T., Oil accumulation and in situ trans/esterification by lipolytic fungal biomass. *Bioresour. Technol.*, 265, 110–118, 2018.
105. Matsuo, T., Terashima, M., Hashimoto, Y., Hasida, W., US Patent 4308 350, 1981.
106. Gierhart, D.L., US Patent 4485, 173, 1984.
107. Glatz, B.A., Hammond, E.G., Hsu, K.H., Baehman, L., Bati, N., Bednarski, W., Brown, D., Floetenmeyer, M., Production and modification of fats and oils by yeast fermentation, in: Ratledge, C., Rattray, J.B.M., Dawson, P.S.S., (Eds.), *Biotechnology for the oils and fats industry*. AOCS Press, Champaign (USA), pp. 163-176, 1984.
108. Montet, D., Ratomahenina, R., Galzy, P., Pina, M., Graille, J., A study of the influence of the growth media on the fatty acid composition in *Candida lipolytica* DIDDENS and LODDER. *Biotechnol. Lett.*, 7, 733-736, 1985.
109. Aggelis, G., Papadiotis, G., Komaitis, M., Microbial fatty acid specificity. *Folia Microbiol.*, 42, 117-120, 1997.
110. Guo, X., Tomonaga, T., Yanagihara, Y., Ota, Y., Screening for yeasts incorporating the exogenous eicosapentaenoic and docosahexaenoic acids from crude fish oil. *J. Biosci. Bioeng.*, 87, 184-188, 1999.
111. Guo, X., Ota, Y., Incorporation of eicosapentaenoic and docosahexaenoic acids by a yeast (FO726A). *J. Appl. Microbiol.*, 89, 107-115, 2000.

112. Kinoshita, H., Ota, Y., Concentration of docosahexaenoic acid from fish oils using *Geotrichum* sp. FO347-2. *Biosci. Biotechnol. Biochem.*, 65, 1022-1026, 2001.
113. Papanikolaou, S., Aggelis, G., Selective uptake of fatty acids by the yeast *Yarrowia lipolytica. Eur. J. Lipid Sci. Technol.*, 105, 651-655, 2003.
114. Najjar, A., Robert, S., Guérin, C., Violet-Asther, M., Carriére, F., Quantitative study of lipase secretion, extracellular lipolysis, and lipid storage in the yeast *Yarrowia lipolytica* grown in the presence of olive oil: analogies with lipolysis in humans. *Appl. Microbiol. Biotechnol.*, 89, 1947-1962, 2011.

5

Overview of Microbial Production of Omega-3-Polyunsaturated Fatty Acid

Farha Deeba, Kukkala Kiran Kumar and Naseem A. Gaur*

International Centre for Genetic Engineering and Biotechnology (ICGEB), New Delhi, India

Abstract

Omega-3 polyunsaturated fatty acids (ω-3 PUFA) are significant in terms of health benefits which led to their higher consumption as dietary supplements. The accelerating demand of ω-3 PUFA has put an additional pressure on global fish stocks. Therefore, alternative and renewable source of ω-3 PUFA production particularly docosahexaenoic acid (DHA) and eicosapentaenoic acid (EPA) have been considered to meet the present demand. Many microorganisms rich in DHA and EPA present a promising source of ω-3 PUFAs. This chapter gives an overview of microbial production platforms for ω-3 fatty acids development. These microbes have been used as bio-factories for ω-3 PUFA commercially in recent years. Current biotechnological and potential metabolic engineering approach for enhanced amount of ω-3 PUFA in microorganisms has also been discussed.

Keywords: Microorganisms, omega-3 polyunsaturated fatty acids, anaerobic polyketide synthase pathway, aerobic desaturase and elongase pathway

5.1 Introduction

Omega-3 polyunsaturated fatty acids (ω-3 PUFA) are extensively distributed in nature and are essential constituents of the membrane that surround each cell in the human body [1]. The three major ω-3 PUFAs are docosahexaenoic acid (DHA), eicosapentaenoic acid (EPA) and α-linolenic acid (ALA) [2]. They are the significant components required for

**Corresponding author:* naseem@icgeb.res.in

Alok Kumar Patel and Leonidas Matsakas (eds.) Nutraceutical Fatty Acids from Oleaginous Microalgae: A Human Health Perspective, (145–164) © 2020 Scrivener Publishing LLC

lipid metabolism and thus are essential for the human system. The ω-3 PUFAs gives energy and plays various significant functions in the body such as in endocrine and immune system, blood vessels, lungs and heart [3, 4]. The human body is unable to synthesize ALA, so it is an essential fatty acid and must be consumed from beverages and foods. ALA is further transformed into EPA and then to DHA in lesser amount which is required at higher concentration for brain and retina function [2]. Therefore, supplementation of these ω-3 fatty acids is essential in the human body.

In this regard, other alternative sources for ω-3 PUFA have been proposed such as transgenic plants, fish oil, microalgae, etc. [3]. Nowadays, microbial production of ω-3 PUFA have been considered a better alternative for commercial applications all over the world [5]. Microbial oil has several advantages of containing phytosterols or squalene which is beneficial to human health, devoid of high amount of cholesterol and unpleasant odour [4]. Also, these microbes are available throughout the year, seasonally independent and are easily cultivated [6–8]. These characteristics make microbial ω-3 PUFA production an attractive and cost-effective strategy for industrial application.

PUFA anabolic pathways have been reported in diverse kinds of microbes and several strategies for their cultivation have been studied. Single cell oils (SCO) production strategy for arachidonic acid (ARA) and DHA as main source of PUFA has also been reported in many literature [9]. The production of very long chain PUFA i.e., ARA and DHA, are commercialized by using various oleaginous microalgae and the oleaginous fungus *Mortierella alpina* [10, 11]. The development of an efficient ω-3-PUFA production system is very much needed to address a major global demand [3]. The metabolic engineering of oleaginous microorganisms for enhanced ω-3-PUFA production will further boost the efforts for industrial application. Recently, Yuan and Alper (2019) has reviewed the metabolic engineering approaches in microbial species with the advancement in ω-3 PUFA production [12]. In this chapter we discuss the overview of ω-3-PUFA production by microorganisms. In brief, the pathway involved in ω-3-PUFA biosynthesis inside microbial cells and the factors that influence their production have been discussed.

5.2 Microbial Sources of ω-3 PUFA

A variety of microorganisms can synthesize different types of PUFAs including algae, fungi, diatom and yeast (Table 5.1). The microalgae *Chlorella minutissima,* diatom *Phaeodactylum tricornutum,* the fungus *Mortieretla*

Table 5.1 Percentage of ω-PUFA in various microorganisms.

PUFA content	Microorganisms	References
EPA (ω-3 PUFA)		
EPA (31.8%)	*Chlorella minutissima* UTEX 2341	[19]
EPA (30.5%)	*Phaeodactylum tricornutum* UTEX 640	[19]
EPA (45%)	*Chlorella minutissima*	[20]
EPA (56.6%)	*Yarrowia lipolytica* Y4184	[21]
EPA (28%)	*Nannochloropsis salina*	[22]
DHA(ω-3 PUFA)		
DHA (61.3–70.5%)	*Aurantiochytrium* sp. ATCC PRA-276	[23]
DHA (31.5%)	*Thraustochytrium* sp. ONC-T18	[24]
DHA (50%)	*Crypthecodinium cohnii*	[18]
DHA (46%)	*Ulkenia sp.*	[10]
EPA/DHA/ARA (ω-3 & ω-6 PUFA)		
ARA + EPA + DHA (56.82% + 3.46% + 4.30%)	*M. alpine CFR-GV15*	[16]
EPA + DHA (36.5% + 23.6%)	*Phaeodactylum tricornutum*	[25]
EPA+DHA (41.5%)	*Pavlova lutheri*	[26]
ALA (ω-3 PUFA)		
ALA (49%)	*Rhodosporidium toruloides*	[27]
ALA (90.5%)	*Candida glabrata* UCP1566	[28]
ALA (75.61%)	*Candida glabrata* UCP1002	
ARA (ω-6 PUFA)		
ARA (35%)	*Yarrowia lipolytica* Y8051	[21]
ARA (50%)	*Mortieretla alpina*	[15]
ARA (25.2%)	*M. alpine NRRL-A-10995*	[17]

alpina, and some Thraustochytrium species are microorganisms that produce high levels of EPA, DHA and ARA [13, 14]. The *M. alpina* is a good producer of ARA [15–17] while *Ulkenia* sp., *Aurantiochytrium* sp., *Thraustochytrium* sp. and *Crypthecodinium cohnii* are good producers of DHA [18].

Pavlova lutheri, *Phaeodactylum tricornutum* and some *M. alpina* species shows higher ARA, EPA or DHA content [16, 25, 26]. The *Candida glabrata* strains (UCP 1556 and 1002) showed increased ALA production of 90.05% and 75.61% when grown on agro-industrial wastes (milk whey and corn steep liquor) [28]. The yeast *Yarrowia lipolytica* has been genetically engineered to produce higher levels of EPA, DGLA and ARA [21]. *Rhodosporidium toruloides* has also been genetically engineered by introducing ω3 desaturase (FAD3) and Δ12 desaturase (FAD2) gene which led to enhanced ALA production by 49% [27].

Some fish oils contains man-made pollutants like dioxins, polychlorinated biphenols (PCBs) and methyl mercury, and also rapid worldwide *decline* of *fish stocks* raised an issue with regard to their utilization for DHA and EPA production [2]. Consequently, microbial oils containing high DHA and EPA content are sustainable, stable, and have simple fatty acid profile as compared to fish oil. Moreover, these microbial oils contain natural antioxidants that prevent them from oxidative damage [2, 29]. Hence, several microbial species could be utilized as producers of special oils with desired specifications. For instance, infant formula particularly rich in DHA was produced from microbial sources of ω-3 oils such as *Crypthecodinium* and *Schizochytrium* species [30]. Breast milk contains long-chain fatty acids (DHA and ARA) and plays an essential role in neonatal development. However, the DHA: EPA: ARA ratio in mother's milk (1.0:0.2:2.0, w/w) is entirely dissimilar to the composition present in fish oil, so fish oils are not suitable for infant formulae in which low EPA and high DHA is favoured [2]. Various microbial oils produced low EPA and are rich in DHA so can be used in combination with other microbial oils rich in ARA to fit the infant formulae requirement. For example, DSM (Dutch State Mines) commercially developed ARA rich oils using *Mortierella alpina* and in collaboration with Martek that produced DHA (40–50%) rich microbial oil from *Crypthecodinium cohnii* [2]. Switzerland Lonza Group commercially produces DHA (45–50%) rich oil from microalgal *Ulkenia* species [2, 10].

Advance techniques such as whole genome mutagenesis have been successfully applied for improving traditional strain with enhanced PUFA production [2]. Shimizu's group performed whole genome mutagenesis of *M. Alpine* strain for commercial Di-homo-γ-linolenic acid (DGLA) rich oil and other fatty acids production [31]. Recently, *Thraustochytrid* strains have been exploited by using efficient techniques such as mutagenesis to effectively trigger mutations and to alter the fatty acid profile [29]. *Thraustochytrid* strains possess distinct features such as elevated PUFA productivity, potential of utilising low-cost carbon sources, and the capability to produce EPA [24].

The traditional methods (EMS mutagenesis, UV mutagenesis, etc.) for strain improvement to enhance the productivity of the desired products have been successful in the past, but these methods are very labour-intensive and time-consuming. Therefore, targeted genome engineering has been a more preferred, fast and efficient method for strain improvement recently [2]. *M. Alpine* was genetically engineered for enhanced ARA production by over expression of elongase (EL2) gene through *Agrobaterium tumefaciens* mediated gene delivery system [31]. DuPont has recently developed a yeast strain (*Yarrowia lipolytica*) by PUFA engineering pathway for commercial production of EPA (55%) rich oils [21]. Genetically engineered oleaginous yeast *Yarrowia lipolytica* has also been developed by Microbia, Inc. using a similar strategy for commercial production of antioxidant (carotenoid) [32]. Another strain *Thraustochytrium* sp. have been reported to produce 80% oil naturally by dry cell weight containing high DHA content (>31%) along with the significant amount of carotenoids such as canthaxanthin, zeaxanthin, astaxanthin and ß-carotene [33]. *Thraustochytrids* species has been genetically engineered by introducing Δ5-desaturase gene for enhanced EPA content indicating it is a promising strategy for generating beneficial PUFAs [34]. Hence, engineered strains with efficient transformation systems have been developed to facilitate high EPA levels, carotenoids, and to enhance DHA productivity in them. Recently, PUFA biosynthetic pathways were explained effectively knocking out FAS and PKS genes in *Schizochytrium* sp. [35]. Overexpression of the carotene synthase (cas) gene in PUFA producing Thraustochytrid (*Aurantiochytrium* sp.) make the process of microalgal lipid extraction easier with extended shelf life of oils [36]. Moreover, in *Schizochytrium* sp. overexpression of this cas gene enhanced antioxidant carotenoid productivity by about 16 times [37].

5.3 ω-3 PUFA Biosynthesis in Microbial Cells

An overview about the ω-3 PUFA biosynthesis in microorganisms is given in Figure 5.1. The continuous supply of NADPH and acetyl-CoA are required for the production of lipid under excess carbon and nutrient

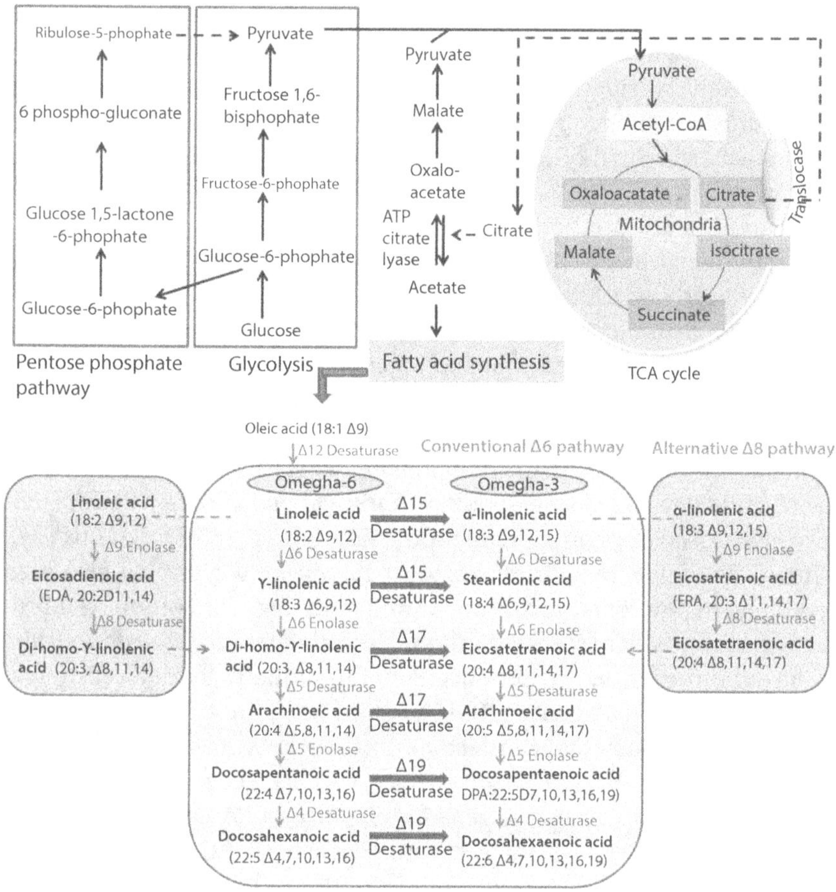

Figure 5.1 Overview of biosynthesis of ω-3 PUFA in microbial cells.

limitation condition [38]. In oleaginous organisms, the acetyl-CoA continuous production is attained by cascade of enzymatic reactions stimulated by nutrient deficient environment which leads to increased pool of citrate inside the mitochondria. An isocitrate dehydrogenase (IDH) enzyme of the kreb cycle is Adenosine mono phosphate (AMP) depended enzyme which catalyzes the oxidative decarboxylation of isocitrate in an oleaginous organism. Under nitrogen limited condition, low levels of AMP inside the mitochondria due to enhanced AMP deaminase activity converts excess AMP to IMP and ammonia. As a result, conversion of isocitrate to citrate by the aconitase enzyme does not occur further and citrate is translocated into the cytoplasm [39]. The ATP:citrate lyase (ACL) enzyme then cleaved citrate to oxalo-acetate (OAA) and acetyl-CoA leading eventually to the continuous acetyl-CoA production for lipid biosynthesis [40]. Reducing power in NADPH form is another parameter apart from acetyl-CoA essential for the fatty acids synthesis. For instance, stearic acid (C18) formation requires 16 moles of NADPH. In some literatures, malic enzyme responsible for decarboxylation of malate to pyruvate is mainly responsible for NADPH supply [38]. However, in some microorganisms, e.g., *Yarrowia lipolytica,* this malic enzyme is involved in NADPH regeneration. An alternative option for NADPH production can be via the pentose phosphate pathway [41]. The end product of lipid biosynthesis in every microorganism is either the palmitic (16:0) or stearic (18:0) acid [4, 42].

The elongation and desaturation for PUFA production occurs through subsequent series of reaction by elongase and desaturase enzymes, respectively. Therefore, the type and amount of PUFA produced depends upon the desaturases and elongases genes present inside the genome of the respective microbe [38].

DHA can either be taken from the diet by humans or can be transformed in lesser concentration from EPA via docosapentaenoic acid (DPA) as an intermediate. This process takes place by an elongation step followed by Δ4-desaturase action. Recently it has been reported that DHA is more likely synthesized via "Sprecher's shunt" by C24 intermediate followed by β-oxidation inside peroxisomes [43]. Therefore, firstly elongation of EPA occurred twice, generating 24:5 ω-3, followed by desaturation to 24:6 ω-3, and then reduced to DHA (22:6 ω-3) via β-oxidation. In microorganisms like fungi, mosses and microalgae DHA biosynthesis occurs through a series of reactions: desaturation catalyzed by desaturase enzyme and elongation catalyzed by elongase enzyme [4]. There are two different pathways such as aerobic desaturase/elongase pathway and anaerobic polyketide synthase (PKS) pathway are reported for biosynthesis of the ω-3 PUFA.

5.3.1 Aerobic Desaturase and Elongase Pathway

This aerobic pathway occurs in eukaryotic and prokaryotic species including bacteria, thraustochytrids, microalgae, fungi, animals, plants and other microbes encoding enzymes required for ω-3 PUFA biosynthesis [4, 44]. The activities of these enzymes can be categorized into two parts: desaturation and chain elongation of fatty acid. The desaturase enzyme catalyses desaturation process by introducing a double bond between the pre-existing double bond and the carboxyl end of a fatty acid substrate. The methyl-end desaturases such as membrane-bound Δ3-, Δ12-, and Δ15-desaturases and front-end desaturases such as Δ4-, Δ5-, Δ6-, and Δ8-desaturases are commonly found in microbial organisms [45, 46]. The fatty acid chain elongation by 2C occurs by the elongation complex containing four distinct subunits: an enoyl-CoA reductase, a hydroxyl acyl-CoA dehydratase, a ketoacyl-CoA reductase and an acyl- CoA elongase [47].

For example, the bio-synthesis of DHA involves the following steps:

1. Desaturation: Δ6 desaturase action at the 6th carbon of ALA to generate stearidonic acid (SDA)
2. Elongation: Δ6 elongase action on SDA to generate eicosatetraenoic acid (ETA)
3. Desaturation: Δ5 desaturase action at the 5th carbon of ETA to generate EPA
4. Elongation: Δ5 elongase action on EPA to generate docosapentaenoic acid (DPA)
5. Desaturation: Δ4 desaturase action at the 4th carbon of DPA to produce DHA [34].

In photosynthetic microalgae, *de novo* fatty acid synthesis occurs in the plastids whereas in non-photosynthetic microbes, it takes place in the cytoplasm. But, the enzymes mainly associated with the endoplasmic reticulum (ER) membrane are involved in fatty acids elongation and desaturation [4]. The end product of fatty acid synthase (FAS) i.e., stearic acid (18:0) is catalyzed by acyl carrier protein (ACP) Δ9-desaturase to oleic acid (18:1Δ9). Some cyanobacteria, many ω-3 PUFA producing microbes and almost all higher plants contains Δ12, and Δ15-/ω3-desaturases to further form linoleic acid (LA; 18:2Δ9,12) and ALA (18:3Δ9,12,15) from oleic acid while humans and animals lack these enzymes [4]. Therefore microbial species deficient in C18-PUFAs in their lipid profile requires the insertion of multiple front-end desaturases, C18- and C20-PUFA-specific elongases [4]. Nowadays, combined approaches of genetic engineering to enhance PUFA using low-cost substrates have been focused on [48].

The ω3-long chain fatty acid synthesis occurs in microorganisms by two different and converging pathways: the conventional Δ6-pathway and the alternative Δ8-pathway. The conventional Δ6-pathway proceeds with catalyzes of LA (n-6) and ALA (n-3) by Δ6-desaturase to form γ-linolenic acid (GLA, 18:3Δ6,9,12) and SDA (18:4Δ6,9,12,15), respectively [49]. Further, Δ6-elongase generates DGLA (20:3Δ8,11,14) and ETA (20:4Δ8,11,14,17), respectively. Further conversion of DGLA and ETA to form ARA (n-6) and EPA (n-3), respectively occurred by Δ5-desaturase. During alternative Δ8-pathway, ω3-very long chain PUFAs biosynthesis initiation occurs by Δ9-elongation of the same substrates as Δ6-pathway, which is catalyzed by a specific Δ9-elongases, generating eicosadienoic acid (EDA, 20:2Δ11,14) and eicosatrienoic acid (ERA, 20:3Δ11,14,17). The Δ8-desaturase further catalyzes these Δ9-elongated fatty acids by desaturation to produce DGLA and ETA, respectively. ARA and EPA are formed later from DGLA and ETA by Δ5-desaturation as in the conventional Δ6-pathway. DHA biosynthesis in DHA-producing microbes involves a specific Δ5-elongation via Δ5/C20-elongases to produce docosapentaenoic acid (DPA, 22:5Δ7,10,13,16,19) from EPA which finally yield DHA (n-3) by Δ4-desaturation. The n-3 and n-6 pathways are interconnected by ω-3 desaturases (Δ15, Δ17, Δ19-desaturase) converting n-6 into n-3 fatty acids [4].

The conventional Δ6-pathway operates in eukaryotic organisms for ω3-very long chain PUFAs-synthesis, whereas the alternative Δ8-pathway has been found in some species of microalgae such as *Isochrysis galbana, Pavlova salina* and in protists such as *Acanthamoeba* spp., *Tetrahymena pyroformis,* and *Euglena gracilis.* All the genes encoding desaturation and elongating activities have been identified from a range of organisms and cloned for desired ω3-long chain PUFAs production. This approach has led to successful assembly of both the alternative Δ8-pathways and the conventional Δ6-pathway in microorganisms [4, 38, 50].

5.3.2 Anaerobic Polyketide Synthase (PKS) Pathway

An alternative PKS pathway for PUFA biosynthesis present in both eukaryotic and prokaryotic organisms does not involve the fatty acid desaturase/elongase system to produce DHA or EPA. Marine bacteria *Shewanella pneumatophori* strain SCRC-2378 was the first microorganism in which PKS pathway has been explained [51]. Later, it was described that in this species *de novo* synthesis occurs for C16 or C18 fatty acids production and undefined aerobic desaturases leads to insertion of double bonds. The PKS-like enzymes are a multi-functional complex that can synthesize EPA without the help of FAS, aerobic elongase and desaturase activities. For the double bonds insertion during extension of fatty acyl chain, the anaerobic PKS pathway involves dehydratase-isomerase module, which is different from aerobic desaturase/elongase pathway that needs oxygen-dependent desaturation for the double bonds insertion during fatty acyl chain elongation [4]. Both the enzymatic system co-exists in *Schizochytrium* [52] and *Thraustochytrium* [53] microbial species. In aerobic desaturase/elongase system, *Schizochytrium* lacks Δ12-desaturase activity and are unable to synthesize very long chain PUFAs from the FAS products. Desaturase genes have been identified both in conventional aerobic pathway and anaerobic PKS-like pathway of *Schizochytrium* and *Thraustochytrium aureum.* These results illustrate that *Schizochytrium* sp. contains partial desaturase/

elongase system [54]. While *T. aureum* have functional Δ12-desaturase that can produce ω3 very long chain PUFAs via the conventional desaturase/elongase pathway and the PKS-like pathway [4]. These observations demonstrate that microbes involve either the anaerobic PKS-like pathway or the conventional aerobic desaturase/elongase pathway or both for the biosynthesis of very long chain PUFAs.

5.4 Factors Affecting ω-3 PUFA Production

5.4.1 Temperature

The ω-3 PUFA production is influenced by temperature. Fermentation at higher temperatures favour increased cell growth while lower temperature favours fatty acid production [55]. Psychrophiles with optimum temperature of <20°C produced negligible amount of PUFA. Hence, PUFA formation is usually rare in cyanobacteria and bacteria [56]. Low growth temperatures are favourable for EPA production. For example, several fungal species of *Mortierella* has been reported to produce higher concentration of EPA at lower temperatures ranging from 12°C-15°C. The shift in temperature from 25°C to 11°C was detected to trigger EPA production but when temperature was increased to 25°C after cultivation at 12°C for 4 days, no further increment in EPA production has been detected [57–59]. Furthermore, EPA (> 85%) produced by *M. hygrophila* at 12°C remains unchanged when temperature shift to 25°C for 5 days. But, EPA production in *M. alpina* rapidly decreased with temperature shift while in *P. tricornutum* optimum temperature for enhanced EPA production was 21.5°C [15, 60]. *Chlorella minutissima* produced more amount of EPA at lower temperature depicting that thermolabile enzymes are involved in desaturation and chain elongation [61]. In *Rhizopus* and *Mucor* species, lower temperature does not show influence on unsaturated fatty acids synthesis [56].

The enhanced unsaturated fatty acids synthesis at lower temperature has been detected in yeast, fungi, eukaryotic algae, bacteria and blue green algae species [56]. This occurs due to increased solubility of oxygen at lower temperature in the medium, as more molecular oxygen will be available for oxygen dependent enzymes which catalyses the desaturation of unsaturated long chain fatty acids [62]. It has been stated that temperature dependent PUFAs production is a process of adaptation to environment [63]. Practically, low temperatures are disadvantageous due to lower growth rate and higher cooling cost. Integration approach of the glucose feeding and temperature shift during the growth may be effective to obtain high EPA within short cultivation period [56].

5.4.2 pH

The PUFA production is also influenced by pH condition of the medium. It is recommended to maintain pH closer to optimal pH required by microbes, otherwise it would waste energy to restore optimal pH [17]. For example, increase in culture pH increases unsaturated fatty acids production in many fungi. Also, in many fungi and algae, an initial optimum pH 6.0-7.6 is required for EPA production while *Thraustochytrium aureum* requires optimum initial pH 6.0 for effective DHA production [59, 60]. Red alga *Porphyridium* and *Mortierella* fungi require optimum initial pH 6-7 while *Mortierella* species (IS-4) requires mild alkaline conditions of pH 8-9 for ARA production [64].

5.4.3 Aeration

The availability of molecular oxygen in the medium also affects the composition of PUFA produced. As the molecular O_2 is required by microbes for the desaturation pathway involved in the biosynthesis of PUFAs, hence presence of oxygen usually determines the degree of unsaturation of fatty acids [65]. For instance, in dinoflagellate *Gyronidium cohnii,* PUFA content elevates with increase in oxygen level in the medium [66]. However, under anaerobic conditions, saturated fatty acids content increases in several species such as during cultivation of *Euglena gracilis* [50].

5.4.4 Media Composition

The factors such as salinity, metal ions, cell density, light intensity and carbon source influences the PUFA composition. For example, in *Mortierella* fungi, EPA production enhances efficiently by using linseed oil as carbon source. The EPA composition increases two-fold when n-octadecane was used as carbon source as compared with glucose in *M. hygrophila* [15]. The maximum EPA production was estimated in *M. elongata* NRRL 5513 during cultivation at 15°C for 8 d in a medium having 2% (w/v) linseed oil [59]. *Mortierella* fungi shows high amount of ARA production on medium containing higher glucose concentration [64]. *Thraustochytrium aureum* effectively produces high DHA when starch, maltose or glucose are used as carbon source [67, 68].

In fungi, bacteria and algae, the proportion of saturated to unsaturated fatty acids gets affected due to the nitrogen concentration and type of nitrogen source provided [69]. *Dunaliella salina, D. bardawal* and *Botryococcus braunii* produces elevated EPA percentage under nitrogen stress condition [70]. Medium containing high C:N ratio increases biomass, lipid content and affects PUFA yields in *Mortierella* species [71].

Numerous metal ions such as manganese deficiency lead to decline in production of EPA in *Euglena* [72]. Higher silica content in the range from 0 to 100 mg l^{-1} leads to reduced growth and EPA productivity in diatom *P. tricornutum* [60]. While the free fatty acids content present in media usually repress lipid synthesis in microorganisms. The productivity of PUFA, especially ARA and EPA increases in *Euglena gracilis* by the supplementation of LA or oleic acid in the cultivation medium [56]. EPA synthesis was inhibited in *P. tricornutum* with the supplementation of oleic acid beyond 2 g/L concentration [19]. Lee *et al.* showed that the C20 PUFA (ARA) content in red alga *Porphyridium cruentum* initially decreases and then enhances with rising NaCl concentrations above 0.8 M NaCl [73].

5.4.5 Incubation Time

Incubation time is also a determining factor for PUFA formation and the type of PUFA to be formed. To attain amplified EPA percentage in lipids by extending the fermentation time, while escalating process costs would decrease the cost and complexity of EPA recovery. Maximum biomass and lipid content was obtained after 20 days of incubation in several fungus species whereas elevated EPA composition was attained after a more prolonged period of 36 days [56].

5.5 Stabilization of ω-3 PUFA

High levels of ω-3 PUFA present in microbial oil are highly susceptible to oxidation that leads to the development of toxic products such as volatile compounds or peroxides causing undesirable off-flavours [2]. The oxidation of Omega-3 PUFA involves complex mechanism which depends on the lipid form and can be catalyzed by other factors such as presence of trace metals, exposure to light or oxygen, and temperature. Hence, to prevent ω-3 PUFA from oxidation; optimum processing, packaging and storage of microbial oil is crucial. Also, to extend the shelf life and improve stability of oil, other parameters are essential to consider [74]. The addition of most common antioxidants such as lactoferrines, gallic acid, propylgallate, ascorbyl palmitate, ascorbic acid, tocopherols etc., are utilized to prevent lipid oxidation [3]. Natural antioxidants has acquired great relevance as an alternative to synthetic compounds such as plant extracts, as rosemary, oregano or olive mill waste waters [3, 75]. Microencapsulation has also been suggested as an effective strategy to control off-flavours and improve the oil stability by retarding lipid oxidation and enzyme hydrolysis [3, 76].

5.6 Conclusions

The ω-3 PUFAs are necessary in human nutrition as they play a significant role in prevention of numerous diseases. Recently, microorganisms have been considered as an important source to attain ω-3 enriched oil. Genetic engineering and technology methods have also been suggested as an alternative strategy for enhanced microbial ω-3 PUFA production. Microencapsulation technologies and adding different types of natural or chemical antioxidants are the most frequent techniques used to prevent ω-3 PUFA oxidation. In this regard, microbial ω-3 PUFA production using naturally producing antioxidants and low-cost substrate are found to be ideal to achieve this aim.

References

1. T.C. Adarme-Vega, D.K.Y. Lim, M. Timmins, F. Vernen, Y. Li, P.M. Schenk, Microalgal biofactories: a promising approach towards sustainable omega-3 fatty acid production, *Microb. Cell Fact.* 11 (2012) 1–10. doi:10.1186/1475-2859-11-96.
2. J.A. Kralovec, S. Zhang, W. Zhang, C.J. Barrow, A review of the progress in enzymatic concentration and microencapsulation of omega-3 rich oil from fish and microbial sources, *Food Chem.* 131 (2012) 639–644. doi:10.1016/j.foodchem.2011.08.085.
3. N. Rubio-Rodríguez, S. Beltrán, I. Jaime, S.M. de Diego, M.T. Sanz, J.R. Carballido, Production of omega-3 polyunsaturated fatty acid concentrates: A review, *Innov. Food Sci. Emerg. Technol.* 11 (2010) 1–12. doi:10.1016/j.ifset.2009.10.006.
4. Y. Gong, X. Wan, M. Jiang, C. Hu, H. Hu, F. Huang, Metabolic engineering of microorganisms to produce omega-3 very long-chain polyunsaturated fatty acids, *Prog. Lipid Res.* 56 (2014) 19–35. doi:10.1016/j.plipres.2014.07.001.
5. O.P. Ward, A. Singh, Omega-3/6 fatty acids: Alternative sources of production, *Process Biochem.* 40 (2005) 3627–3652. doi:10.1016/j.procbio.2005.02.020.

6. F. Deeba, V. Pruthi, Y.S. Negi, Converting paper mill sludge into neutral lipids by oleaginous yeast *Cryptococcus vishniaccii* for biodiesel production, *Bioresour. Technol.* 213 (2016) 96–102. doi:10.1016/j.biortech.2016.02.105.
7. F. Deeba, A. Patel, N. Arora, V. Pruthi, P.A. Pruthi, Y.S. Negi, Amaranth seeds (*Amaranthus palmeri* L.) as novel feedstock for biodiesel production by oleaginous yeast, *Env. Sci Pollut Res.* 25 (2017) 353–362. doi:10.1007/s11356-017-0444-x.
8. F. Deeba, V. Pruthi, Y.S. Negi, Aromatic hydrocarbon biodegradation activates neutral lipid biosynthesis in oleaginous yeast, *Bioresour. Technol.* 255 (2018) 273–280. doi:10.1016/j.biortech.2018.01.096.
9. R.E. Armenta, M.C. Valentine, Single-cell oils as a source of omega-3 fatty acids: An overview of recent advances, *JAOCS, J. Am. Oil Chem. Soc.* 90 (2013) 167–182. doi:10.1007/s11746-012-2154-3.
10. T. Kiy, M. Rüsing, D. Fabritius, Production of docosahexaenoic acid by the marine microalga, Ulkenia sp., in: *Single Cell Oils*, AOCS Publishing, 2005: pp. 104–111.
11. C. Ratledge, Fatty acid biosynthesis in microorganisms being used for Single Cell Oil production, *Biochimie.* 86 (2004) 807–815. doi:10.1016/j.biochi.2004.09.017.
12. S.-F. Yuan, H.S. Alper, Metabolic engineering of microbial cell factories for production of nutraceuticals, *Cell Fact.* 18 (2019) 46. doi:10.1186/s12934-019-1096-y.
13. W. Amjad Khan, H. Chun-Mei, N. Khan, A. Iqbal, S.W. Lyu, F. Shah, Bioengineered Plants Can Be a Useful Source of Omega-3 Fatty Acids, *Biomed Res. Int.* 2017 (2017) 7348919. doi:10.1155/2017/7348919.
14. E. Abedi, M.A. Sahari, Long-chain polyunsaturated fatty acid sources and evaluation of their nutritional and functional properties, *Food Sci. Nutr.* 2 (2014) 443–463. doi:10.1002/fsn3.121.
15. S. Shimizu, H. Kawashima, Y. Shinmen, K. Akimoto, H. Yamada, Production of Eicosapentaenoic acid by *Mortierella* fungi, JAOCS, *J. Am. Oil Chem. Soc.* 65 (1988) 1455–1459. doi:10.1007/BF02932833.
16. G. Vadivelan, G. Venkateswaran, Production and Enhancement of Omega-3 Fatty Acid from *Mortierella alpina* CFR-GV15: Its Food and Therapeutic Application, *Biomed Res. Int.* 2014 (2014) 1–9. doi:10.1155/2014/657414.
17. A.A. Mironov, V.A. Nemashkalov, N.N. Stepanova, S. V. Kamzolova, W. Rymowicz, I.G. Morgunov, The effect of ph and temperature on arachidonic acid production by glycerol-Grown *Mortierella alpina* NRRL-A-10995, *Fermentation.* 4 (2018) 1–12. doi:10.3390/fermentation4010017.
18. A. Mendes, T. Lopes Da Silva, A. Reis, DHA Concentration and Purification from the Marine Heterotrophic Microalga *Crypthecodinium cohnii* CCMP 316 by Winterization and Urea Complexation, *Food Technol. Biotechnol.* 45 (2007) 38–44.

19. W. Yongmanitchai, O.P. Ward, Screening of algae for potential alternative sources of eicosapentaenoic acid, *Phytochemistry*. 30 (1991) 2963–2967. doi:10.1016/S0031-9422(00)98231-1.
20. A. Seto, H.L. Wang, C.W. Hesseltine, Culture conditions affect eicosapentaenoic acid content of *Chlorella minutissima*, *J. Am. Oil Chem. Soc.* 61 (1984) 892–894. doi:10.1007/BF02542159.
21. Z. Xue, P.L. Sharpe, S.P. Hong, N.S. Yadav, D. Xie, D.R. Short, H.G. Damude, R.A. Rupert, J.E. Seip, J. Wang, D.W. Pollak, M.W. Bostick, M.D. Bosak, D.J. Macool, D.H. Hollerbach, H. Zhang, D.M. Arcilla, S.A. Bledsoe, K. Croker, E.F. McCord, B.D. Tyreus, E.N. Jackson, Q. Zhu, Production of omega-3 eicosapentaenoic acid by metabolic engineering of *Yarrowia lipolytica*, *Nat. Biotechnol.* 31 (2013) 734–740. doi:10.1038/nbt.2622.
22. J. Van Wagenen, T.W. Miller, S. Hobbs, P. Hook, B. Crowe, M. Huesemann, Effects of light and temperature on fatty acid production in *Nannochloropsis salina*, *Energies*. 5 (2012) 731–740. doi:10.3390/en5030731.
23. V. Júnior, M. Furlan, V. Maus, I. Batista, N.M. Bandarra, Production of docosahexaenoic acid by Aurantiochytrium sp. ATCC PRA-276, *Brazilian J. Microbiol.* 48 (2017) 359–365. doi:10.1016/j.bjm.2017.01.001.
24. A.M. Burja Helia Radianingtyas Anthony Windust Colin J Barrow, Isolation and characterization of polyunsaturated fatty acid producing *Thraustochytrium* species: screening of strains and optimization of omega-3 production, *Appl Microbiol Biotechnol.* 72 (2006) 1161–1169. doi:10.1007/s00253-006-0419-1.
25. M.L. Hamilton, S. Powers, J.A. Napier, O. Sayanova, Heterotrophic Production of Omega-3 Long-Chain Polyunsaturated Fatty Acids by Trophically Converted Marine Diatom *Phaeodactylum tricornutum*, *Mar. Drugs*. 14 (2016) 1–10. doi:10.3390/md14030053.
26. F. Guihéneuf, V. Mimouni, L. Ulmann, G. Tremblin, Combined effects of irradiance level and carbon source on fatty acid and lipid class composition in the microalga *Pavlova lutheri* commonly used in mariculture, *J. Exp. Mar. Bio. Ecol.* 369 (2009) 136–143. doi:10.1016/j.jembe.2008.11.009.
27. Y. Liu, C.M. Koh, L. Ji, Methods for efficient production of polyunsaturated fatty acids (PUFA) in *Rhodosporidium* and *Rhodotorula* species, WO 2016/039685 Al, 2016.
28. R.A. Lima, R.F.S. Andrade, D.R. Ribeaux, P.N. Santos, C.D.C. Albuquerque, G.M. Campos-Takaki, Production of Very Long Chain Polyunsaturated Omega 3 and Omega 6 Fatty Acids by *Candida glabrata* Strains, 2015. http://www.ijcmas.com (accessed November 5, 2019).
29. R.E. Armenta, M.C. Valentine, Single-cell oils as a source of omega-3 fatty acids: An overview of recent advances, *JAOCS, J. Am. Oil Chem. Soc.* 90 (2013) 167–182. doi:10.1007/s11746-012-2154-3.
30. M. Yeiser, C.L. Harris, A.L. Kirchoff, A.C. Patterson, J.L. Wampler, E.N. Zissman, C.L. Berseth, Growth and tolerance of infants fed formula with a new algal source of docosahexaenoic acid: Double-blind, randomized,

controlled trial, *Prostaglandins Leukot. Essent. Fat. Acids.* 115 (2016) 89–96. doi:10.1016/j.plefa.2016.09.001.

31. E. Sakuradani, S. Shimizu, Single cell oil production by *Mortierella alpina*, *J. Biotechnol.* 144 (2009) 31–36. doi:10.1016/j.jbiotec.2009.04.012.
32. D.G. Grenfell-Lee, A Fermentation-Based Process for Carotenoid Synthesis in Yarrowia lipolytica, in: *SIM Annu. Meet. Proceedings. Ind. Microbiol. Biotechnol.* Canada, 2009: p. 74.
33. A.M. Burja, H. Radianingtyas, A. Windust, C.J. Barrow, Isolation and characterization of polyunsaturated fatty acid producing *Thraustochytrium* species: Screening of strains and optimization of omega-3 production, *Appl. Microbiol. Biotechnol.* 72 (2006) 1161–1169. doi:10.1007/s00253-006-0419-1.
34. T. Kobayashi, K. Sakaguchi, T. Matsuda, E. Abe, Y. Hama, M. Hayashi, D. Honda, Y. Okita, S. Sugimoto, N. Okino, M. Ito, Increase of Eicosapentaenoic Acid in Thraustochytrids through Thraustochytrid Ubiquitin Promoter-Driven Expression of a Fatty Acid 5 Desaturase Gene, *Appl. Environ. Microbiol.* 77 (2011) 3870–3876. doi:10.1128/AEM.02664-10.
35. Y. Liang, Y. Liu, J. Tang, J. Ma, J.J. Cheng, M. Daroch, Transcriptomic Profiling and Gene Disruption Revealed that Two Genes Related to PUFAs/DHA Biosynthesis May be Essential for Cell Growth of *Aurantiochytrium* sp., *Mar. Drugs.* 16 (2018) 1–15. doi:10.3390/md16090310.
36. H. Iwasaka, R. Koyanagi, R. Satoh, A. Nagano, K. Watanabe, K. Hisata, N. Satoh, T. Aki, A possible trifunctional β-carotene synthase gene identified in the draft genome of *Aurantiochytrium* sp. Strain KH105, *Genes* (Basel). 9 (2018) 1–14. doi:10.3390/genes9040200.
37. C.A. Weaver, J.G. Metz, J.M. Kuner, F.H. Overton, Jr., Carotene synthase gene and uses therefor, U.S. Patent 7585659, 2009. https://patentimages.storage.googleapis.com/c9/e4/4f/e068cdc005e8c5/US7585659.pdf.
38. K. Ochsenreither, C. Glück, T. Stressler, L. Fischer, C. Syldatk, Production strategies and applications of microbial single cell oils, *Front. Microbiol.* 7 (2016) 1–26. doi:10.3389/fmicb.2016.01539.
39. A. Patel, N. Arora, K. Sartaj, V. Pruthi, P.A. Pruthi, Sustainable biodiesel production from oleaginous yeasts utilizing hydrolysates of various non-edible lignocellulosic biomasses, *Renew. Sustain. Energy Rev.* 62 (2016) 836–855. doi:10.1016/j.rser.2016.05.014.
40. A. Patel, L. Matsakas, K. Hrůzová, U. Rova, P. Christakopoulos, Biosynthesis of Nutraceutical Fatty Acids by the Oleaginous Marine Microalgae *Phaeodactylum tricornutum* Utilizing Hydrolysates from Organosolv-Pretreated Birch and Spruce Biomass, *Mar. Drugs.* 17 (2019) 1–17. doi:10.3390/md17020119.
41. H. Zhang, L. Zhang, H. Chen, Y.Q. Chen, C. Ratledge, Y. Song, W. Chen, Regulatory properties of malic enzyme in the oleaginous yeast, *Yarrowia lipolytica*, and its non-involvement in lipid accumulation, *Biotechnol. Lett.* 35 (2013) 2091–2098. doi:10.1007/s10529-013-1302-7.

42. D. Alves Martins, L. sa Custódio, L. sa Barreira, H. Pereira, R. Ben-Hamadou, J. Varela, K.M. Abu-Salah, Alternative Sources of n-3 Long-Chain Polyunsaturated Fatty Acids in Marine Microalgae, *Mar. Drugs.* 11 (2013) 2259–2281. doi:10.3390/md11072259.
43. R. De Caterina, G. Basta, n-3 Fatty acids and the inflammatory response-biological background, 2001. https://academic.oup.com/eurheartjsupp/article-abstract/3/suppl_D/D42/369530.
44. S.L. Pereira, A.E. Leonard, P. Mukerji, Recent advances in the study of fatty acid desaturases from animals and lower eukaryotes, *Prostaglandins Leukot. Essent. Fat. Acids.* 68 (2003) 97–106. doi:10.1016/S0952-3278(02)00259-4.
45. J. Shanklin, E.B. Cahoon, Desaturation and related modifications of fatty acids, *Annu. Rev. Plant Physiol. Plant Mol. Biol.* 49 (1998) 611–641. doi:10.1146/annurev.arplant.49.1.611.
46. P. Sperling, P. Ternes, T.K. Zank, E. Heinz, The evolution of desaturases, *Prostaglandins Leukot. Essent. Fat. Acids.* 68 (2003) 73–95. doi:10.1016/S0952-3278(02)00258-2.
47. A.E. Leonard, S.L. Pereira, H. Sprecher, Y.-S. Huang, Elongation of long-chain fatty acids, *Prog. Lipid Res.* 43 (2004) 36–54. doi:10.1016/S0163-7827(03)00040-7.
48. A. Beopoulus, J.-M. Nicaud, Yeast: A new oil producer?, *OCL.* 19 (2012) 22–28. doi:10.1684/ocl.2012.0426.
49. L. V Michaelson, C.M. Lazarus, G. Griffiths, J.A. Napier, A.K. Stobart, Isolation of a 5-Fatty Acid Desaturase Gene from *Mortierella alpina*, *J. Biol. Chem.* 273 (1998) 19055–19059. http://www.jbc.org/.
50. O.P. Ward, A. Singh, Omega-3/6 fatty acids: Alternative sources of production, *Process Biochem.* 40 (2005) 3627–3652. doi:10.1016/j.procbio.2005.02.020.
51. K. Yazawa, Production of eicosapentaenoic acid from marine bacteria, *Lipids.* 31 (1996) S297–S300. doi:10.1007/bf02637095.
52. J.G. Metz, P. Roessler, D. Facciotti, C. Levering, F. Dittrich, M. Lassner, R. Valentine, K. Lardizabal, F. Domergue, A. Yamada, K. Yazawa, V. Knauf, J. Browse, Production of polyunsaturated fatty acids by potyketide synthases in both prokaryotes and eukaryotes, *Science* (80-.). 293 (2001) 290–293. doi:10.1126/science.1059593.
53. X. Qiu, H. Hong, S.L. MacKenzie, Identification of a 4 Fatty Acid Desaturase from *Thraustochytrium* sp. Involved in the Biosynthesis of Docosahexanoic Acid by Heterologous Expression in *Saccharomyces cerevisiae* and *Brassica juncea*, *J. Biol. Chem.* 276 (2001) 31561–31566. doi:10.1074/jbc.M102971200.
54. J.C. Lippmeier, K.S. Crawford, C.B. Owen, A.A. Rivas, J.G. Metz, K.E. Apt, Characterization of both polyunsaturated fatty acid biosynthetic pathways in schizochytrium sp., *Lipids.* 44 (2009) 621–630. doi:10.1007/s11745-009-3311-9.
55. R.J. Winwood, Recent developments in the commercial production of DHA and EPA rich oils from micro-algae, *OCL.* 20 (2013) 1–5. doi:10.1051/ocl/2013030.

56. P. Bajpai, P.K. Bajpai, Eicosapentaenoic acid (EPA) production from microorganisms: a review, *J. Biotechnol.* 30 (1993) 161–183. doi:10.1016/0168-1656(93)90111-Y.
57. P.K. Bajpai, P. Bajpai, O.P. Ward, Production of arachidonic acid by *Mortierella alpina* ATCC 32222, *J. Ind. Microbiol.* 8 (1991) 179–185. doi:10.1007/BF01575851.
58. P. Bajpai, P.K. Bajpai, O.P. Ward, Eicosapentaenoic acid (EPA) formation: comparative studies with *Mortierella* strains and production by *Mortierella elongata, Mycol. Res.* 95 (1991) 1294–1298. doi:10.1016/S0953-7562(09)80577-7.
59. P.K. Bajpai, P. Bajpai, O.P. Ward, Optimisation of culture conditions for production of eicosapentaenoic acid by *Mortierella elongate* NRRL 5513, *J. Ind. Microbiol.* 9 (1992) 11–17. doi:10.1007/BF01576363.
60. W. Yongmanitchai, O.P. Ward, Growth of and Omega-3 Fatty Acid Production by *Phaeodactylum tricornutum* under Different Culture Conditions, *Appl. Environ. Microbiol.* 57 (1991) 419–425. http://aem.asm.org/.
61. A. Seto, H.L. Wang, C.W. Hesseltine, Culture conditions affect eicosapentaenoic acid content of *Chlorella minutissima, J. Am. Oil Chem. Soc.* 61 (1984) 892–894. doi:10.1007/BF02542159.
62. C.M. Brown, A.H. Rose, Fatty-Acid Composition of *Candida utilis* as Affected by Growth Temperature and Dissolved-Oxygen Tension, *J. Bacteriol.* 99 (1969) 371–378. http://jb.asm.org/.
63. H. Keller, G. Reinhardt, N. Rettenmaier, A. Schorb, M. Dittrich Heidelberg, Environmental assessment of algae-based polyunsaturated fatty acid production, 2017. www.ifeu.de.
64. P. Bajpai, P. Bajpai, Arachidonic acid production by microorganisms, *Biotechnol. Appl. Biochem.* 15 (1992) 1–10. doi:10.1111/j.1470-8744.1992.tb00194.x.
65. J.A. Erwin, *Lipids and biomembranes of eukaryotic microorganisms*, Academic Press, 1973.
66. G.W. Harrington, G.G. Holz, The monoenoic and docosahexaenoic fatty acids of a heterotrophic dinoflagellate, *Biochim. Biophys. Acta - Lipids Lipid Metab.* 164 (1968) 137–139. doi:10.1016/0005-2760(68)90083-0.
67. P. Bajpai, P.K. Bajpai, O. Ward, Production of docosahexaenoic acid by *Thraustochytrium aureum, Appl. Microbiol. Biotechnol.* 35 (1991) 706–710. doi:10.1007/BF00169881.
68. K. Wu, L. Ding, P. Zhu, S. Li, S. He, L. Dak, S. Yip, Y. Chin, K. Li, Application of the Response Surface Methodology to Optimize the Fermentation Parameters for Enhanced Docosahexaenoic Acid (DHA) Production by Thraustochytrium sp. ATCC 26185, *Molecules* 23 (2018) 1–11. doi:10.3390/molecules23040974.
69. M. Tossavainen, U. Ilyass, V. Ollilainen, K. Valkonen, A. Ojala, M. Romantschuk, Influence of long term nitrogen limitation on lipid, protein and pigment production of *Euglena gracilis* in photoheterotrophic cultures, *Peer J.* 6624 (2019) 1–19. doi:10.7717/peerj.6624.

70. A. Ben-Amotz, T.G. Tornabene, W.H. Thomas, Chemical profile of selected species of microalgae with emphasis on lipids, *J. Phycol.* 21 (2004) 72–81. doi:10.1111/j.0022-3646.1985.00072.x.
71. L. Grantina-Ievina, A. Berzina, V. Nikolajeva, P. Mekss, I. Muiznieks, Production of fatty acids by *Mortierella* and *Umbelopsis* species isolated from temperate climate soils, 2014. http://eeb.lu.lv/EEB/201403/EEB_12_Grantina-Ievina.pdf.
72. G. Constantopoulos, J. Bryant, Lipid Metabolism of Manganese-deficient Algae, 1970. https://www.ncbi.nlm.nih.gov/pmc/articles/PMC396358/pdf/plntphys00193-0083.pdf.
73. Y.-K. Lee, H.-M. Tan, C.-S. Low, Effect of salinity of medium on cellular fatty acid composition of marine alga *Porphyridium cruentum* (Rhodophyceae), *J. Appl. Phycol.* 1 (1989) 19–23. doi:10.1007/BF00003531.
74. A. Kamal-Eldin, N. V. Yanishlieva, N-3 fatty acids for human nutrition: stability considerations, *Eur. J. Lipid Sci. Technol.* 104 (2002) 825–836. doi:10.1002/1438-9312(200212)104:12<825::AID-EJLT825>3.0.CO;2-N.
75. D. Jimenez-Alvarez, F. Giuffrida, P.A. Golay, C. Cotting, A. Lardeau, B.J. Keely, Antioxidant Activity of Oregano, Parsley, and Olive Mill Wastewaters in Bulk Oils and Oil-in-Water Emulsions Enriched in Fish Oil, *J. Agric. Food Chem.* 56 (2008) 7151–7159. doi:10.1021/jf801154r.
76. R. Matsuno, S. Adachi, Lipid encapsulation technology - techniques and applications to food, *Trends Food Sci. Technol.* 4 (1993) 256–261. doi:10.1016/0924-2244(93)90141-V.

6

Autotrophic Cultivation of Microalgae for the Production of Polyunsaturated Fatty Acid

Pallavi Saxena, Mukesh Kumar and Harish*

Department of Botany, Mohanlal Sukhadia University, Udaipur, Rajasthan, India-313001

Abstract

PUFAs are one of the essential components for the growth and development of higher hierarchy organisms. Bioactive components belonging to PUFA provide immense health benefits to humans. It is estimated that PUFA production will increase by 13% from 2014 to 2020 and that could lead to an increase in the pressure on fish feed-stocks. Therefore, there is an alarming need of other alternatives such as microalgae. Microalgae serve the purpose behind the production of such a diverse variety of compounds to ensure survival and to adapt to the stringent and varying environmental conditions in their habitat. The fatty acids which some microalgae tend to accumulate serve purposes that allow the microalgae to survive in adverse conditions when cultivated under different cultivation systems, namely autotrophic, heterotrophic and mixotrophic. In autotrophic cultivation the chances of contamination are less, production cost is low and no negative impact on environmental conditions is imposed. Cellular lipid contents were found higher under autotrophic cultivation; however, production is also found low in many algal species.

Keywords: Algae, abiotic stress, PUFAs, lipid, metabolomics

6.1 Introduction

Algae are one of the primary natural resource of polyunsaturated fatty acids (PUFAs) bioactive molecules. They are considered to be the most

**Corresponding author*: harish.botany1979@gmail.com

Alok Kumar Patel and Leonidas Matsakas (eds.) Nutraceutical Fatty Acids from Oleaginous Microalgae: A Human Health Perspective, (165–186) © 2020 Scrivener Publishing LLC

promising vegetative and non-polluted reserve for bioengineering of PUFA production as substitute to fish oil [1]. Algal flora comprising different morpho-physiological characteristic, supports approximately 40% of the photosynthesis at global level [2]. Microalgae having a size of 3-10 μm (length or diameter) confer the ability to produce a number of bioactive compounds in order to survive [3]. In the last decade, the research focus has shifted to algae as they could be a potential resource for the production of biofuels but also a plethora of high-value compounds, i.e., unsaturated fatty acids, carotenoids, pigments, etc. [4–5]. Their high photosynthetic efficacy, no requirement of arable land and carbon source, have made algae a competent resource for bioactive compounds and bioenergy purposes [6]. Algae have been validated as a potential source of phenolic and other antioxidant compounds [7]. The extracted compounds exhibited potential applications in the pharmaceutical, healthcare and food industries [8–9]. Commercial production of PUFAs from algae has emerged as an alternative to fish oils, as fish feedstocks solely were found unable to meet the global market demands [10]. Moreover, metal contaminants present in fishes could be biomagnified if that stock would be used for PUFAs production (Figure 6.1) [11].

Only a limited number of species have been optimized to fulfill market demands [12]. But by using different cultivation techniques and leading engineering technologies, production of the desired compound can be enhanced in different algal species [13]. Exploration of suitable strains for

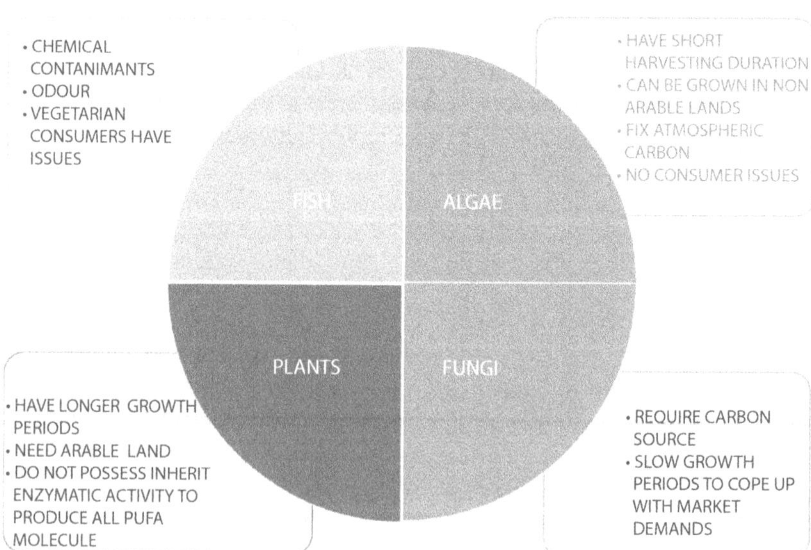

Figure 6.1 Various resources of PUFAs biomolecules.

Table 6.1 Various fatty acid molecules derived from algae.

Fatty acid molecule	Chemical formulae	Molecular weight	Structure
(EPA)	20:5 n-3$C_{20}H_{30}O_2$ (5Z,8Z,11Z,14Z,17Z)-5,8,11,14,17-icosapentaenoic acid	302.458 g/mol	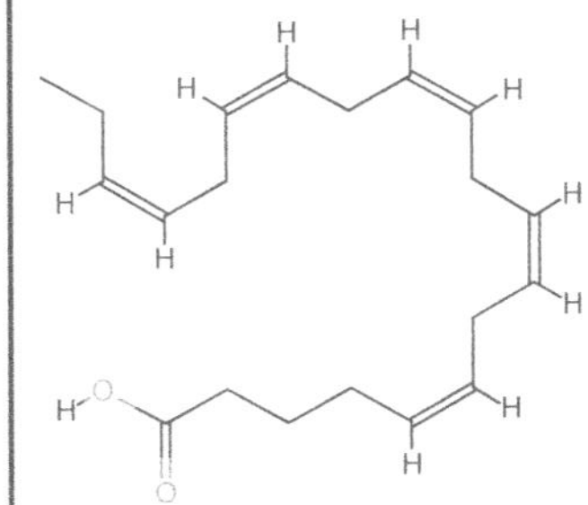 PubChem CID: 446284
(DHA)	22:6 n-3$C_{22}H_{32}O_2$ all-cis-docosa-4,7,10,13,16,19-hexa-enoic acid	328.496 g/mol	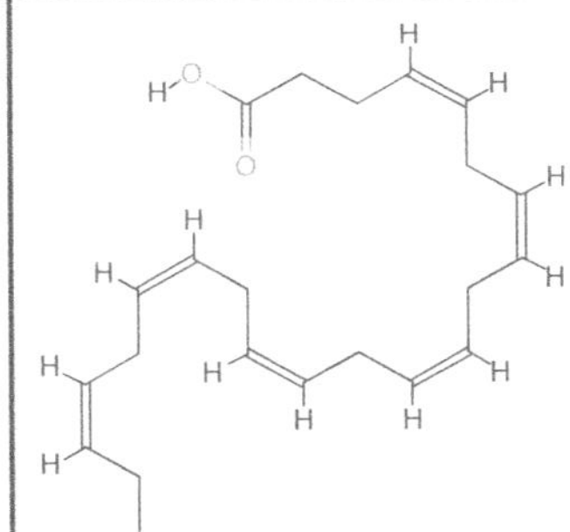 PubChem CID: 445580

(*Continued*)

Table 6.1 Various fatty acid molecules derived from algae. (*Continued*)

Fatty acid molecule	Chemical formulae	Molecular weight	Structure
Arachidonic acid (ARA)	20:4(n-6)$C_{20}H_{32}O_2$ (5Z,8Z,11Z,14Z)-5,8,11,14,-eicosatetraenoic acid	304.474 g/mol	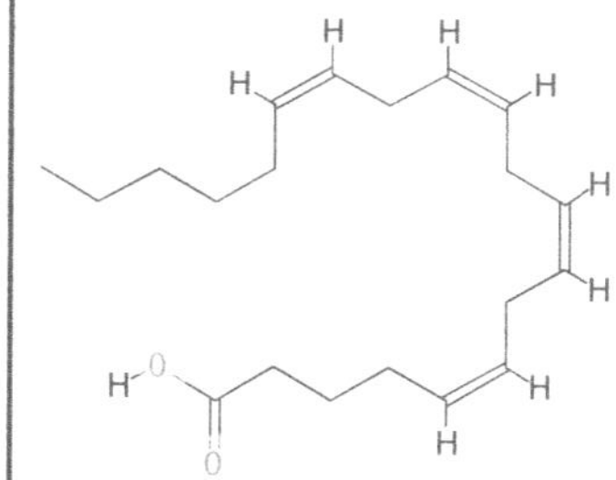 PubChem CID: 444899
γ-linolenic acid (GLA)	$C_{18}H_{30}O_2$18:3(n-6) All-cis-6,9,12-octadecatrienoic acid	278.436 g/mol	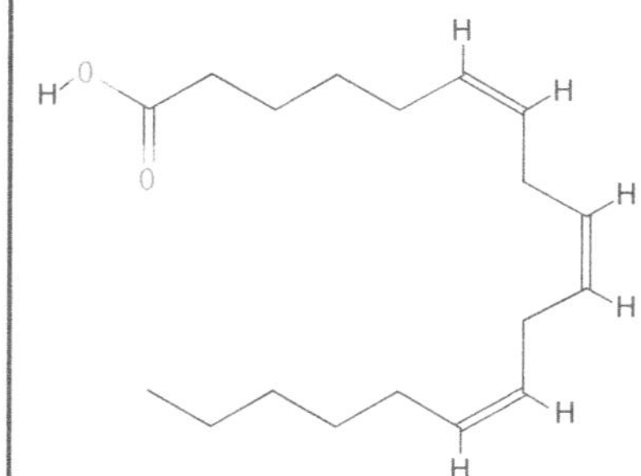 PubChem CID: 5280933

(*Continued*)

Table 6.1 Various fatty acid molecules derived from algae. (*Continued*)

Fatty acid molecule	Chemical formulae	Molecular weight	Structure
Linoleic acid (LA)	$C_{18}H_{32}O_2$18:3 n-3	280.4472 g/mol	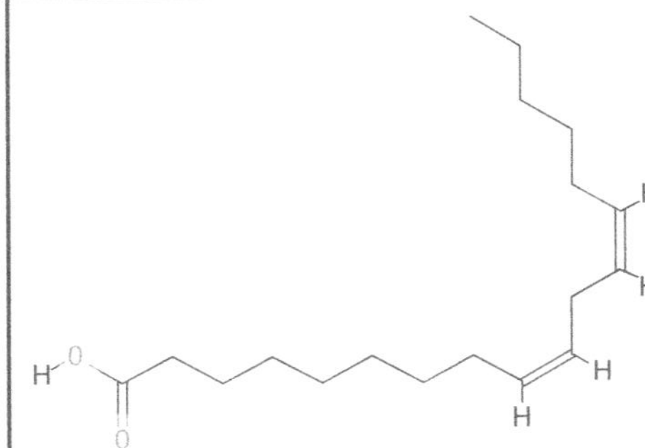 PubChem CID: 5280450

robust production of preferable compound with the implementation of successful extraction technique is an ongoing process [14]. Various fatty acid molecules derived from algae are shown in the Table 6.1. This chapter discusses recent developments of autotrophic cultivation technique.

6.2 Importance of PUFAs

PUFAs tend to have a lot of benefits as they are an important component of the membranes inside human cells [15]. They are predominantly found as a component forming the structure of the neuronal cells in the human body [16]. PUFAs tend to show a positive effect as supplements and help prevent several cardiovascular conditions in the nervous system and act as anti-inflammatory compounds [17]. Regular consumption of omega-3 fatty acids also tends to reduce the chances of several illnesses effectively reducing the chances of hypertension, thrombosis as well as the chances of cardiac arrhythmias [18]. The reason behind this health benefit that accompanies the consumption of the omega-3 fatty acid is because they tend to increase the ratio of high-density lipoprotein (HDL) to low-density lipoprotein (LDL), while at the same time they tend to reduce the overall cholesterol to HDL. These are only a few examples of health benefits that omega-3 are known to have for the human body, and exactly because of these benefits, currently, the microalgae are in an exploding demand in the pharmaceutical industry [19]. Nutraceutical companies using PUFAs produced by the microalgae are getting attention and as a result market demand has risen over the last few years [20, 21].

Several face care products, such as anti-aging creams, regenerative care products tend to make use of the microalgal extract which enhances the collagen synthesis in skin [22]. This process tends to bring significant improvement in skin regeneration as well as cause an overall reduction of wrinkled skin. Another similar microalgal extract that the skincare industry uses is protulines, which is extracted from *Arthrospira* and tends to protect the skin from early aging and slows down the process of formation of wrinkles on the skin [23]. Additionally, the consumption of omega-3 also tends to enhance brain function, with a considerable improvement often observed in the nervous system [24]. It has been observed that in the case of pregnant women, the consumption of EPA and DHA in appropriate amounts is vital for an overall development of the foetus brain [25]. The consumption of the omega-6 fatty acid (arachidonic acid) is also considered beneficial for infants, since it is of great importance for their normal growth and development [26]. Periodical consumption of omega-3 fatty acids demonstrate also immuno-modulatory effects in tests treating people suffering from rheumatoid arthritis, ulcerative colitis, psoriasis, asthma and several other diseases [27, 28]. Interestingly, the consumption of omega-3 acids also tends to help reduce depression in an individual [29]. Although the process itself is unclear, the consumption of EPA and DHA has been observed to have caused a decrement in pain and overall improvement in the condition of patients who suffer from rheumatoid arthritis [30].

6.3 Biosynthesis of PUFA in Autotrophic Algae

Under stress conditions, microalgae grow slowly and their growth tends to be reflected in the fatty acids that have been accumulated by them. Besides fatty acids, microalgae tend to produce several other compounds. The purpose behind the production of such a diverse variety of compounds is to ensure survival and adaptation to the stringent and varying environmental conditions [13]. There are different factors that influence autotrophic cultivation for PUFA production (Figure 6.2). There are several microalgal strains that tend to have 10-50% w/w oil content, while there are some other strains that can accumulate as high as 70% w/w [31]. *Phaeodactylum*, *Nannochloropsis*, *Thraustochytrium* and *Schizochytrium* are examples that demonstrate a significantly high accumulation of either EPA or DHA, or both [32–36]. Some strains, such as *Phaeodactylum tricornutum* and *Nannochloropsis* sp., accumulate as high as 39% w/w of EPA in their total lipids. On the opposite side, strains of *Thraustochytrium* and *Schizochytrium limacinum* tended to demonstrate a 30-40 % w/w accumulation of DHA in their total lipids. In order to adopt a cost-effective method for biomolecule production usually three types of cultivation methods are in use: autotrophic, heterotrophic

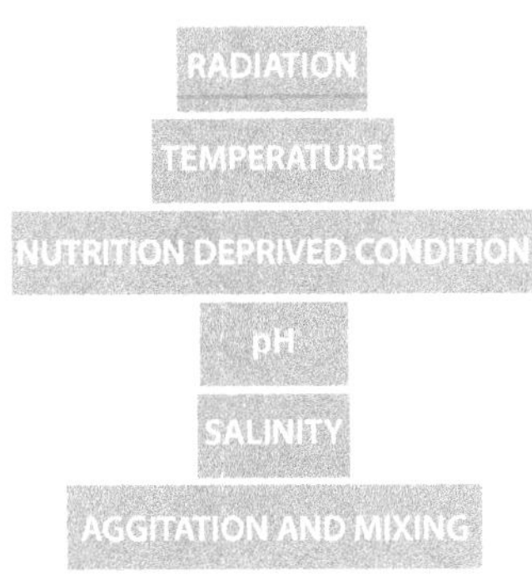

Figure 6.2 Factors affecting autotrophic cultivation for PUFA production.

and mixotrophic. In the autotrophic approach microalgae absorb light from natural resource and fix CO_2 in order to produce natural organic matter. In heterotrophic, an external organic source is usually provided to the algae, whereas mixing of both these cultivation system techniques is known as mixotrophic cultivation. Autotrophic process of cultivation is contingent to the lowest risk of contamination, therefore, it is a more prominent method to transfer growing algae from a closed system to an open system [37]. It is the most cost-effective method too. Even, Borowitzka [38] concluded that for the large-scale production of algae like *Scenedesmus* sp., and *Chlorella* sp., *Spirulina* sp., *Dunaliella sp.* the outdoor autotrophic cultivation method is the best to attain food grade algal biomass production. It is reported that open autotrophic-based microalgal cultivation requires minimum energy requirements, and therefore has better net energy ratio than the closed systems [39]. In autotrophic cultivation, high biomass content and EPA as well as DHA levels which are considered acceptable can be achieved in controlled pH conditions as well as regulated temperature levels where nitrogen and the carbon contents in the environment are appropriate [40, 41]. If these factors are controlled in an ideal manner, high cell densities allowing for microalgal bloom and increased DHA productivities can be achieved [42]. The change in the lipid content of microalgae actually affect the dietary composition of several levels in the food chain, such as the zooplanktons, and some other fish as well [43]. If there is a change in the accumulation of the EPA and DHA content in the food chain, the same becomes evident in the organisms which are superior in the food chain [9]. Ruangsomboon *et al.* [44] worked with microalga strain, *Botryococcus braunii* KMITL 2 and reported that the biomass growth in autotrophic cultivation is influenced by the light intensity and photoperiods. Under 24:0 light cycle reported biomass concentration was found four times higher than the 12:12 light/dark cycle. However, the highest lipid content was detected under 16:8 light/dark cycle. Contrastingly, Liang *et al.* [45] reported the higher cellular lipid content (38%) under autotrophic mode of cultivation in *Chlorella vulgaris* in comparison to the heterotrophic and mixotrophic mode but the lipid productivity was found lower in autotrophic in comparison to others. Santos *et al.* [46] reported the increase in microalgae *Chlorella protothecoides* biomass and lipid productivity 94% and 87% in comparison to their control under autotrophic conditions when aerated with the outlet gas stream from the yeast *Rhodosporidium toruloides* fermenter. Vidyashankar *et al.* [47] investigated 12 indigenous algal strains regarding PUFA productivity under autotrophic conditions. Results showed that the fresh water algae such as *Chlorococcum* sp. and *Scenedesmus dimorphus* were best for ALA production as on an average it produced 3 mgL^{-1} day^{-1} of it and *Nannochloropsis* sp. and *Chlorella* sp. were observed to be best strains for EPA production. Whereas, strains like *Tetraselmis theli* significantly found affected by growth medium composition in terms of biomass production. Similarly, Guedes *et al.* [48] investigated 16 microalgal species under autotrophic mode of cultivation and reported that among them, *Nannochloropsis* sp. was most suitable for ALA production. Eustigmatophyceae members followed by Chlorophyceae, Rhodophyceae, Prymnesiophyceae were listed in descending order as the best producers of ALA. Campenni *et al.* [49] concluded that *Chlorella protothecoides* under autotrophic conditions had lower biomass and lipid productivity but it produced higher amount of carotenoid in comparison to the heterotrophic conditions. Study reported autotrophic cultivation cost-effective and eco-friendly approach. Chang *et al.* [50] concluded that the increase in light intensity and temperature leads to increase in the amount of trans-fatty acid and ARA in *Porphyridium purpureum.* In scale up cultivation, the conversion pathway from C18:2 to ARA was found to be faster in comparison to the flask cultivation, leading to higher ARA accumulation. Moreover, limiting phosphate also resulted in elevated lipid and PUFA synthesis. Light spectra were also found to play an important role in some algal strains regarding the lipid and biomass production [51]. Nascimento *et al.* [52] reported that under nutrient deficient conditions, lipid content increased by 43% in *Botryococcus terribilis.* Sato *et al.* [53] worked with the red algae *Cyanidioschyzon merolae* and *Porphyridium purpureum* and revealed that *C. merolae* does not accumulate PUFAs and therefore could be the ideal strain for biodiesel production. Lin *et al.* [54] stated that a local strain in china, HDMA-11, has the potential of high fatty acid synthesis and especially ALA when grown autotrophically, resulting to nearly 39.2% out of the total lipids ALA content. Hu *et al.* [55] reported that the alga *Tisochrysis lutea* under the addition of carbon source was found to accumulate lower amount of PUFAs in mixotrophic condition than autotrophic. Another study conducted by López *et al.* [40] concluded that mixotrophic and heterotrophic cultivation condition were best suited for biomass and PUFAs production whereas for higher total lipid accumulation, autotrophic cultivation is better in *Galdieria* sp.

6.4 Harvesting of Algae and Extraction of Fatty Acids

Harvesting of microalgae from the culture systems is one of the most intriguing and difficult tasks. Dewatering or harvesting adds an additional cost of about 30% to the total production costs [56] and therefore a lot of R&D efforts have been targeted towards achieving low cost, eco-friendly, fast and efficient methods. Many methods have been developed for harvesting of algae, which can be categorized as physical, chemical and biological methods [57, 58]. Among the physical methods, the different approaches used are centrifugation, filtration, flotation, etc. Chemical approaches like inorganic flocculants, organic flocculants, electro-flocculation and pH modulation, are also applied in different studies [59]. Use of algal flocculants, bacteria and fungi has also been reported in a few studies. For proper extraction of desired compound from harvested algae, many methods have been used to assist like use of high pressure homogenization [60], microwave-assisted disruption [61], pulsed electric field [62] and ultrasound [63] and biochemical like alkali and acid treatment, enzymatic disruption methods etc. [64, 65].

In one study, four marine microalgae (*Nannochloropsis oculata, P. tricornutum, Thalassiosira pseudonana,* and *Pavlova lutheri*) were studied for the production of long chain (LC) PUFAs. Algal cells were harvested by centrifugation at 13,000 rpm for 15 min. For extraction of LC-PUFA from the interior of algal cell, chemical methods were employed in which first the pellet, after centrifugation, was mixed with chloroform methanol and frozen in liquid nitrogen. After this treatment, cell debris was separated by centrifugation and KCl is added to the total lipid supernatant. Further lipids were extracted using hexane and this represented the total lipid extract [66]. Likewise, in another study, twelve different microalgae were grown in a bioreactor for production of PUFAs, algal cells were harvested using centrifugation and lipids were extracted using the Bligh and Dyer method [67]. These studies focused more on screening of microalgae for potential production of PUFA and not much focus was given on developing an energy-efficient, fast method for harvesting of algae and extraction of desired compounds. However, in recent years, focus has shifted to make the entire bioprocess engineering and post-production process more energy efficient, easy, eco-friendly and less time consuming.

In one recent study aimed towards production of omega-3 fatty acid, *N. gaditana* was harvested from the culture medium after changing the pH of the media to about 10 pH by using NaOH [68]. Lipid extraction from harvested algae was done by acetone in combination with ultrasonic physical disruption of the cells [69]. In another study extraction of EPA from *N. gaditana* revealed that optimum yield is obtained at 236.54°C temperature, 13.95 minutes of time duration, and 60.50 g/L of algal biomass in subcritical water extraction (SWE) reactor [70]. In one unique study conducted recently, high lipid production and harvesting in *Nannochloropsis oceanica* is achieved by bio-flocculation with the oleaginous fungi *Mortierella elongata* [71]. In a recently published study, a novel approach is used where ultrasonication-microwave treatment is applied on wet biomass of yeast for lipid extraction [72]. This study, although performed on yeast biomass, can be useful if applied on microalgae. Recently developed techniques for extraction of desired compound from algae have been reviewed [9, 73, 74]. Likewise, energy efficient passive methods for extraction of lipids from algae are also reviewed [75]. In view of the above challenges, a concept of bio-refinery has been proposed (Figure 6.3).

6.5 Metabolic Engineering Towards Increasing Production of PUFA's by Algae

Microalgae possess an inherent nature to accumulate high-value compounds under stress conditions and undoubtedly have become one of the most significant unconventional sources for PUFAs [76]. Metabolic

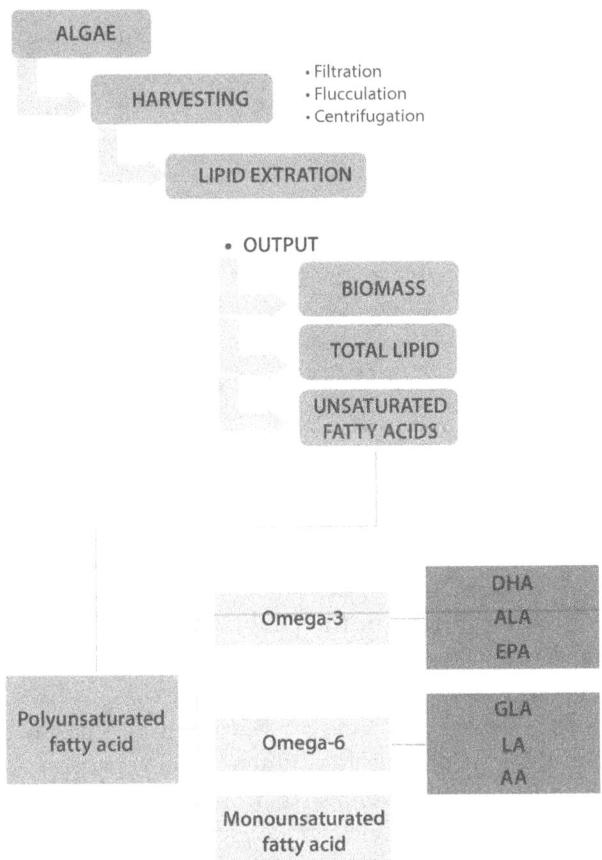

Figure 6.3 Concept of bio-refinery proposed nowadays for economical sustenance by the researchers.

engineering has played a vital role for increasing the efficacy of microalgae. Nowadays, researchers are focused on developing sustainable bio-refinery approaches for the production of PUFAs along with other lipids and biomolecules in a cost-effective manner [77, 78]. Onset of metabolic engineering for PUFA production started with *Chlamydomonas reinhardtii* after the genes encoding fatty acids synthesis were identified [79]. From that time, *C. reinhardtii* served as a model organism for protein expressions, especially considering the availability of molecular toolkits [80]. In earlier days, *Thalassiosira psuedonana* and *Phaeodactylum triconutum* were also used for studying metabolic pathway for fatty acid synthesis [66]. Earlier studies revealed information mostly about fatty acid synthesis, carboxylation and chain elongation mechanism at that point. That knowledge permitted us to make changes in enzyme regulations which lead to better protein expressions, and in turn higher production of desired biomolecules. Recently, it was reported that the marine diatom *P. tricornutum* was efficient for large-scale production of omega-3, in which an increase in DHA content was also observed after overexpressing the heterologous enzyme Δ6-desaturase and Δ5-elongase activities [32, 81]. Later on, the same strain was reported to accumulate a high amount of the omega-3 EPA and DHA by using a recombinant protein phytase [32]. Similarly, Xue *et al.* [82] reported that the total lipid content in *P. tricornutum* increased in transgenic cells by 2.5-fold and the dry weight by 57.8% of biomass as a result of malic enzyme overexpression. By altering the expression capability increase in pyruvate metabolism and carbon fixation tended to be enhanced, resulting in higher lipid accumulation in algal cells. Another study using *Nannochloropsis oceanica* was performed by Kaye [83] under nitrogen starvation conditions by overexpressing NoD12 under the controller of the stress-inducible endogenous lipid droplet surface protein (LDSP) promoter, which significantly increases the 18:2 proportion in phosphatidylcholine and TAG in the algae. Similarly, under silicon starvation conditions alteration of the elongation steps in the pathways had been made by Cook and Hildebrand [84]. Results indicated that the alterations lead into the significant increase in DHA content only, whereas the EPA and DHA per cell were found maximal. Su *et al.* [85] reported that limiting phosphate concentration promoted unsaturated fatty acids and arachidonic acid in *Porphyridium purpureum.* Their study concluded that lower concentrations of phosphate elevated Δ6-desaturase enzymatic activities which played a vital role in catalyzing the conversion of C16:0 to C18:2 within *Porphyridium purpureum.* Dong *et al.* [86] constructed the recombinant vectors Syd6D, Syd15D and Syd6Dd15D to study the expression of Δ15 and Δ6 desaturase genes in the cyanobacterial strain *Synechococcus* sp. study reported that by overexpressing Δ15 desaturase gene in *Synechococcus* sp. accumulation of α-linolenic acid enhanced by >5 times in comparison to their wild type. Moreover, overexpression of Δ6 desaturase gene enhanced GLA and SDA content with in the algal cells. Wang *et al.* [87] reported the overexpression of MCAT and PtD5b resulted into elevated lipid accumulation in engineered *P. tricornutum*. Study reported the highest accumulation 85.35 μg/mg of EPA along with 18.98 μg/mg and 9.15 μg/mg (dry weight) of arachidonic acid and DHA in their total lipid content. To achieve this hyper-accumulation, they coordinately expressed malonyl CoA-acyl carrier protein transacylase. 1-acyl-sn-glycerol-3-phosphate acyltransferase overexpression could elevate the TAG biosynthesis in *P. tricornutum* by 1.81 folds and simultaneously increase EPA and DHA content with in algal cells as reported by Balamurugan *et al.* [88]. Another study reported by Rengel *et al.* [89] worked with *C. reinhardtii* by overexpressing acetyl-CoA synthetase. Transformed cells showed 2.4-folds higher accumulation of TAG in algal cells. In another study conducted by Santos Merino *et al.* [90] in which overexpression of fabF and deletion of fadD factors in combination with desaturase genes were engineered within *Synechococcus elongates* in order to improve fatty acid production. Results revealed alpha-linolenic acid production was improved significantly. Shi *et al.* [91] reported that Δ6 fatty acid desaturase (FADS6) in microalgae catalytically enhance the ALA production in *T. pseudonana*. A similar enzyme was overexpressed to increase the production of EPA in *D. salina* by 343.8-fold and 25-fold higher than those in wild-type *D. salina*, respectively. In a recent study conducted by Norashikin *et al.* [92], agrobacterium-mediated transformed algae cells of *Chlorella vulgaris* had an upregulated expressing of the FAD gene expressions, produced higher content of α-linolenic acid in transgenic *chlorella vulgaris*. Similarly in another transformed line expression of several other enzymes like β-ketoacyl ACP synthase I, stearoyl-ACP desaturase and omega-6 desaturase genes were also found upregulated in transgenic *Chlorella vulgaris*. Pudney *et al.* [33] recently used two different recombinant proteins one from *Aspergillus niger* and another from *Escherichia coli*. *P. tricornutum* engineered using these two proteins and found to be able to accumulate higher levels of EPA and DHA. Overexpression of the promoters Pt202 and Pt667 at the same time in *P. tricornutum* under the stress conditions. Two essential lipogenic genes, malic enzyme and 5-desaturase were found overexpressed in algal cells and neutral lipid was found to increase more than twofold in the study performed by Zou *et al.* [93].

6.6 Conclusion

Undoubtedly, microalgae have emerged as a potential feedstock for high-value compounds like DHA, EPA, omega-3 and others due to the high cell growth and high efficacy of photosynthesis. They do not require any carbon source to grow as we need in the case of yeast or bacteria. Moreover, production of high-value compound could be coupled with biofuel production. This makes the process economically feasible for both biofuel as well as high-value compounds production. Furthermore, in comparison to other higher order organisms like fish it is easier to alter metabolic pathways of algal strains in order to enhance the production of desired compounds. More algal strains need to be further explored to identify appropriate candidates for biomolecules production as well as to improve product extraction. Advancement in current technologies to improve the cost-effective growth of microalgae is still required.

6.7 Acknowledgement

The contribution of Pallavi Saxena to this study was financially supported by University Grants Commission (UGC), New Delhi, India, in the form of BSR Meritorious Fellowship [F.25-a/2013-14(BSR)/7-125/2007(BSR)].

References

1. Khozin-Goldberg, I., Iskandarov, U., Cohen, Z., LC-PUFA from photosynthetic microalgae: occurrence, biosynthesis, and prospects in biotechnology. *Appl. Microbiol. Biotechnol.*, 91(4), 905, 2011.
2. Moreno-Garrido, I., Microalgae immobilization: current techniques and uses. *Bioresour., Technol.*, 99(10), 3949-3964, 2008.
3. Kumar, B.R., Deviram, G., Mathimani, T., Duc, P.A., Pugazhendhi, A., Microalgae as rich source of polyunsaturated fatty acids. *Biocat. Agricul. Biotechnol.*, 2019.
4. Biller, P., Friedman, C., Ross, A.B., Hydrothermal microwave processing of microalgae as a pre-treatment and extraction technique for bio-fuels and bio-products. *Bioresour. Technol.*, *136*, 188-195, 2013.
5. Mathimani, T., Baldinelli, A., Rajendran, K., Prabakar, D., Matheswaran, M., van Leeuwen, R.P., Pugazhendhi, A., Review on cultivation and thermochemical conversion of microalgae to fuels and chemicals: process evaluation and knowledge gaps. *J. Clean. Produc.*, 2018.
6. Mathimani, T., Kumar, T. S., Chandrasekar, M., Uma, L., Prabaharan, D., Assessment of fuel properties, engine performance and emission

characteristics of outdoor grown marine *Chlorella vulgaris* BDUG 91771 biodiesel. *Renew. Ener.*, 105, 637-646, 2017.
7. Morowvat, M. H., Ghasemi, Y., Evaluation of antioxidant properties of some naturally isolated microalgae: Identification and characterization of the most efficient strain. *Biocat. Agri. Biotechnol.*, 8, 263-269, 2016.
8. Sahu, A., Pancha, I., Jain, D., Paliwal, C., Ghosh, T., Patidar, S., ... Mishra, S., Fatty acids as biomarkers of microalgae. *Phytochemistry*, 89, 53-58, 2013.
9. Kumar, V., Arora, N., Nanda, M., & Pruthi, V., Different cell disruption and lipid extraction methods from microalgae for biodiesel production, in: *Microalgae biotechnology for development of biofuel and wastewater treatment*, Alam, Md. Asraful, Wang, Zhongming (Eds.), pp. 265-292, Springer, Singapore, 2019.
10. Szulinska, M., Gibas-Dorna, M., Miller-Kasprzak, E., Suliburska, J., Miczke, A., Walczak-Gałezewska, M., ... Bogdanski, P., *Spirulina maxima* improves insulin sensitivity, lipid profile, and total antioxidant status in obese patients with well-treated hypertension: A randomized double-blind placebo-controlled study. *Eur Rev Med Pharmacol. Sci.*, 21(10), 2473-2481, 2017.
11. Abyor, N., Dessy, A., Hady, H., Potential Production of Polyunsaturated Fatty Acids from Microalgae. *Int. J. Sci. Eng.*, *2*(1), 13-16, 2011.
12. Bhalamurugan, G.L., Valerie, O., Mark, L., Valuable bioproducts obtained from microalgal biomass and their commercial applications: A review. *Environ. Eng.*, *23*(3), 229-241, 2018.
13. Jahan, A., Ahmad, I.Z., Fatima, N., Ansari, V.A., Akhtar, J., Algal bioactive compounds in the cosmeceutical industry: a review. *Phycologia*, 56(4), 410-422, 2017.
14. Steinrücken, P., Erga, S.R., Mjøs, S.A., Kleivdal, H., Prestegard, S. K., Bioprospecting north atlantic microalgae with fast growth and high polyunsaturated fatty acid (PUFA) content for microalgae-based technologies. *Algal Res.*, 26, 392-401, 2017.
15. Ruxton, C., Reed, S., Simpson, M., Millington, K., The health benefits of omega-3 polyunsaturated fatty acids: a review of the evidence. *J. Hum. Nutri. Diet.*, 20(3), 275-285, 2007.
16. Sharon, R., Bar-Joseph, I., Frosch, M.P., Walsh, D.M., Hamilton, J.A., Selkoe, D.J., The formation of highly soluble oligomers of α-synuclein is regulated by fatty acids and enhanced in Parkinson's disease. *Neuron*, 37(4), 583-595, 2003.
17. Kris-Etherton, P.M., Grieger, J.A., Etherton, T.D., Dietary reference intakes for DHA and EPA. *Prostaglandins, Leukotrienes Essent. Fatty Acids*, 81(2-3), 99-104, 2009.
18. Swanson, D., Block, R., Mousa, S.A., Omega-3 fatty acids EPA and DHA: health benefits throughout life. *Adv. Nutri.*, 3(1), 1-7, 2012.
19. Luo, X., Su, P., Zhang, W., Advances in microalgae-derived phytosterols for functional food and pharmaceutical applications. *Mar. Drugs*, 13(7), 4231-4254, 2015.
20. Matos, J., Cardoso, C., Bandarra, N.M., Afonso, C., Microalgae as healthy ingredients for functional food: a review. *Food Func.*, 8(8), 2672-2685, 2017.

21. Li, X., Liu, J., Chen, G., Zhang, J., Wang, C., Liu, B., Extraction and purification of eicosapentaenoic acid and docosahexaenoic acid from microalgae: A critical review. *Algal Res.*, 43, 101619, 2019.
22. CODIF Recherche & Nature: DERMOCHLORELLA D - DERMOCHLORELLA DP skin restructuring, in: *St Malo cedex*, France: CODIF Recherche & Nature (Ed.), 2006.
23. Berthon, J.Y., Nachat-Kappes, R., Bey, M., Cadoret, J.P., Renimel, I., Filaire, E., Marine algae as attractive source to skin care. *Free Rad. Res.*, *51*(6), 555-567, 2017.
24. Dyall, S.C., Long-chain omega-3 fatty acids and the brain: a review of the independent and shared effects of EPA, DPA and DHA. *Front. Aging Neurosci.*, 7, 52, 2015.
25. Makrides, M., ω-3 Fatty Acids in Pregnancy: Time for Action. *J. Nutri.*, 149(4), 549-550, 2019.
26. Oken, E., Berghella, V., Barss, V.A., Fish consumption and marine n-3 long. chain polyunsaturated fatty acid supplementation in pregnancy. *UpToDate. UpToDate Inc, Waltham* (*https://www. uptodate.com;* accessed on April 16, 2019).
27. Siriwardhana, N., Kalupahana, N.S., Moustaid-Moussa, N., Health benefits of n-3 polyunsaturated fatty acids: eicosapentaenoic acid and docosahexaenoic acid, in: *Advances in food and nutrition research*, Se-Kwon Kim (Ed.), Vol. 65, pp. 211-222, Academic Press, 2012.
28. Sun, G.Y., Simonyi, A., Fritsche, K.L., Chuang, D.Y., Hannink, M., Gu, Z., ... Beversdorf, D.Q., Docosahexaenoic acid (DHA): An essential nutrient and a nutraceutical for brain health and diseases. *Prostaglandins, Leukotrienes Essen. Fatty Acids*, 136, 3-13, 2018.
29. Hibbeln, J. R., Salem Jr, N., Dietary polyunsaturated fatty acids and depression: when cholesterol does not satisfy. *Am. J. Clin. Nutr., 62*(1), 1-9, 1995.
30. Dawczynski, C., Dittrich, M., Neumann, T., Goetze, K., Welzel, A., Oelzner, P., ... Pace, S., Docosahexaenoic acid in the treatment of rheumatoid arthritis: A double-blind, placebo-controlled, randomized cross-over study with microalgae vs. sunflower oil. *Cli. Nutri.*, 37(2), 494-504, 2018.
31. Ward, O. P., Singh, A., Omega-3/6 fatty acids: alternative sources of production. *Proc. Biochem.*, 40(12), 3627-3652, 2005.
32. Hamilton, M. L. *et al.*, Towards the industrial production of omega-3 long chain polyunsaturated fatty acids from a genetically modifed diatom *Phaeodactylum tricornutum.* Plos One 10, https://doi.org/10.1371/journal.pone.0144054 (2015).
33. Pudney, A., Gandini, C., Economou, C.K., Smith, R., Goddard, P., Napier, J.A., ... Sayanova, O., Multifunctionalizing the marine diatom *Phaeodactylum tricornutum* for sustainable co-production of omega-3 long chain polyunsaturated fatty acids and recombinant phytase. *Sci. Rep.*, 9(1), 1-10, 2019.
34. Poliner, E., Pulman, J.A., Zienkiewicz, K., Childs, K., Benning, C., Farré, E.M., A toolkit for *Nannochloropsis oceanica* CCMP 1779 enables gene stacking and genetic engineering of the eicosapentaenoic acid pathway for enhanced

long-chain polyunsaturated fatty acid production. *Plant Biotechnol. J.*, 16(1), 298-309, 2018.

35. Xie, X., Meesapyodsuk, D., Qiu, X., Enhancing oil production in *Arabidopsis* through expression of a ketoacyl-ACP synthase domain of the PUFA synthase from *Thraustochytrium*. *Biotechnol. Biofuels*, 12(1), 172, 2019.
36. Li, Z., Meng, T., Ling, X., Li, J., Zheng, C., Shi, Y., ... He, N., Overexpression of malonyl-CoA: ACP transacylase in *Schizochytrium* sp. to improve polyunsaturated fatty acid production. *J. Agri. Food Che.*, 66(21), 5382-5391, 2018.
37. Razzak, S.A., Ali, S. A.M., Hossain, M.M., deLasa, H., Biological CO_2 fixation with production of microalgae in wastewater–a review. *Renew. Sust. Energy Reviews*, *76*, 379-390, 2017.
38. Borowitzka, M.A., Commercial production of microalgae: ponds, tanks, and fermenters, in: *Progress in industrial microbiology*, Osinga R., Tramper J., Burgess J.G., Wijffels R.H., (Eds.), pp. 313-321, Elsevier, 1999.
39. Jorquera, O., Kiperstok, A., Sales, E.A., Embirucu, M., Ghirardi, M.L., Comparative energy life-cycle analyses of microalgal biomass production in open ponds and photobioreactors. *Bioresour. Technol., 101*(4), 1406-1413. 2010.
40. López, G., Yate, C., Ramos, F.A., Cala, M.P., Restrepo, S., Baena, S., Production of polyunsaturated fatty acids and lipids from autotrophic, mixotrophic and heterotrophic cultivation of *Galdieria* sp. strain USBA-GBX-832. *Sci. Rep.*, 9(1), 10791, 2019.
41. Mehta, P., Singh, K., Gupta, R.P., Shankar, A., Cultivation of microalgae for the production of biomolecules and bioproducts at an industrial level. *Sustain. Downstream Proc. Microalgae Ind. App.*, 2019.
42. Yin, F.W., Guo, D.S., Ren, L.J., Ji, X.J., Huang, H., Development of a method for the valorization of fermentation wastewater and algal-residue extract in docosahexaenoic acid production by *Schizochytrium* sp. *Bioresour. Technol.*, 266, 482-487, 2018.
43. Nguyen, Q.V., Malau-Aduli, B.S., Cavalieri, J., Malau-Aduli, A.E., Nichols, P.D., Enhancing omega 3 long-chain polyunsaturated fatty acid content of dairy-derived foods for human consumption. *Nutrients*, 11(4), 743, 2019.
44. Rismani, S., Shariati, M., Changes of the total lipid and omega-3 fatty acid contents in two microalgae *Dunaliella salina* and *Chlorella vulgaris* under salt stress. *Braz. Arch. Biol. Technol.*, 60, 2017.
45. Liang, Y., Sarkany, N., & Cui, Y., Biomass and lipid productivities of Chlorella vulgaris under autotrophic, heterotrophic and mixotrophic growth conditions. *Biotechnol. Let.*, *31*(7), 1043-1049, 2009.
46. Santos, C.A., Caldeira, M.L., da Silva, T.L., Novais, J.M., Reis, A., Enhanced lipidic algae biomass production using gas transfer from a fermentative *Rhodosporidium toruloides* culture to an autotrophic *Chlorella protothecoides* culture. *Bioresour. Technol.*, *138*, 48-54, 2013.
47. Vidyashankar, S., Sireesha, E., Chauhan, V. S., & Sarada, R., Evaluation of microalgae as vegetarian source of dietary polyunsaturated fatty acids

under autotrophic growth conditions. *J. Food Sci. Technol.*, *52*(11), 7070-7080, 2015.
48. Guedes, A. C., Amaro, H. M., Barbosa, C. R., Pereira, R. D., & Malcata, F. X., Fatty acid composition of several wild microalgae and cyanobacteria, with a focus on eicosapentaenoic, docosahexaenoic and α-linolenic acids for eventual dietary uses. *Food Res Int.*, *44*(9), 2721-2729, 2011.
49. Campenni, L., Nobre, B. P., Santos, C. A., Oliveira, A. C., Aires-Barros, M. R., Palavra, A. M. F., & Gouveia, L., Carotenoid and lipid production by the autotrophic microalga Chlorella protothecoides under nutritional, salinity, and luminosity stress conditions. *Appl. Microbial Biotechnol.*, *97*(3), 1383-1393, 2013.
50. Chang, J., Le, K., Song, X., Jiao, K., Zeng, X., Ling, X., ... & Lin, L., Scale-up cultivation enhanced arachidonic acid accumulation by red microalgae *Porphyridium purpureum*. *Biopro. Biosyst. Eng.*, *40*(12), 1763-1773, 2017.
51. Vadiveloo, A., Moheimani, N.R., Cosgrove, J.J., Parlevliet, D., Bahri, P.A., Effects of different light spectra on the growth, productivity and photosynthesis of two acclimated strains of Nannochloropsis sp. *J Appl. Phycol.*, *29*(4), 1765-1774, 2017.
52. Nascimento, I.A., Cabanelas, I.T.D., dos Santos, J.N., Nascimento, M.A., Sousa, L., Sansone, G., Biodiesel yields and fuel quality as criteria for algal-feedstock selection: Effects of CO2-supplementation and nutrient levels in cultures. *Algal Res.*, *8*, 53-60, 2015.
53. Sato, N., Moriyama, T., Mori, N., Toyoshima, M., Lipid metabolism and potentials of biofuel and high added-value oil production in red algae. *World J. Microbiol. Biotechnol.*, 33(4), 74, 2017.
54. Lin, Y., Ge, J., Ling, H., Zhang, Y., Yan, X., Ping, W., Isolation of a novel strain of *Monoraphidium* sp. and characterization of its potential for α-linolenic acid and biodiesel production. *Bioresour. Technol.*, 267, 466-472, 2018.
55. Hu, H., Ma, L.L., Shen, X.F., Li, J.Y., Wang, H.F., Zeng, R.J., Effect of cultivation mode on the production of docosahexaenoic acid by *Tisochrysis lutea*. *Amb Express*, 8(1), 50, 2018.
56. Tork, M.B., Khalilzadeh, R., Kouchakzadeh, H., Efficient harvesting of marine *Chlorella vulgaris* microalgae utilizing cationic starch nanoparticles by response surface methodology. *Bioresour. Technol.*, 243, 583-588, 2017.
57. Sathe, S., & Durand, P.M., A low cost, non-toxic biological method for harvesting algal biomass. *Algal Res.*, *11*, 169-172, 2015.
58. Bhattacharya, A., Mathur, M., Kumar, P., Prajapati, S. K., Malik, A., A rapid method for fungal assisted algal flocculation: critical parameters & mechanism insights. *Algal Res.*, *21*, 42-51, 2017.
59. Branyikova, I., Prochazkova, G., Potocar, T., Jezkova, Z., & Branyik, T., Harvesting of microalgae by flocculation. *Fermentation*, *4*(4), 93, 2018.
60. Samarasinghe, N., Fernando, S., Lacey, R., & Faulkner, W.B., Algal cell rupture using high pressure homogenization as a prelude to oil extraction. *Renew.*, *48*, 300-308, 2012.

61. Ali, M., & Watson, I.A., Microwave treatment of wet algal paste for enhanced solvent extraction of lipids for biodiesel production. *Renew. Ener.*, *76*, 470-477, 2015.
62. Yodsuwan, N., Kamonpatana, P., Chisti, Y., & Sirisansaneeyakul, S., Ohmic heating pretreatment of algal slurry for production of biodiesel. *J. Biotechnol.*, *267*, 71-78, 2018.
63. Mittal, R., Tavanandi, H. A., Mantri, V. A., Raghavarao, K.S.M.S., Ultrasound assisted methods for enhanced extraction of phycobiliproteins from marine macro-algae, *Gelidium pusillum* (Rhodophyta). *Ultrason sonochem*, *38*, 92-103, 2017.
64. Zhang, S., He, Y., Sen, B., Chen, X., Xie, Y., Keasling, J.D., Wang, G., Alleviation of reactive oxygen species enhances PUFA accumulation in *Schizochytrium* sp. through regulating genes involved in lipid metabolism. *Metabolic Eng. Communi.*, 6, 39-48, 2018.
65. Phasey, J., Vandamme, D., Fallowfield, H.J., Harvesting of algae in municipal wastewater treatment by calcium phosphate precipitation mediated by photosynthesis, sodium hydroxide and lime. *Algal Res.*, *27*, 115-120, 2017.
66. Tonon, T., Harvey, D., Larson, T.R., Graham, I.A., Long chain polyunsaturated fatty acid production and partitioning to triacylglycerols in four microalgae. *Phytoche.*, 61(1), 15-24, 2002.
67. Patil, V., Källqvist, T., Olsen, E., Vogt, G., Gislerød, H.R., Fatty acid composition of 12 microalgae for possible use in aquaculture feed. *Aquacul. Int.*, 15(1), 1-9, 2007.
68. Camacho-Rodríguez, J., González-Céspedes, A.M., Cerón-García, M.C., Fernández-Sevilla, J.M., Acién-Fernández, F.G., Molina-Grima, E., A quantitative study of eicosapentaenoic acid (EPA) production by *Nannochloropsis gaditana* for aquaculture as a function of dilution rate, temperature and average irradiance. *Appl. Microbial Biotechnol.*, *98*(6), 2429-2440, 2014.
69. Abirami, S., Murugesan, S., Sivamurugan, V., Sivaswamy, S. N., Screening and optimization of culture conditions of Nannochloropsis gaditana for omega 3 fatty acid production. *J. Appl. Biol. Biotechnol.*, 5, 13-17, 2017.
70. Ho, B.C.H., Kamal, S. M. M., Danquah, M.K., Harun, R., Optimization of subcritical water extraction (SWE) of lipid and eicosapentaenoic Acid (EPA) from *Nannochloropsis gaditana*. *BioMed Res. Int.*, 2018.
71. Du, Z.Y., Alvaro, J., Hyden, B., Zienkiewicz, K., Benning, N., Zienkiewicz, A., ... Benning, C., Enhancing oil production and harvest by combining the marine alga *Nannochloropsis oceanica* and the oleaginous fungus *Mortierella elongata*. *Biotechnol. Biof.*, 11(1), 174, 2018.
72. Patel, A., Arora, N., Pruthi, V., & Pruthi, P.A., A novel rapid ultrasonication-microwave treatment for total lipid extraction from wet oleaginous yeast biomass for sustainable biodiesel production. *Ultrasonic. Sonoche.*, 51, 504-516, 2019.

73. Salinas-Salazar, C., Garcia-Perez, J.S., Chandra, R., Castillo-Zacarias, C., Iqbal, H.M., Parra-Saldívar, R., Methods for Extraction of Valuable Products from Microalgae Biomass. in: *Microalgae biotechnology for development of biofuel and wastewater treatment*, Alam, Md. Asraful, Wang, Zhongming (Eds.), pp. 245-263, Springer, Singapore, 2019.
74. Zhang, R., Parniakov, O., Grimi, N., Lebovka, N., Marchal, L., & Vorobiev, E., Emerging techniques for cell disruption and extraction of valuable biomolecules of microalgae *Nannochloropsis* sp. *Bioproc. Biosys. Eng.*, 42(2), 173-186, 2019.
75. Nagappan, S., Devendran, S., Tsai, P.C., Dinakaran, S., Dahms, H.U., Ponnusamy, V.K., Passive cell disruption lipid extraction methods of microalgae for biofuel production–A review. *Fuel*, 252, 699-709, 2019.
76. Stengel, D.B., Connan, S., & Popper, Z.A., Algal chemodiversity and bioactivity: sources of natural variability and implications for commercial application. *Biotechnol Adv.*, *29*(5), 483-501, 2011.
77. Adarme-Vega, T.C., Lim, D.K., Timmins, M., Vernen, F., Li, Y., Schenk, P.M., Microalgal biofactories: a promising approach towards sustainable omega-3 fatty acid production. *Microbial Cell Factories*, *11*(1), 96, 2012.
78. Mayfield, S., Golden, S.S., *Photosynthetic bio-manufacturing: food, fuel, and medicine for the 21st century*, 2015.
79. Chi, X., Zhang, X., Guan, X., Ding, L., Li, Y., Wang, M., ... & Qin, S., Fatty acid biosynthesis in eukaryotic photosynthetic microalgae: identification of a microsomal delta 12 desaturase in Chlamydomonas reinhardtii. *J. Microbiol.*, *46*(2), 189-201, 2008.
80. Scranton, M.A., Ostrand, J.T., Fields, F.J., Mayfeld, S.P., *Chlamydomonas* as a model for biofuels and bio-products production. *Plant J.* 82, 523–531, 2015.
81. Hamilton, M. L., Haslam, R. P., Napier, J. A., & Sayanova, O., Metabolic engineering of *Phaeodactylum tricornutum* for the enhanced accumulation of omega-3 long chain polyunsaturated fatty acids. *Metabolic Eng.*, 22, 3-9, 2014.
82. Xue, J., Niu, Y. F., Huang, T., Yang, W. D., Liu, J. S., & Li, H. Y., Genetic improvement of the microalga *Phaeodactylum tricornutum* for boosting neutral lipid accumulation. *Metabolic Eng.*, 27, 1-9, 2015.
83. Kaye, Y., Grundman, O., Leu, S., Zarka, A., Zorin, B., Didi-Cohen, S., ... & Boussiba, S., Metabolic engineering toward enhanced LC-PUFA biosynthesis in *Nannochloropsis oceanica*: Overexpression of endogenous Δ12 desaturase driven by stress-inducible promoter leads to enhanced deposition of polyunsaturated fatty acids in TAG. *Algal Res.*, *11*, 387-398, 2015.
84. Cook, O., Hildebrand, M., Enhancing LC-PUFA production in *Thalassiosira pseudonana* by overexpressing the endogenous fatty acid elongase genes. *J. App. Phycol.*, *28*(2), 897-905, 2016.
85. Su, G., Jiao, K., Li, Z., Guo, X., Chang, J., Ndikubwimana, T., ... & Lin, L., Phosphate limitation promotes unsaturated fatty acids and arachidonic acid biosynthesis by microalgae *Porphyridium purpureum*. *Bioproc. Biosy. Eng.*, 39(7), 1129-1136, 2016.

86. Dong, X., He, Q., Peng, Z., Yu, J., Bian, F., Li, Y., & Bi, Y., Production of γ-linolenic acid and stearidonic acid by *Synechococcus* sp. PCC7002 containing cyanobacterial fatty acid desaturase genes. *Chi. J. Oceanol. Limnol.*, 34(4), 772-780, 2016.
87. Wang, F., Bi, Y., Diao, J., Lv, M., Cui, J., Chen, L., & Zhang, W., Metabolic engineering to enhance biosynthesis of both docosahexaenoic acid and odd-chain fatty acids in *Schizochytrium* sp. S31. *Biotechnol. Biofuels*, 12(1), 141, 2019.
88. Balamurugan, S., Wang, X., Wang, H. L., An, C. J., Li, H., Li, D. W., ... & Li, H. Y., Occurrence of plastidial triacylglycerol synthesis and the potential regulatory role of AGPAT in the model diatom *Phaeodactylum tricornutum*. *Biotechnol. Biof.*, *10*(1), 97, 2017.
89. Rengel, R., Smith, R. T., Haslam, R. P., Sayanova, O., Vila, M., & León, R., Overexpression of acetyl-CoA synthetase (ACS) enhances the biosynthesis of neutral lipids and starch in the green microalga *Chlamydomonas reinhardtii*. *Algal Res.*, 31, 183-193, 2018.
90. Santos-Merino, M., Garcillán-Barcia, M. P., & de la Cruz, F., Engineering the fatty acid synthesis pathway in *Synechococcus elongatus* PCC 7942 improves omega-3 fatty acid production. *Biotechnol. Biof.*, 11(1), 239, 2018.
91. Shi, H., Luo, X., Wu, R., & Yue, X., Production of eicosapentaenoic acid by application of a delta-6 desaturase with the highest ALA catalytic activity in algae. *Micro. Cell Factories*, 17(1), 7, 2018.
92. Norashikin, M. N., Loh, S. H., Aziz, A., & San Cha, T., Metabolic engineering of fatty acid biosynthesis in *Chlorella vulgaris* using an endogenous omega-3 fatty acid desaturase gene with its promoter. *Algal Res.*, 31, 262-275, 2018.
93. Zou, L. G., Balamurugan, S., Zhou, T. B., Chen, J. W., Li, D. W., Yang, W. D., ... & Li, H. Y., Potentiation of concurrent expression of lipogenic genes by novel strong promoters in the oleaginous microalga *Phaeodactylum tricornutum*. *Biotechnology and Bioengineering*, 2019.

7

Production of Omega-3 and Omega-6 PUFA from Food Crops and Fishes

Km Sartaj and R. Prasad*

Department of Biotechnology, Indian Institute of Technology Roorkee, Roorkee, Uttarakhand

Abstract

A highly stressed and busy life schedule demands a balanced diet with all nutritive components. Thus, among all other nutrition, intake of essential fatty acids (EFAs) is highly recommended due to the inability of the human body to synthesize these essential fats. Fishes, algae, oilseed crops, aquatic foods and vertebrates are the key sources of polyunsaturated fatty acids (PUFAs), mainly eicosapentaenoic acid (EPA) and docosahexaenoic acid (DHA). However, demand for these nutraceuticals is burgeoning at a tremendous rate which has led to enormous pressure on the current market. So, there should be constant efforts to advance the current production potential by upgrading the technologies in obtaining these nutraceuticals from aquatic and agricultural sources. The current chapter mainly explores the prospective of fish and food crops for n-3 and n-6 PUFA production and also discusses the variation in fatty acid profiles in interspecific and intraspecific association of species. Furthermore, this article also describes the variation of lipid profile with different factors like geographical locations, season, temperature, salinity, farmed or wild fish.

Keywords: Polyunsaturated fatty acids, eicosapentaenoic *acid* (EPA), *docosahexaenoic acid* aquaculture, geographical locations

**Corresponding author*: rapdyfbs@iitr.ac.in

Alok Kumar Patel and Leonidas Matsakas (eds.) Nutraceutical Fatty Acids from Oleaginous Microalgae: A Human Health Perspective, (187–208) © 2020 Scrivener Publishing LLC

7.1 Introduction

Ever rising population, technology advancement and modernization play important roles in the day-to-day lives of human beings. To accomplish an upgraded and advanced lifestyle they fit themselves into a hectic schedule that affects their health directly. Therefore, numerous campaigns related to health awareness gain people's attention, which makes them more concerned about maintaining their health [1]. Consequently, the trend of utilizing compounds having high nutritive value has increased the pharmaceutical industry's interest in them. Among several healthy and nutritious foods, polyunsaturated fatty acids (PUFAs) are one of the basic and vital nutrients for humans [2]. Principally it is a category of fatty acids having 18-22 carbons with two or more than two double bonds in the existing carbon series [2, 3]. Omega-3 and 6 fatty acids (ω-3 and ω-6) are the major families of PUFA that include α linoleic acid (ALA) and linolenic acid respectively [3]. These are considered as essential fatty acids due to the inability of humans to synthesize these by their own metabolism because of lack of an enzyme desaturase, which is mainly responsible for insertion of double bond at different positions. Therefore, dietary sources are the major options to fulfil the requirement of these fatty acids [4, 5]. Due to high nutritional value of these compounds some dietary organizations and government agencies recommend daily intake level of PUFA ranging from 0.2-0.3g/day [6]. Several countries like Canada and Scandinavian countries have now established daily recommended dietary intake values for essential fat [7]. Thus, the worldwide consumption of PUFAs are rapidly increasing day by day; however, the rising demand can't be fulfilled by the traditional dietary sources of omega-3 fatty acids. This is the major factor that supports its commercial production [8]. Although it has some major challenges such as availability of appropriate PUFA containing resources, that makes them more costly. Thus, a large number of the population cannot afford these supplements. Therefore, it is necessary to escalate research in the field of production and purification of PUFA so that everyone can fulfil the demand of their daily nutritional requirement in spite of their busy schedules. Considering the importance of PUFA, many researchers have tried to increase the yield of these supplements by modifying the natural tendency of lipid accumulation in plants and animals via two different approaches; genetic engineering and designing or managing food systems to maximize the production of fatty acids by utilizing the natural resources. In addition to this, the scientific communities are also striving to produce long chain fatty acids through newer sources [9].

The aim of the current chapter is to discuss PUFA production strategies or processes from different food crops and fishes, and concurrently enlighten its role in terms of nutritional compounds and applications in public health. The literature also reviews challenges and opportunities in terms of long chain fatty acid production.

7.2 PUFA as a Dietary Supplement

According to the Brazilian Society of Cardiology, American College of Cardiology and American Heart Association, daily oral supplementation with ω-3 PUFAs are approximately 2-4g [10] is an effective treatment to prevent heart failure and also atherosclerosis along with cardioprotective benefits [11, 12]. Deficiencies of these fatty acids not only adversely affect human health but are also responsible for major disorders such as hyperactivity disorder (ADHD), autism, dyspraxia, etc. [13, 14].

7.2.1 Omega-3 (n-3) Fatty Acids

Omega-3 fatty acids are one of the family of PUFAs that can be depicted by various terminology like ω-3 fatty acids or n-3 fatty acids, etc., which basically represent the position of double bonds located in between third and fourth carbon if counting starts from methyl group (CH_3) [15]. Nowadays these n-3 fatty acids are gaining more attention towards commercialization and their utilization as an important dietary supplement for preventing the risk of cardiovascular disease, arteriosclerosis, etc. In addition, they also serve as essential components for membrane lipids [16]. Two most important categories of n-3 fatty acids are Docosahexaenoic acid (DHA, 22:6) and Eicosapentaenoic acid (EPA, 20:5) having 22 carbon atoms with six double bonds and 20 carbons with five double bonds, respectively. This degree of unsaturation is inversely proportional to the melting point of fatty acids that further linked to the fluidity. Synthesis of EPA and DHA involves a n-3 precursor; α-linolenic acid (ALA; 18:3) that is an essential component for human diet and cannot be synthesized endogenously [15]. Both of these essential fats play a significant role in larval development of fish, molluscs and shrimp [7]. Chloroplast of green leafy vegetables, walnuts, flax seeds chia seeds, flax, rape and perilla has enough amount of unsaturated fats [11].

7.2.2 Omega-6 (n-6) Fatty Acids

This category of PUFA is mainly known for its applicability in precursor compounds that are metabolized by numerous enzymes and produce a wide range of biologically as well as clinically important eicosanoids hormones, including leukotrienes, thromboxanes and prostaglandins, etc., some of which play substantial roles in combating or preventing deadly diseases [17]. Arachidonic acid (ARA) and γ-linoleic acid (GLA) are typical ω-6 fatty acids that are present in breast milk [17]. Apart from this, animal tissues such as liver and adrenal glands contain ω-6 ARA whereas plants belonging to Borage, Oenothera and other species are the primary and traditional source of ω-6 GLA extraction [18]. Seeds of most plants except a few of these like cocoa, coconut and palm are commonly rich in ω-6 GLA (Figure 7.1) [11].

7.2.3 Health Aspects and Physiological Functions of PUFA

The long chain (LC) fatty acids are gaining people's attention due to their health promotive and disease reductive qualities. Fatty acids also play a key role in maintaining blood lipid profile, lipid composition, gene expression and cell signalling cascade, etc. [19]. Between countless benefits, these are essential elements of all kinds of life forms. Formation of cell membrane

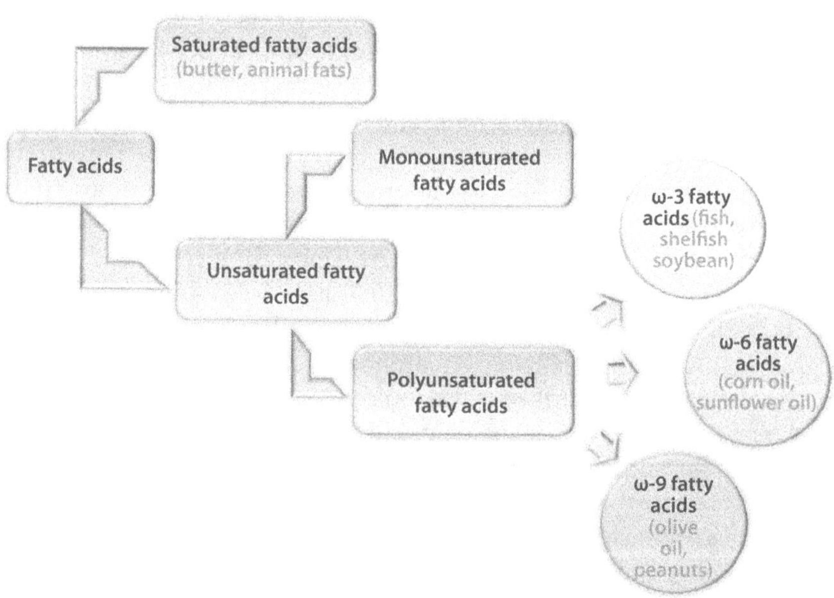

Figure 7.1 Overview of fatty acids.

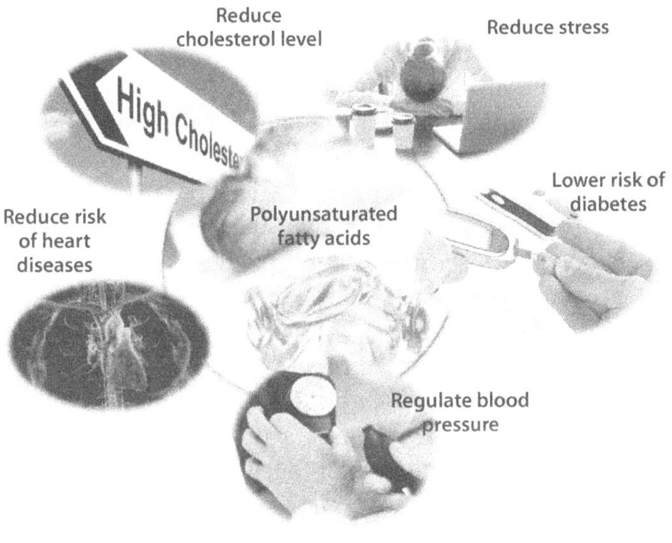

Figure 7.2 Health benefits of PUFA.

structure, source of energy, hormone precursor [20] and modulators of immune response [21] are some unique properties that cannot be replaced by any other food component (Figure 7.2). Therefore, consuming fatty acids is increasing day by day and people are using supplements rich in PUFA, especially ω-3 long chain fatty acids, in their diet on a daily basis. According to Lovo-Martins *et al.*, fish oil having n-3 unsaturated fat is helpful in controlling inflammation that is necessary for the host defence system against several infectious diseases like Chagas, etc. [11]. Past clinical trials in human patients and experimental data have shown that LC n-3 PUFA rich supplements act as powerful anti-inflammatory and immunomodulatory agents for rheumatoid arthritis [22], inflammatory bowel disease [23], autoimmune [24] and even infectious diseases [11, 25, 26]. In addition to this, a study on pregnant and lactating women reveals that DHA oil is good for infants' eye and brain function. Inhibition of blood platelet aggregation, decreased fat accumulation in liver and improved functions of vascular endothelial cell are some more health benefits of PUFA [27].

7.3 Biosynthesis and Metabolism of PUFA

Although essential fatty acids are directly or indirectly associated with physiological health of all vertebrates, their metabolic system is not, however, suitable to synthesize PUFA from monounsaturated fats since its synthesis depends on endogenous capacity of species turning from PUFA (C_{18}) to the biologically active highly unsaturated fatty acids (HUFA): EPA, DHA and ARA [28]. Moreover, quantitative and qualitative requirements of essential fatty acids also vary among different animal species. Therefore, it is essential to gather deep insight into its biosynthetic part, which requires intense knowledge about fatty acid chemistry and their biochemical mechanisms.

The process of PUFA synthesis follows two main routes: aerobic and anaerobic pathways. Where an aerobic pathway involves a chain of distinct desaturation and elongation reactions, an anaerobic pathway is accomplished by "polyketide synthase," a multi-enzyme complex that performs desaturation and elongation (2-carbon) by using malonyl coenzyme A (malonyl-CoA) and acetyl coenzyme A (acetyl-CoA) [29]. Generally fatty acid synthase synthesizes saturated fatty acid, palmitic (C16:0) or stearic (C18:0) acid that act as a precursor for PUFA production. These precursors go through several modifications and lead to formation of malonyl-CoA and acetyl-CoA that is considered as a primary stage of fatty acid synthesis. The whole process is accomplished in the presence of two main leading enzymes, desaturase and elongase. Both the enzymes participate in terms of increasing the degree of unsaturation by adding extra double bonds to the carboxyl end of the fatty acids and increasing the chain length, respectively. However, these enzymes are absent (or present in slight amounts) in some mammals including Homo sapiens [29]. The process of PUFA synthesis varies slightly among different organisms from plants to mammals. Occurrence of Δ12 and n3 desaturase activities in plants assist synthesis of linoleic (18:2n6) and linolenic acid (18:2n6). Whereas on the other side, mammals complete their requirement of these essential fatty acids from the diet directly [29]. Further reactions (desaturation and elongation) in endoplasmic reticulum convert these essential fatty acids into ARA (20:4n6) and DHA (22:6n3). Unlike plants and mammals, the free-living nematode *Caenorhabditis elegans* have capacity to produce ARA (20:4n6) and EPA (20:5n3), but are unable to elongate C20 PUFAs further [29].

Current studies in this field reveal a novel biosynthetic pathway (polyketide syntheses) for PUFA in both prokaryotes and eukaryotic microorganisms. Although it follows the same reaction as FAS, the uniqueness of the pathway is that it uses a polyketide synthase rather than elongase/desaturase structure to produce 20:5n3 and 22:6n3 [30]. Moreover, the pathway runs in presence of six different enzymes: 3-ketoacyl synthase (KS), 3-ketoacyl-ACP-reductase (KR), dehydrase (DH), enoyl reductase (ER), dehydratase/ 2-trans, 3-cis isomerase (DH/2,3I), dehydratase/2-trans and 2-cis isomerase (DH/2,2I). Irrespective of desaturase pathway it inserts double bond to nascent acyl chains that makes this pathway energetically more proficient [28].

7.4 Potential Commodities for PUFA Production

Diseases such as diabetes, heart attacks, arthritis, etc., are intensely connected to the daily life of humans in which their food habits play a major role. More consumption of processed food in children also causes obesity, fatty liver, etc.; therefore different government agencies are trying to increase awareness [3]. Consequently, regular visits to dieticians also enhances people's knowledge about daily requirements of the body, like vitamin, calcium, fatty acid, supplements: PUFA, etc. Exclusively the current generation is becoming more advanced and playing a major role in increasing consumption of food and supplements that provide full nourishment to their body along with good mental health. So, increasing demand of PUFA can only be fulfilled by focusing research more toward exploring the feedstock for PUFA production along with advanced techniques that can enhance the production up to a large scale. At present, sea foods including fishes, most of aquatic animals and food crops are the major source of highly unsaturated fatty acid production (HUFA). The present chapter describes the potential of some traditional sources for PUFA production.

7.4.1 Food Crops

Apart from the advanced farming of fishes, rapid depletion has been noticed in aquatic life forms [31]. This is due to the increased globalization of some fish oil-based supplements and pharmaceutical products. Subsequently, increasing concern towards marine pollution and over-fishing are the major key points that force an evaluation of potential resources for LC PUFA production that can meet any increased demand. Moreover, plants have also been explored as potential sources for production of EFAs usually produced by fish for example, linoleic and linolenic acids along with ARA, DHA and EPA. Inability of land plants towards synthesis of LC-PUFAs above C18 is a major demerit but considerable research in public and private sectors has made it possible to produce ω-3 fatty acids in a wide range of oilseed crops [32]. For example, seeds of borage (*Borago oficinalis*), evening primrose (*Oenothera biennis*) and black current (*Ribes nigrum*) comprised of 24-25%, 8-10% and 16-17% (w/w) GLA, respectively [8]. Hence, PUFA production from plants and food crops is an appealing approach for renewable supply of ω-3 fatty acids. Table 7.1 shows the different content of PUFA in food crops and plants.

Table 7.1 PUFA contents in food crops and plants.

Sources	Category	Types of PUFA	Content (%)/g	References
Oenothera	Herbaceous flowering plants	GLA	–	[2]
Borago	Herbs			
Ribes	Flowering plants	ALA	–	
Coconut oil	Food crop	Lauric acid (C12:0)	49	[33]
		Myristic acid (C14:0)	8	
		Palmitic acid (C16:0)	8	
		Stearic acid (C18:0)	2	
		Oleic acid (C18:1)	6	
		Linoleic acid (C18:2)	2	
Palm kernel oil	Food crop	Lauric acid (C12:0)	15.17	[34]
Palm oil	Food crop	Capric acid (C 10:0)	49.18	

(*Continued*)

Table 7.1 PUFA contents in food crops and plants. (*Continued*)

Sources	Category	Types of PUFA	Content (%)/g	References
		Palmitic acid (C16:0)	8.47	
		Lauric acid (C12:0)	0.36	
		Palmitic acid (C16:0)	44.57	
		Oleic acid (C18:1)	55.05	
Sunflower	Plant	Saturated fatty acids (SFA)	12	[3]
		Monounsaturated fatty acids (MFA)	20.5	
		Polyunsaturated fatty acids (PUFA)	67.5	
Corn	Food crop	Saturated fatty acids (SFA)	14.5	[3]
		Monounsaturated fatty acids (MFA)	29.9	
		Polyunsaturated fatty acids (PUFA)	55.6	

(*Continued*)

Table 7.1 PUFA contents in food crops and plants. (*Continued*)

Sources	Category	Types of PUFA	Content (%)/g	References
Soya	Food crops	Saturated fatty acids (SFA)	15.6	[3]
		Monounsaturated fatty acids (MFA)	21.2	
		Polyunsaturated fatty acids (PUFA)	63.2	
Flax seed	Seeds	α-Linolenic acid	22.8 g/100g	[3, 35]
Walnuts, black	Nuts	α-Linolenic acid	3.3g/100g	
Leeks (freeze- dried)	Vegetables	α-Linolenic acid	0.7g/100g	

Food crops, mainly oilseed crops, are majorly grown for oil in their seeds and vary in terms of oil content, composition and quality. Oil obtained from these crops is rich in long chain fatty acids (C14 to C24), EPA, DHA (10-15%) and can be utilized in the form of nutraceuticals like fish oil capsules or as feed for animals and aquaculture that provide a better option to get ω-3 LC-PUFA enriched eggs, meat, milk and fish, etc., directly from consuming these products [36]. High production capacity of approximately 2.5 million hectares (2% of total world acreage) and relatively low cost makes these crops an excellent and sustainable alternative for ω-3 fatty acid production [36]. The vegetarian nature of food crops also reduces pressure on marine aquaculture and satisfies the requirement for an important sector of the global market. Soybean, rapeseed, groundnut, sesame seed, linseed, mustard seed and safflower are some worldwide known oilseed crops. Seeds like flaxseed oil and walnut are known for ω-3 fatty acids (DHA/EPA), whereas safflower oil, corn oil and soybean oil are popular for ω-6 fatty acids [3]. However, linseed oil provides both ω-3 and 6 fatty acids. Experimental data suggest that some plant breeders improved traditional crops in an efficient way so that these could produce nutritionally favourable products in higher amounts. Sunflower (NusunTM), rapeseed (MonolaTM, canola) soybean and linseed (LinolaTM, solin) are some examples of plant species which have been bred with increased oleic acid content [32, 36].

7.4.1.1 Soybean Seeds

The polyunsaturated fatty acids in traditional soybean oil varies from 4-11% linolenic acid (C18:3) and 48-59% linoleic acid (C18:2). The global production of soybean oil in 2002 was 180Mt, almost 50% of the total oilseed production. While seed oil content in soybean is approximately 20% lower as compared to other crops, that can be overcome by applying genetic engineering approaches for further improvement in fatty acid prolife [3, 36].

7.4.1.2 Rapeseed

Higher proportion of oleic (C18:1), linoleic (C18:2), and linolenic (C18:3) fatty acids enhances nutritional value of current canola oil as compared to traditionally present oil that was rich in erucic acid and saturated fatty acids. Although this is rated in fifth position with 33 Mt production worldwide, there has been a continual decline in its production. Along with France, Canada and Germany, India is also a major producer of canola oil [3, 36].

7.4.1.3 *Safflower*

Fatty acid profile of safflower oil depicted that it contains high oleic acid (70-84%) as compared to traditional one (68-83% linoleic acid and 8-22% oleic acid). However, early cultivars of safflower were not suitable for commercial production because of low oil content 30%. Since several breeding programmes have increased more lipid than was previously the case, which improved its production to industrial level [3, 36].

7.4.1.4 *Sesame and Linseed*

Sesame is an important crop that is not only rich in PUFA but also shows antioxidant properties that have excellent health benefits. It holds 25% of protein but oil content is much higher (approximately 50%) than protein. Oleic acid (40%) and linoleic acid (45%) are majorly found fatty acids. Canada, China, USA and India are major producers of linseed. Growing conditions of linseed majorly affects its production level. LinolaTM 989 is linseed oil containing 34% protein and 46% oil (dry basis) [3, 36].

7.4.1.5 *Sunflower*

The sunflower oil constitutes low linolenic acid (2-17%) and high oleic acid (75-91%) with higher levels of alpha tocopherol (Vitamin E). Some mid oleic cultivars such as NuSun™ are also available with 8% saturated fatty acid and 20-30% linoleic acid [3, 36].

7.4.2 Transgenic Plants

Intensive research has also resulted in the introduction of various transgenic plants having good amount of very long chain of polyunsaturated fatty acids (VLC-PUFAs) in their vegetative tissues. This approach mainly targets expression of gene from zebra fish and algae to seeds of plants, for example, transgenic plant of soybean refined to get increased DHA approximately 3% [37, 38].

7.4.3 Fishes

Terrestrial and aquatic (marine or freshwater) ecosystems are both awidespread as a traditional source for large-scale PUFA production. Fishes and shellfish are considered a nutritionally high-quality food for humans and also a key source of ω-3 fatty acids [9]. Therefore, global demand for fishes has increased, resulting in overfishing of seas, for example; during the past 20 years, production and consumption of salmonids has dramatically increased. Where dieticians recommend adding oily marine fish such as tuna, herring, mackerel or salmon in the diet twice a week, a large number of the population is not able to access these fishes [39]. Therefore, formulation and processing of fish oil supplements rich in PUFAs is becoming mandated, which requires proper knowledge about biochemical constituents of fishes along with its lipid profile.

7.4.3.1 *Fish Bioecology and Lipid Content*

Production of EFAs from fish is directly or indirectly affected by different habitat, generally their aquatic environment (cold, warm, marine and freshwater) and considerably with climate, diet, maturity, age and type of species, sexual maturity, migration and spawning period, etc. [40, 41]. Water (60-90%), mineral (0.5-5%), protein (10-22%) and fats (1-20%) are the main composite of fishes in which triacylglycerol and phosphoglycerides are major form of lipid. That are rich in long chain fatty acids, especially EPA and DHA. Analytical studies from tropical climate (lower lipid) to the Arctic region show a clear difference in total amount of lipid. Additionally, most fish from Indian water have different lipid composition, from myristic, palmitic and stearic acids to monounsaturated and polyunsaturated group like palmitoleic and oleic acids. Fishes preferably found in Western countries like salmon (*Salmo salar*), tuna (*Thunnus thynnus*) and cod (*Gadus morhua*) not only provide DHA but also contain tryptophan (precursor of serotonin) and vitamin D [42]. Where freshwater fish rich in linoleic (C18:2 n-6), linolenic (C18:3 n-3) and EPA [40] LA and ALA [41, 43] marine fish constitute with EPA and DHA [44, 45]. Thus, marine fish are considered a source of n-3 PUFA whereas freshwater fish as n-6 PUFA; this is because of different feeding habits, as freshwater fishes mainly depend on vegetation and plant materials while marine fishes feed on zooplanktons; PUFA rich in microscopic organisms. Fish species from the Northern Hemisphere, herring, sardines and menhaden, etc., are known for 18:12 oil, that is the most common fish oil supplement which follows different composition of 180mg (18%) EPA and 120 mg (12%) of DHA per 1000 mg of oil. One of the major restrictions in consumption of fish rich in LC PUFA like, etc., is the bioaccumulation of marine pollutants like heavy metals and polychlorinated biphenyl (PCB) toxins, etc. [46].

7.5 Alternate Sources of PUFA

Fish oil have certain demerits like bad taste, high cost, fishy smell, stability and sustainability problems, etc. Further, the quality of product is also dependent on season, location and ocean pollution. Low yield of oilseed crops, and requirement of a large land area are the major hurdles with PUFA production [47]. Therefore, it is equally important to explore new alternatives of LC-PUFAs that could eliminate these problems and discard the shortcomings of a fish oil-based process. Thus, many researchers screened lower and higher plant, algae and microbes for their potential to produce n-3 and n-6 PUFA. According to current dogma, bacteria and marine microalgae have all the necessary enzymes involved in *de novo* PUFA synthesis and these are consider as a primary source of oil enriched with DHA (up to 40%) [47]. This statement is also supported by the fact that fish also get unsaturated fats by eating microalgae. Protein (12-35%), carbohydrate (4.6-23%) and fats (7.2-23%) (% dry weight) are the basic constituents of microalgae and nutraceuticals obtained from these can be available in the form of capsules, powders, tablets and concentrates [47].

Phaeodactylum tricornutum (EPA 29.8%), *Nitzchia* and *Nannochoropsis* are some examples of marine microalgae that are traditionally used as EPA producers [10, 48]. Apart from these advantages, large-scale production of PUFA seems impossible due to some hindrances like slow growth rate and low accumulated biomass of microalgae. Therefore, intervention of various advanced techniques like metabolic engineering and omega-3 biotechnology in algae and algae like microorganisms for n-3 fatty acids enhanced its production level. PUFA in most of oleaginous fungi belongs to Zygomycetes store in form of triacylglycerol that is economically more valuable. Recently E.I. DuPont Company is utilizing metabolically engineered yeast strain of *Yarrowia lipolytica* for commercial production of EPA [49]. As a result of progressive research in this field, Suntory, Martek, Gist Brocades and Lion have developed processes for fungal PUFAs. Omega-Tech, Scotia Pharmaceutical, Heliosynthese, Sagami and Nestle examined new approaches for microalgal and bacterial synthesis of PUFAs, respectively [8, 18, 46, 47]. Table 7.2 shows different sources of PUFA production.

7.6 Future Avenues

The nutritional value of unsaturated fatty acids is an accepted fact worldwide since their role in cardiovascular and neuronal disease prove they are a most prized component of diet. Marine aquaculture and genetically

Table 7.2 Different sources of PUFA production.

PUFA	Type of fatty acid	Conventional sources	Microbial sources			References
			Fungi	Algae	Bacteria	
DHA	ω-3	**Fish:** cod, tuna, herring, salmon sardine and menhaden ***Shellfish:*** blue crab, mussel oyster and lobster	*Thraustochytrium aureum*	*microalgae MK8805, Gyrodinium nelsoni*	*Vibrio spp., Rhodopseudomonas spp.*	[50, 51]
EPA	ω-3	**Fish:** menhaden and herring **Shellfish:** blue crab, mussel lobster and oyster	*Mortierella elongata, Pythium irregular*	*Chlorella minutissima, Porhyridium cruentum*	*Shewanella putrefaciens*	[8, 52]
GLA	ω-6	**Plants:** Evening primrose, Blackcurrant and Borage	*Mucor mucedo, Cunninghamella elegans*	*Spirulina platensis, Chlorela vulgaris*	–	[8, 37, 53]

(*Continued*)

Table 7.2 Different sources of PUFA production. (*Continued*)

PUFA	Type of fatty acid	Conventional sources	Microbial sources			References
			Fungi	Algae	Bacteria	
DPA (Docosapentaenoic acid)	ω-3	**Fish**	*Schyzochytrium sp.*	–	–	[54]
AA (Arachidonic acid)	ω-6	**Animal tissues** **Fish:** *Brevoortia* **Mosses:** *Ctenidium molluscum*	*Pythium ultimum*	*Euglena gracilis* *Sargassum salicifolium*	–	[8, 55]
ETA (Eicosatetranoic acid)	ω-3	**Animal tissues**	–	–	–	[56]
DGLA (Dihomo-γ-linoleic acid)	ω-6	**Human milk, Animal tissues, Fish, Mosses:** *Pogonatum urnigerum*	*Saprolegnia ferax*	–	–	[56]
MA (Mead acid)	ω-9	**Animal tissues**	*Mortierella alpina*	–	–	[57]

modified oilseed crops are considered as origin of n-3 and n-6 LC PUFAs and must meet the demand of a burgeoning human population. Since consumption of these resources will increase in the near future, progressive research toward evaluating alternative sources of PUFA production is a much-needed task. More advanced techniques related to metabolic engineering of microorganisms for tailoring lipid will also help in overcoming the pressure on existing species. Hence, it becomes essential to resolve all the glitches linked with large-scale production and further improvement is also required to augment the EPA and DHA level in farmed fish and aquatic ecosystem without any negative impact on feed efficiency, fish growth and health. Price implications, negative perceptions about genetically modified (GM) derived oil and hurdles in regulatory approvals are some major drawbacks which need to be resolved on an urgent basis so that a balance between supply and demand can be maintained in the coming years [58].

7.7 Conclusion

The present study summarised all the interesting aspects related to PUFA from fishes and food crops, whether it is about health benefits or existing resources for PUFA production. Advanced fish farming and cultivation of oilseed crops are effectually fulfilling all the requirements of essential fatty acids, although much remains to be investigated about alternative for PUFA production due to high demand of EPA and DHA. However, the biggest challenge is to continue discovering a new way to introduce these compounds into human diet in a safe and therapeutic manner.

References

1. Kaczorowski, J., Chambers, L. W., Dolovich, L., Paterson, J. M., Karwalajtys, T., Gierman, T., Zagorski, B., Improving cardiovascular health at population level: 39 community cluster randomised trial of Cardiovascular Health Awareness Program (CHAP). *Bmj.*, 342, 442, 2011.
2. Alonso, D. López., F. García Maroto., Plants as 'chemical factories' for the production of polyunsaturated fatty acids. *Biotechnol Adv.*, 18, 6, 481-497, 2000.
3. Amjad Khan, W., Chun-Mei, H., Khan, N., Iqbal, A., Lyu, S. W., Shah, F., Bioengineered plants can be a useful source of omega-3 fatty acids. *Biomed Res. Int.*, 2017.
4. Simopoulos, A., An increase in the omega-6/omega-3 fatty acid ratio increases the risk for obesity. *Nutrients*, 8, 128, 2016.

5. Isabel Lovo-Martins, M., Cardoso Martins-Pinge, M., Pinge-Filho, P., Fish Oil and Inflammation: A Perspective on the challenges of evaluating efficacy in *Trypanosoma cruzi* infection. *Biology of Trypanosoma cruzi IntechOpen.*, 2019.
6. Kalogeropoulos, N., Chiou, A., Gavala, E., Christea, M., Andrikopoulos, N.K., Nutritional evaluation and bioactive microconstituents (carotenoids, tocopherols, sterols and squalene) of raw and roasted chicken fed on DHA-rich microalgae. *Food Res. Int.*, 43, 2006–2013, 2010.
7. Barclay, W. R., Meager, K.M., Abril, J.R., Heterotrophic production of long chain omega-3 fatty acids utilizing algae and algae-like microorganisms. *J. Appl. Phycol.*, 6, 123–129, 1994.
8. Certik, M., Shimizu, S., Biosynthesis and regulation of microbial polyunsaturated fatty acid production. *J. Biosci. Bioeng.*, 87, 1–14, 1999.
9. Patel, A., Matsakas, L., Hrůzová, K., Rova, U., Christakopoulos, P., Biosynthesis of nutraceutical fatty acids by the oleaginous marine microalgae *Phaeodactylum tricornutum* utilizing hydrolysates from organosolv-pretreated birch and spruce biomass. *Mar. Drugs*, 17, 119, 2019.
10. Scheben, A. Edwards, D., Bottlenecks for genome-edited crops on the road from lab to farm. *Genome Biol.*, 19, 178, 2018.
11. Lovo-Martins, M.I., Malvezi, A.D., da Silva, R.V., Zanluqui, N.G., Tatakihara, V.L., Câmara, N.O., de Oliveira, A.P.L., Peron, J.P., Martins-Pinge, M.C., Fritsche, K.L., Pinge-Filho, P., Fish oil supplementation benefits the murine host during the acute phase of a parasitic infection from *Trypanosoma cruzi. Nutr. Res.*, 41, 73–85, 2017.
12. Yancy, C.W. *et al.* ACCF/AHA Guideline for the management of heart failure. *J. Am. Coll. Cardiol.*, 62, 147–239, 2013.
13. Pote, S., Bhadekar, R., Statistical approach for production of PUFA from *Kocuria* sp. BRI 35 isolated from marine water sample. *BioMed Res. Int.*, 2014, 1–9, 2014.
14. Schuchardt, J. P., Huss, M., Stauss-Grabo, M., Hahn, A., Significance of long-chain polyunsaturated fatty acids (PUFAs) for the development and behaviour of children. *Eur. J. Pediatr.*, 169, 149–164, 2010.
15. Health benefit of omega-3 polyunsaturated fatty acids. https://onlinelibrary.wiley.com/action/showCitFormats, 2019.
16. Hayashi, S., Satoh, Y., Ujihara, T., Takata, Y., Dairi, T., Enhanced production of polyunsaturated fatty acids by enzyme engineering of tandem acyl carrier proteins. *Sci. Rep.*, 6, 35441, 2016.
17. Kapoor, R., Huang, Y.S., Gamma linolenic acid: an antiinflammatory omega-6 fatty acid. *Curr. Pharm. Biotechno.*, 7(6), 531-534, 2006.
18. Ji, X.J., Huang, H., Engineering microbes to produce polyunsaturated fatty acids. *Trends Biotechnol.*, 37, 2018.
19. Shahidi, F., Wanasundara, N., Concentrates: nutritional aspects and production technologies. *Trends Food Sci. Technol.*, 11, 1998.
20. Turchini, G.M., Nichols, P.D., Barrow, C., Sinclair, A.J., Jumping on the omega-3 bandwagon: Distinguishing the role of long-chain and short-chain omega-3 fatty acids. *Crit. Rev. Food Sci. Nutr.*, 52, 795–803, 2012.

21. Fritsche, K., Fatty Acids as modulators of the immune response. *Annu. Rev. Nutr.*, 26, 45–73, 2006.
22. Woo, S.J., Lim, K., Park, S.Y., Jung, M.Y., Lim, H.S., Jeon, M.G., Lee, S.I., Park, B.H., Endogenous conversion of n-6 to n-3 polyunsaturated fatty acids attenuates K/BxN serum-transfer arthritis in fat-1 mice. *J. Nutr. Biochem.*, 26, 713–720, 2015.
23. Belluzzi, A., Boschi, S., Brignola, C., Munarini, A., Cariani, G., Miglio, F., Polyunsaturated fatty acids and inflammatory bowel disease. *Am. J. Clin. Nutr.*, 71(1), 339-342, 2000.
24. Pestka, J.J., n-3 Polyunsaturated fatty acids and autoimmune-mediated glomerulonephritis. *Prostaglandins Leukot. Essent. Fat. Acids.*, 82, 251–258, 2010.
25. Irons, R., Anderson, M.J., Zhang, M., Fritsche, K.L., Dietary fish oil impairs primary host resistance against *Listeria monocytogenes* more than the immunological memory response. *J. Nutr.*, 133, 1163–1169, 2003.
26. Blok, W. L., Vogels, M. T., Curfs, J. H., Eling, W. M., Buurman, W. A., van der Meer, J. W., Dietary fish-oil supplementation in experimental gram-negative infection and in cerebral malaria in mice. *J. Infect.*, 165(5), 898-903, 1992.
27. Zárate, R., Jaber-Vazdekis, N., Tejera, N., Pérez, J.A., Rodríguez, C., Significance of long chain polyunsaturated fatty acids in human health. *Clin. Transl. Med.*, 6, 25, 2017.
28. Bell, M.V., Tocher, D.R., Biosynthesis of polyunsaturated fatty acids in aquatic ecosystems: general pathways and new directions. *Lipids in Aquatic Ecosystems*, 211–236, 2009.
29. Wallis, J.G., Watts, J.L., Browse, J., Polyunsaturated fatty acid synthesis: what will they think of next? *Trends Biochem. Sci.*, 27, 467, 2002.
30. Metz, J.G., Production of polyunsaturated fatty acids by polyketide synthases in both prokaryotes and eukaryotes. *Science*, 293, 290–293, 2001.
31. Bhattacharya, A., Sarkar, S.K., Impact of overexploitation of shellfish: Northeastern coast of India. *Ambio.*, 70-75, 2003.
32. Hirschi, K.D., Nutrient biofortification of food crops. *Annu. Rev. Nutr.*, 29, 401–421, 2009.
33. Boateng, L., Ansong, R., Owusu, W., Steiner-Asiedu, M., Coconut oil and palm oil's role in nutrition, health and national development. *Ghana Medical Journal.*, 50(3), 189-196, 2016.
34. Ejiofor, E., Shirley, E., Obike, A., Onyedikachi, B., Okechukwu, A., Obinna, A., Michael, K., Omeh, N., Fatty acids composition profile evaluation of palm oil in crude oil polluted environment. *Asian J Agri & Biol.*, 6(3), 373-378, 2018.
35. Patil, V., Källqvist, T., Olsen, E., Vogt, G., Gislerød, H.R., Fatty acid composition of 12 microalgae for possible use in aquaculture feed. *Aquac Int.*, 15(1), 1-9, 2007.
36. Hbw, P., Jb, T., Tm, R., Cl, W., 380 OILSEEDS, OVERVIEW. 7
37. Damude, H.G., Kinney, A.J., Enhancing plant seed oils for human nutrition. *Plant Physiol.*, 147, 962–968, 2008.

38. Tucker, G., Nutritional enhancement of plants. *Curr. Opin. Biotechnol.*, 14, 221–225, 2003.
39. Strobel, C., Jahreis, G., Kuhnt, K., Survey of n-3 and n-6 polyunsaturated fatty acids in fish and fish products. *Lipids Health Dis.*, 11, 144, 2012.
40. Sahena, F., Zaidul, I.S.M., Jinap, S., Saari, N., Jahurul, H.A., Abbas, K.A., Norulaini, N.A., PUFAs in Fish: Extraction, Fractionation, Importance in Health. *Compr. Rev.* ***Food*** *Sci.* ***Food*** *Saf.*, 8, 59-74, 2009.
41. Taşbozan, O., Gökçe, M.A., Fatty acids in fish. *Fat. Acids*, 2017.
42. Mohanty, B.P., Sankar, T.V., Ganguly, S., Mahanty, A., Anandan, R., Chakraborty, K., Paul, B.N., Sarma, D., Dayal, J.S., Mathew, S., Asha, K.K., Micronutrient composition of 35 food fishes from India and their significance in human nutrition. *Biol. Trace Elem. Res.*, 174, 448–458, 2016.
43. Tocher, D.R., Metabolism and functions of lipids and fatty acids in Teleost fish. *Rev. Fish. Sci.*, 11, 107–184, 2003.
44. Steffens, W., Wirth, M., Freshwater fish-an important source of n-3 polyunsaturated fatty acids: a review. *Archives of Polish Fisheries*, 13, 5, 2005.
45. Grahl-Nielsen, O., Averina, E., Pronin, N., Radnaeva, L., Käkelä, R., Fatty acid profiles in different fish species in Lake Baikal. *Aquat. Biol.*, 13, 1–10, 2011.
46. Walsh, T.A., Metz, J.G., Producing the omega-3 fatty acids DHA and EPA in oilseed crops. *Lipid Technol.*, 25, 103-105, 2013.
47. Handayania, N.A., Ariyantib, D., Potential production of polyunsaturated fatty acids from microalgae. *J. Bioprocess. Biotech.*, 01, 2012.
48. Ward, O.P., Singh, A., Omega-3/6 fatty acids: Alternative sources of production. *Process Biochem.*, 40, 3627–3652, 2005.
49. Ledesma-Amaro, R., Nicaud, J.M., *Yarrowia lipolytica* as a biotechnological chassis to produce usual and unusual fatty acids. *Prog. Lipid Res.*, 61, 40–50, 2016.
50. Vazhappilly, R., Chen, F., Eicosapentaenoic acid and docosahexaenoic acid production potential of microalgae and their heterotrophic growth. *J. Am. Oil Chem. Soc.*, 75, 393–397, 1998.
51. Singh, A., Ward, O.P., Microbial production of docosahexaenoic acid (DHA, C22:6). *Adv Appl Microbiol.*, 45, 271–312, 1997.
52. Bajpai, P.K., Bajpai, P., Ward, O.P., Optimisation of culture conditions for production of eicosapentaenoic acid by *Mortierella elongata* NRRL 5513. *J. Ind. Microbiol.*, 9, 11-17, 1992.
53. Cuellar-Bermudez, S.P., Aguilar-Hernandez, I., Cardenas-Chavez, D.L., Ornelas-Soto, N., Romero-Ogawa, M.A., Parra-Saldivar, R., Extraction and purification of high-value metabolites from microalgae: essential lipids, astaxanthin and phycobiliproteins: High-value metabolites form algae. *Microb. Biotechnol.*, 8, 190–209, 2015.
54. Wang, Q., Ye, H., Sen, B., Xie, Y., He, Y., Park, S., & Wang, G., Improved production of docosahexaenoic acid in batch fermentation by newly-isolated strains of *Schizochytrium* sp. and *Thraustochytriidae* sp. through bioprocess optimization. *Synth Syst Biotechnol.*, 3(2), 121-129, 2018.

55. Gandhi, S.R., Weete, J.D., Production of the polyunsaturated fatty acids arachidonic acid and eicosapentaenoic acid by the fungus *Pythium ultimum*. *J. Gen. Microbiol.*, 137, 1825–1830, 1991.
56. Abedi, E., Sahari, M.A., Long-chain polyunsaturated fatty acid sources and evaluation of their nutritional and functional properties. *Food Sci. Nutr.*, 2, 443–463, 2014.
57. Sakuradani, E., Advances in the production of various polyunsaturated fatty acids through oleaginous fungus *Mortierella alpina* breeding. *Biosci. Biotechnol. Biochem.*, 74, 908–917, 2010.
58. Tocher, D.R., Betancor, M.B., Sprague, M., Olsen, R.E., Napier, J.A., Omega-3 long-chain polyunsaturated fatty acids, EPA and DHA: Bridging the gap between supply and demand. *Nutrients*, 11, 89, 2019.

8

The Role of Metabolic Engineering for Enhancing PUFA Production in Microalgae

Neha Arora

Department of Chemical Engineering, Indian Institute of Technology Bombay, Powai, Mumbai, India

Abstract

Polyunsaturated fatty acids (PUFA) are vital nutritional ingredients for normal human development and growth. Since the human body cannot meet the daily nutritional requirements of omega-3 and omega-6 PUFAs on its own, they need to be consumed as food supplements or nutritive drinks. This has led to an upsurge in the demand and a rapid expansion of PUFA-based nutraceuticals. However, the current fish farming cannot meet the demand and supply chain, thereby necessitating the development of a sustainable alternative. In this regard, microalgae have emerged as potential producers for omega-3 and omega-6 fatty acids. The present chapter focuses on the importance of microalgae for enhanced PUFA production. A brief overview of the major pathways operating in different microalgae for the biosynthesis of PUFA has been discussed. The chapter also reviews the characterization of novel PUFA enzymes from different microalgal strains which can be engineered in microalgae to improve the overall PUFA productivity. A catalogue of metabolic engineering studies carried out to date in various microalgae to enhance the PUFA production has been listed.

Keywords: PUFA, microalgae, omega-3, omega-6, metabolic engineering

8.1 Introduction

Lipids, particularly triacylglycerol (TAGs), are considered as universal energy storage molecules, playing a vital role in the cellular metabolism [1].

Email: nehaarorajiit@gmail.com

Alok Kumar Patel and Leonidas Matsakas (eds.) Nutraceutical Fatty Acids from Oleaginous Microalgae: A Human Health Perspective, (209–226) © 2020 Scrivener Publishing LLC

Depending on the fatty acid composition (carbon number, chain length and degree of unsaturation) of TAGs, they can be utilized as nutritional supplement or biofuels. Among TAGs, polyunsaturated fatty acids (PUFAs) are essential components of human diet and are responsible for maintaining normal development and health of an individual [2]. The long-chain PUFAs (LC-PUFAs) are categorized according to the number of hydrocarbon acyl chains (18-22 carbons) with several cis-double bonds which are positioned either on 3rd carbon (termed as omega-3 fatty acids) or on 6th carbon (termed as omega-6 fatty acids) being closest to the methyl (omega) end [3]. LC-PUFAs have higher stability and bioavailability than free fatty acids or ethyl ester lipids [1]. Multiple PUFAs are recognized as essential fatty acids including linoleic acid (C18:2n6), γ-linolenic acid (C18:3n6), α linolenic acid (C18:3n3), octadecatetraenoic acid (C18:4n3), dihomo-γ-linolenic acid (C20:3n6), arachidonic acid (C20:4n6), eicosapentaenoic acid (C20:5n3) and docosahexaenoic acid (C22:6n3) [4]. Due to the unique structural and functional characteristics of LC-PUFAs, they are involved in the regulation of the design, dynamics, permeability and phase transition of cell membranes [5]. They are responsible for modulating the behaviour of membrane receptors, ATPases and ion channels. Deficiency of PUFAs in the human body can lead to skin abnormalities, abnormal development of central nervous, immune, inflammatory, cardiovascular, respiratory and reproductive systems [5].

PUFAs, particularly eicosapentaenoic acid (EPA) and docosahexaenoic acid (DHA), are crucial for the growth and development of humans [6]. EPA is a key component of the glycolipids and phospholipids which maintain and adjust the membrane fluidity [7]. It is also a precursor for various biological active regulators such as hormones, signalling molecules etc., which are involved in cellular metabolism [7]. DHA is abundantly present in the brain tissue and is essential to maintain and improve the brain function [8]. It is imperative for the overall neurodevelopment including visual and cognitive maturity. Regular consumption of EPA and DHA in diet has been linked to lower cholesterol levels, alleviation of inflammation, proper functioning of circulatory system, putative role in curing certain cancers, prevention and treatment of chronic diseases, arthritis, asthma, coronary heart disease, Alzheimer's, depression and autoimmune diseases [2, 4, 7]. Indeed, humans can synthesize DHA and EPA on their own, however, only ~8% of dietary α-linolenic acid (ALA) is converted into EPA while ~4% to DHA, which is not sufficient to meet the daily nutritional requirements of humans [9]. Thus, to meet the daily nutritional requirements, EPA and DHA are required to be consumed as dietary supplements.

Currently, marine fishes are the main sources for EPA and DHA which are non-sustainable due to the rapid depletion of fish stocks, contamination of the aquatic ecosystems particularly by heavy metals, inadequate quality, instability in the supply chain and high purification cost of the desired product [6, 7]. This necessitates the development of alternative sources for the production of DHA and EPA, which are sustainable, cost-effective and non-toxic. In this regard, microalgae are the primary producers of the LC-PUFAs in aquatic ecosystems and thus offer an advantageous feedstock for DHA and EPA production. Indeed, fishes don't synthesize LC-PUFA but eat microalgae which leads to accumulation of DHA and EPA in them. Microalgal lipids are an excellent source of PUFA as they are rich in both omega-3 (ω-3) including ALA, EPA and DHA and omega-6 (ω-6) fatty acids such as γ-linolenic acid (GLA) [10]. The main advantages of utilizing microalgae as potential feedstocks for the production PUFAs include [10]:

(i) Oleaginous marine microalgae can accumulate ~ 30-40% of intracellular lipids/dry cell weight.
(ii) In comparison to marine fishes, microalgae have rapid growth rates and capability to survive under various physiological conditions inducing stress which in turn increases the lipid.
(iii) Microalgal farming is unaffected by the seasonal/climatic alterations and could be done all year round.
(iv) High-quality PUFA production can be achieved, since lipid accumulation and profile can be manipulated easily.
(v) Numerous microalgae species are available which have unique enzymes for PUFA synthesis, providing greater flexibility to improve the yield and properties.
(vi) Plethora of existing platforms for genetic engineering of microalgal strains to enhance the PUFA productivity.
(vii) Microalgae biomass after the extraction of PUFA can be utilized for other valuable product generation such as pigments, carotenoids, vitamins, etc., or as biofertilizer, as biofuel feedstocks, thus establishing a biorefinery concept.

Keeping the above view in mind, in the present chapter the importance of microalgae as a potential source for PUFAs synthesis is discussed. The overview of the metabolic pathway for the synthesis of ω-3 and ω-6 fatty acids is provided. Further, the identification and characterization of key enzymes in various microalgal strains has been detailed. The chapter also reviews the key metabolic engineering strategies that have been carried out in different microalgal strains for increasing the DHA and EPA synthesis.

8.2 LC-PUFA Biosynthesis in Microalgae

LC-PUFA biosynthesis in microalgae occurs via two major pathways: the conventional aerobic pathway which is prevalent in all eukaryotic microalgae, plants, fungi and bacteria and the unique anaerobic pathway which operates in a few eukaryotic microalgae and prokaryotic bacteria. This distinction is made in terms of the utilization of molecular oxygen by the desaturase enzymes for the synthesis of EPA and DHA via the aerobic pathway while the anaerobic pathway eliminates the need for oxygen and requires less reducing power, NADPH [11]. A brief overview of the two pathways is discussed in the section below.

8.2.1 Conventional Aerobic Pathway

The conventional aerobic PUFA biosynthesis is initiated in the endoplasmic reticulum (ER) with conversion of palmitic acid (C16:0) to stearic acid (C18:0), which is then desaturated to oleic acid ($C18{:}1^{\Delta 9}$) by Δ9 desaturase (Figure 8.1). Oleic acid is then converted to linoleic acid (LA, $C18{:}2^{\Delta 9,12}$) commencing the ω-6 PUFA pathway, which is catalyzed by Δ12 desaturase. The subsequent desaturation of linoleic acid to ALA ($C18{:}3^{\Delta 9,12,15}$), thereby committing it to the ω-3 PUFA pathway [12, 13]. The ω-3 fatty acid synthesis begins by the desaturation of ALA to stearidonic acid (SDA, $C18{:}4^{\Delta 6,9,12,15}$) catalyzed by Δ6 desaturase, followed by elongation of SDA to eicosatetraenoic acid (ETA, $C20{:}4^{\Delta 8,11,14,17}$), mediated by Δ6 elongase which is then desaturated to EPA via Δ5 desaturase (Figure 8.1). Alternating Δ5 elongase and Δ5 desaturase then finally convert EPA to DHA, respectively. Similarly, for the ω-6 fatty acid synthesis, a series of desaturases and elongases catalyze the complete pathway resulting in the biosynthesis of GLA ($C18{:}3^{\Delta 6,9,12}$), di-homo-γ -linoleic acid (DGLA, $C20{:}3^{\Delta 8,11,14}$), arachidonic acid (ARA, $C20{:}4^{\Delta 5,8,11,14}$) and docosatetraenoic (DTA, $C22{:}4^{\Delta 7,10,13,16}$) and docosapentaenoic acid (DPA, $C22{:}5^{\Delta 4,7,10,13,16}$) (Figure 8.1).

However, an alternative pathway also exists, which is present in a few microalgae including *Isochrysis galbana* and *Pavlova salina* [12]. This pathway bypasses the Δ6 desaturase enzyme and is a Δ9 specific elongation of LA or ALA to eicosadienoic acid (EDA, $C20{:}2^{\Delta 11,14}$) or ETA, respectively [13]. The respective products EDA and ETA are desaturated via Δ8 desaturase and then subsequently converted to ARA and EPA, thus integrating in the common pathway of either ω-6 or ω-3 fatty acid synthesis respectively (Figure 8.1).

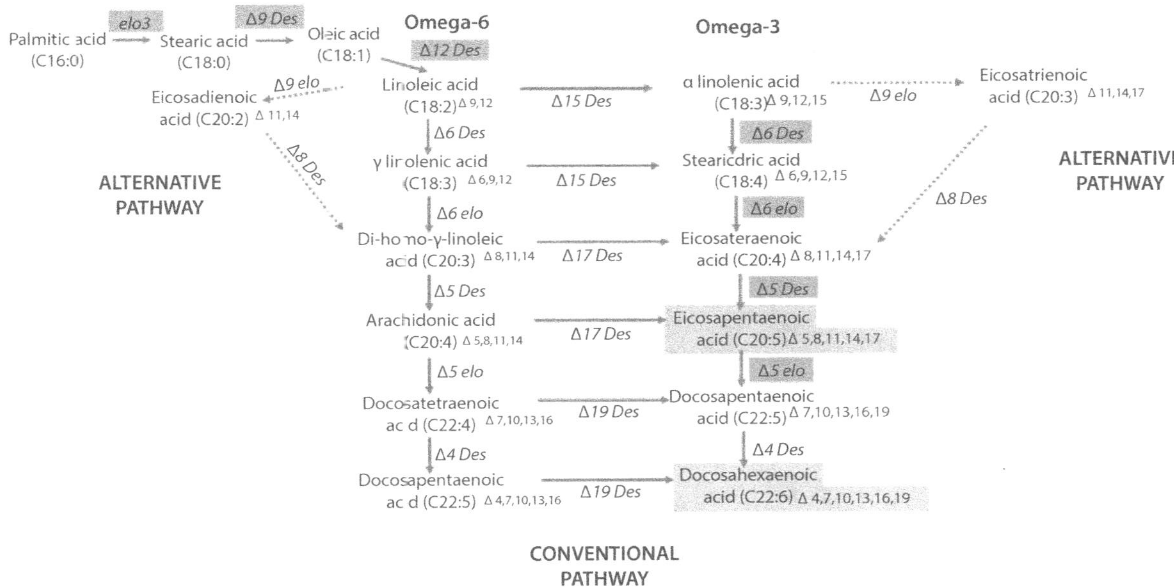

Figure 8.1 Schematic of aerobic conventional pathway and alternative pathway for biosynthesis of omega-3 and omega-6 fatty acids in microalgae. Adapted from [12, 13].

8.2.2 Anaerobic Pathway

The DHA synthesis in thraustochyrids is initiated via bacterial like polyketide synthase (PKS) or PUFA synthase which excludes the *in-situ* reduction of the intermediates like the conventional aerobic pathway [14]. The PUFA synthase is composed of multiple domains (08) which carry out the PUFA synthesis in *Schizochytrium* sp. [11, 14]. These domains include 3-ketoacylsynthase (KS) which is a condensing enzyme, acyl-carrier protein (ACP), malonyl-CoA:ACP acyltransferase (MAT), 3-ketoacyl-ACP reductase (KR), acyltransferase (AT), chain length factor (CLF), enoyl reductase (ER) and dehydrase/isomerase (DH) [14]. The synthesis of DHA is initiated by the successive elongation with malonyl-CoA to DHA (Figure 8.2).

8.3 Identification and Characterization of Enzymes Involved in PUFA Synthesis

The growing importance of microalgae-based LC-PUFA synthesis has led to a keen interest in the identification and characterization of novel and efficient PUFA genes which can potentially increase the metabolic flux towards DHA and EPA production. Microalgae are a diverse group of organisms with every species having a variable PUFA accumulating capacity, attributed to the presence of a unique and different set of PUFA genes. These genes encode enzymes with different substrate preference (here substrate refers to the fatty acid precursor of ω-6 or ω-3 PUFA biosynthesis pathway) and conversion efficiency,

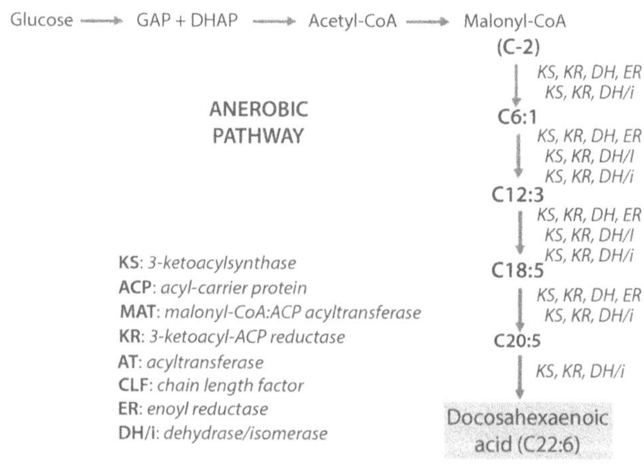

Figure 8.2 Overview of anaerobic biosynthesis of DHA in microalgae. GAP: glyceraldehyde 3-phosphate, DHAP: dihydroxyacetone phosphate. Adapted from [14].

resulting in the formation of EPA and DHA. Identification and characterization of different PUFA genes can potentially leap towards enhancing the EPA and DHA productivity in microalgae via genetic engineering approaches. To date, various desaturases and elongases have been identified, characterized and expressed in yeast from different microalgae to study their substrate affinity and conversion rate as listed in Table 8.1. Fatty acid desaturases majorly belong to two main classes, namely soluble enzymes and transmembrane enzymes [15]. The soluble enzymes insert double bond to an acyl-ACP substrate and exist in the stroma of chloroplast using ferredoxin (Fd) as an electron acceptor [15]. On the other hand, transmembrane enzymes, add the double bond to acyl-glycerolipids with three electron acceptors namely Fd (chloroplast desaturase), cyctochrome b5 (Cytb5) for ER desaturase and Cytb5 fusion which is located in the ER or plastid of the cell [15]. A maximum conversion (54%) of ALA to SDA was reported by Δ6 desaturase isolated from *Ostreococcus* sp. RCC 809 [16]. The Δ6 elongase isolated from *Myrmecia incisa Reisigl H4301* showed maximum conversion (41%) of SDA to ETA [17]. Further, the Δ5 desaturase derived from *Pavlova salina* exhibited ~ 11% conversion of ETA to EPA [18]. The subsequent conversion of EPA to DPA was reported to be 75% when catalysed by *Pyamimonas cordata* Δ5 elongase [19]. The final step of the desaturation of DPA to DHA was maximum by Δ4 desaturase isolated from *Pavlova lutheri* [20]. Interestingly, the authors also reported an equal affinity of the enzyme towards ω-6 fatty acid substrate (DTA) and its desaturation to DPAn-6. The alternative pathway enzymes Δ8 desaturase and Δ9 elongase have also been characterized indicating the feasibility of engineering two pathways in microalgae for further enhancing the PUFA production (Table 8.1). Similarly, a few of the desaturase had a higher affinity for ω-6 fatty acid substrates including Δ6 elongase (*Parietochloris incsia*) and Δ5 elongase (*Pavlova salina*) (Table 8.1). Indeed, more enzymes need to be identified and characterized to strengthen the molecular tool box for engineering microalgae.

8.4 Metabolic Engineering for Enhancing the LC-PUFA Production in Microalgae

To date, only a few microalgal strains have been genetically engineered to enhance DHA and EPA production. The model diatom *P. tricornutum* under optimum growth conditions accumulates ~ 35% of EPA content with trace levels of DHA [6]. To increase the DHA content, heterologous overexpression of phospholipid dependent Δ6- desaturase from the picoalga *Ostreococcus tauri* in *P. tricornutum*-UTEX-646 lead to only 1.8% of DHA in total lipids as

Table 8.1 List of desaturases and elongases isolated and characterized from different microalgae with their substrate specificity and conversion efficiency.

Microalgae	Enzyme	Substrate specificity	Product	Conversion efficiency (%)	Reference
Parietochloris incsia	Δ6 desaturase	–	C18:2, C20:3, C18:3, C20:4	–	[33]
	Δ5 desaturase	–		–	
	Δ6 elongase	C18:3 (GLA)	C20:3 (DGLA)	16.4	[34]
		C18:4 (STA)	C20:4 (ETA)	24.6	
Ostreococcus sp. RCC809	Δ6 desaturase	C18:3 (ALA)	C18:4 (SDA)	54	[16]
	Δ4 desaturase	C22:5 (DPA)	C22:6 (DHA)	15	
Ostreococcus lucimarinus	Δ6 desaturase	C18:3 (ALA)	C18:4 (SDA)	38.8	[19]
		C18:2 (LA)	C18:3 (GLA)	6.6	
Pyamimonas cordata	Δ6 elongase	C18:4 (SDA)	C20:5 (EPA)	65.6	
	Δ5 elongase	C20:5 (EPA)	C22:5 (DPA)	75	

(Continued)

Table 8.1 List of desaturases and elongases isolated and characterized from different microalgae with their substrate specificity and conversion efficiency. (*Continued*)

Microalgae	Enzyme	Substrate specificity	Product	Conversion efficiency (%)	Reference
Isochrysis glabana	Δ9 elongase	C18:3 (ALA)	C20:3 (ETA)	45	[35]
		C18:2 (LA)	C20:2 (EDA)	45	
				57.6	[17]
		C18:3 (ALA)	C20:4 (ETA)	56.1	
Phaeodactylum tricornutum	Δ9 desaturase	C16:0 (PA)	C16:1	–	[15]
	Δ12 desaturase (ω6)	C18:1 (OA)	C18:2 (LA)		
	Δ15 desaturase (ω3)	C18:3 (ALA)	C18:4 (SDA)		
	Δ5 desaturase (cis)	C20:4 (ETA)	C20:5 (EPA)		
	Δ5 desaturase (cis)	C22:5 (DPA)	C22:6 (DHA)		
	Δ12 desaturase (cis)	C16:1	C16:2		
	Δ6 desaturase (cis)	C16:2	C16:3		

(*Continued*)

Table 8.1 List of desaturases and elongases isolated and characterized from different microalgae with their substrate specificity and conversion efficiency. (*Continued*)

Microalgae	Enzyme	Substrate specificity	Product	Conversion efficiency (%)	Reference
	Δ3 desaturase (trans)	C16:0 (PA)	C16:1		
	ω3desaturase (cis)	C16:3	C16:4		
	Δ9 desaturase (cis)	C18:0 (SA)	C18:1 (OA)		
	Δ11 elongase	C20:5 (EPA)	C22:5 (DPA)	93.1	[36]
		C20:4 (ARA)	C22:4 (DTA)	79.4	
Thalassiosira pseudonana	Δ11 desaturase	C16:0	C16:1	–	[37]
Pavlova lutheri	Δ4 desaturase	C22:4 (DTA)	C22:5 (DPA)	35	[20]
		C22:5 (DPA)	C22:6 (DHA)	35	

(*Continued*)

Table 8.1 List of desaturases and elongases isolated and characterized from different microalgae with their substrate specificity and conversion efficiency. (*Continued*)

Microalgae	Enzyme	Substrate specificity	Product	Conversion efficiency (%)	Reference
Pavlova salina	Δ4 desaturase	C22:4 (DTA)	C22:5 (DPA)	3	[18]
		C22:5 (DPA)	C22:6 (DHA)	2.4	
	Δ5 desaturase	C20:4 (ETA)	C20:5 (EPA)	11.3	
	Δ8 desaturase	C20:3 (ERA)	C20:4 (ETA)	4.8	
	Δ5 elongase	C20:5 (EPA)	C22:5 (DPA)	42.4	[38]
		C20:4 (AA)	C22:4 (DTA)	42.3	
Glossomatrix chrysoplasta	Δ6 desaturase	C18:2 (LA)	C18:3 (GLA)	7	[39]
		C18:3 (ALA)	C18:4 (SDA)	6	
Nannochloropsis oculata		C18:2 (LA)	C18:3 (GLA)	–	[40]
		C18:3 (ALA)	C18:4 (SDA)	–	
Myrmecia incisa Reisigl H4301	Δ6 elongase	C18:3 (GLA)	C20:3 (DGLA)	24	[17]
		C18:4 (SDA)	C20:4 (ETA)	41	

compared to 1.3% of DHA in the wild type strain [21]. The EPA content was also marginally increased in the transgenic strain. These results indicated that overexpressing the first committed step towards EPA/DHA synthesis was not sufficient to induce DHA accumulation in this diatom. However, when the authors overexpressed *O. tauri* derived Δ5- elongase there was an 8-fold increase in DHA content as compared to wild type. Such an increase was attributed to the activity of C-20 Δ5- elongase which is responsible for the C-2 elongation of EPA-CoA to DHA-CoA, a precursor of DHA. Contrary to the above findings, the endogenous overexpression of Δ6- desaturase gene led to ~47% increase in EPA production with a total accumulation of 33.035 mg/g in the engineered diatom as compared to the wild type (25.803 mg/g) [22]. One possible reason for such a difference in EPA synthesis could be the origin of the overexpressed Δ6- desaturase gene, i.e., the gene obtained from *P. tricornutum* was more efficient to redirect the flux towards the EPA production in comparison to the *O. tauri* derived desaturase gene. Thus, emphasizing on the identification, characterization and selection of efficient genes (desaturases and elongases) that can drive the EPA and DHA production in microalgae (detailed in section 3). Furthermore, the endogenous overexpression of Δ5- desaturase (rate limiting enzyme in PUFA synthesis-depicted in Figure 8.1) in *P. tricornutum* resulted in ~65% increase in the total neutral lipid content as compared to the wild type with EPA contributing to 38.9 mg/g of dry cell weight [23]. Further, the authors reported accumulation of 0.44 mg/g DHA in the engineered strains which was not detected in the wild type. Another study reported that the heterologous overexpression of malonyl-CoA acyl carrier protein transacylase (MCAT) obtained from diatom *Fragilariopsis cylindrus* along with homologous overexpression of Δ5- desaturase resulted in enhanced EPA content (85.35 mg/g) and DHA content (9.15 mg/g) in *P. tricornutum* [24]. MCAT catalyzes the formation of malonyl-ACP required for the initiation of type II fatty acid synthesis.

Another model diatom, *Thalassiosira pseudonana* is reported to accumulate ~15-20% EPA and DHA with a ratio of ~ 9.6-11.6: 1 respectively [8]. *T. pseudonana* encodes three genes for fatty acid elongases and 20 genes for fatty acid desaturases [8]. The overexpression of three endogenous elongases genes (elo1, elo2 and elo3) was accomplished in *T. pseudonana* to augment the PUFA content [8]. The elo1 gene has stronger preference for Δ6-elongase, elo2 is analogous to Δ5- elongase and elo3 is specific to elongases that add two carbon unit to saturated and monosaturaed fatty acids as illustrated in Figure 8.1 [8]. The authors reported that overexpression of elo1, elo2 or elo3 resulted in 2.3 to 4.3-fold in all the three transgenic lines as compared to the wild type when subjected to 24 h silicon (Si) starvation. However, no alteration in the EPA content was recorded in either of the engineered strain. The overexpression of Diglyceride acyltransferase (DGAT-2) in *T. pseudonana* under Si starvation led to enhanced 1.5-fold EPA and 1.2- fold DHA content in the transgenic lines as compared the wild type strain [25]. DGAT's catalase the final and committed step towards TAG synthesis by converting diacylglycerol (DAG) to TAG [1].

The unicellular marine microalga *Nannochloropsis* sp. accumulates high EPA content and has been an industrial established strain to produce EPA [26]. *Nannochloropsis* sp. synthesize ω-3 LC-PUFA majorly as a part of chloroplast and membrane galactolipids [27]. The PUFA genes have been putatively annotated including Δ12, Δ6, Δ5 – desaturases and Δ6 elongase [27]. The overexpression of Δ12 desaturase (obtained from *Nannochloropsis oceanica*) under nitrogen starvation led to substantial increase in the arachidonic acid (3.1% of total lipids) as compared to wild type (2% of total lipids) [27]. However, the authors reported no significant increase in the EPA content. Poliner *et al.* (2018) overexpressed endogenous Δ9, Δ12 and Δ5 desaturases individually and in combination to evaluate the alterations in the EPA content of the generated transgenic lines of *N. oceanica* [28]. Among the individual expressed desaturases, overexpression of Δ5 desaturase resulted in maximum EPA content (30.21% of total lipids) followed by Δ12 desaturase and then Δ9 desaturase [28]. The overexpression of either Δ12 or Δ5 desaturase resulted in ~25% increase in EPA content as compared to the wild type. Interestingly, on combined overexpression of Δ12 and Δ9 desaturases or all three desaturases (Δ9, Δ12 and Δ5) did not have an additive effect and the EPA accumulation was similar to the individual overexpression of genes indicating the limitation of PUFA accumulation in *N. oceanica* [28]. The common halotolerant unicellular wall-less microalga, *Dunaliella salina* is known to accumulate high ALA but limited EPA synthesis occurs in the organism [29]. In order to reroute the flux towards EPA synthesis in *D. salina*, Δ6 desaturase isolated from *T. pseudonana* was overexpressed [29]. The authors reported a drastic increase in the EPA content (28.12% of total lipids) as compared to the wild type (1.91% of total lipids).

The DHA dietary supplement obtained from thraustochytrids is already commercialized by various companies [30]. Thraustochytrids are heterotrophic marine microalgae which accumulate ~35-40% of total lipids as DHA [31]. Such a high biosynthesis of DHA is attributed to the alternative pathway for DHA synthesis which utilizes polypeptide synthase enzyme complex (Figure 8.2; section 8.2: LC-PUFA biosynthesis in microalgae). Among the thraustochytrids, *Schizochytrium* sp. has been genetically engineered to augment the DHA production. Heterologous overexpression of codon optimized elongase 3 gene obtained from *Morbierella alpina*, catalyzing the conversion of palmitic acid to oleic acid resulted in increase in overall lipid content indicating redirection of carbon flux towards lipid synthesis [32]. Further, the co-overexpression of malic enzyme derived from *Chrypthecodinium cohnii* enhanced the DHA content to 27.70% of total fatty acids as compared to wild type (19.25% of total fatty acids). Hayashi *et al.* (2016), synthetically constructed PUFA synthase derivates with 11 active acyl carrier proteins (ACP) domains in OrfA of *Schizochytrium* sp. (native species have 9 active ACP domains) which resulted in 1.8-fold enhanced DHA accumulation [11].

8.5 Conclusion and Future Perspective

The growing awareness of the health benefits of LC-PUFAs over the past few years in society has led to a dramatic expansion of ω3 and ω6 fatty acids. The ω3 fatty acid industry is expected to grow at a CAGR (Compound annual growth rate) of 13.1% from 2019, reaching USD8.5 billion market sales by 2025 (https://www.marketsandmarkets.com/Market-Reports/omega-3-omega-6-227.html). This necessitates the urgent need to search and explore novel sources for the sustainable production of LC-PUFAs. In the last decade, microalgae has been a central organism which can potentially be exploited for various value-added products including LC-PUFA and biofuels. Indeed, the inherent capability of various microalgae to accumulate high amounts of LC-PUFA intracellularly, led to a commercialization of algal-based ω3 fatty acids, particularly DHA. The marketed algal-based PUFA include NUTROVA (DHA derived from marine algae), Unived OVEGHA (DHA), AlgaPrime™ (DHA), Solgar Omega-3 (DHA) and Nordic naturals algae omega.

However, there is still a lacuna in terms of isolation of novel strains that can accumulate high amounts of essential LC-PUFA such as ARA, DHA and EPA. The high PUFA accumulating strains could be scanned for efficient desaturase and elongates followed by their characterization leading to the development of a library of PUFA genes. These genes can then be genetically engineered in microalgal strains to enhance the targeted PUFA content and productivity. Further, there is an urgent need to develop a library of constitutive native promoters, ribosome binding sites and terminators which can efficiently drive the PUFA synthesis in engineered organisms. These genetic elements can be identified using OMICS approaches particularly transcriptomics, aiding in understanding of the mRNA level response under a set of abiotic or biotic conditions. Indeed, abiotic stresses such as temperature, light, salinity, etc., play a vital role in augmenting the PUFA content in microalgae, thus warranting a detailed investigation and understanding at molecular level using OMICS studies. Ultimately, to realise the feasibility of economical and sustainable algal-derived PUFA production, development of the entire chain is essential. This starts by bioprospecting of high-producing strains, understanding optimum growth conditions for enhanced PUFA accumulation, characterization of efficient enzymes, establishing the genetic tool box and downstream processing of the PUFA.

References

1. Xin, Y., Shen, C., She, Y., Chen, H., Wang, C., Wei, L., Yoon, K., Han, D., Hu, Q., Xu, J. Biosynthesis of Triacylglycerol Molecules with a Tailored PUFA Profile in Industrial Microalgae. *Mol. Plant.* 12, 474–488, 2016.
2. Hoffmann, M., Wagner, M., Abbadi, A., Fulda, M., Feussner, I. Metabolic Engineering of ω3-Very Long Chain Polyunsaturated Fatty Acid Production by an Exclusively acyl-CoA-dependent pathway. *J Biol Chem.* 283, 22352–22362, 2008.
3. Poliner, E., Farré, E.M., Benning, C., Benning, C. Advanced genetic tools enable synthetic biology in the oleaginous microalgae Nannochloropsis sp. *Plant Cell Rep.* 37,1383-1399, 2018.
4. Alonso, D.L., Maroto, F.G. Plants as 'chemical factories' for the production of polyunsaturated fatty acids. *Biotechnol Adv.* 18, 481–497, 2000.
5. Shimizu, S. Biosynthesis and Regulation of Microbial Polyunsaturated Fatty Acid Production. *J Biosci Bioeng.* 87, 1–14,1999.
6. Hamilton, M.L., Warwick, J., Terry, A., Allen, M.J., Napier, J.A., Sayanova, O. Towards the industrial production of omega-3 long chain polyunsaturated fatty acids from a genetically modified diatom phaeodactylum tricornutum *PLoS One.* 10, 1–15, 2015.
7. Cao, Y., Cao, Y., Zhao, M. Biotechnological production of eicosapentaenoic acid: From a metabolic engineering point of view. *Process Biochem.* 47, 1320–1326, 2012.
8. Cook, O., Hildebrand, M. Enhancing LC-PUFA production in Thalassiosira pseudonana by overexpressing the endogenous fatty acid elongase genes. *Journal of Applied Phycology*, 28, 897-905, 2015.
9. Ruiz-lopez, N., Usher, S., Sayanova, O. V., Napier, J.A., Haslam, R.P. Modifying the lipid content and composition of plant seeds: engineering the production of LC-PUFA. 143–154, 2015.
10. Ferreira, G.F., Pinto, L.F.R., Filho, R.M., Leonardo, V. Microalgal Biomass as a Source of Polyunsaturated Fatty Acids for Industrial Application: a Mini-Review. *Chemical Engineering Transactions.* 74, 163–168, 2019.
11. Hayashi, S., Satoh, Y., Ujihara, T., Takata, Y., Dairi, T. Enhanced production of polyunsaturated fatty acids by enzyme engineering of tandem acyl carrier proteins. *Nat. Publ. Gr.* 1–6, 2016.

12. Khozin-Goldberg, I., Iskandarov, U., Cohen, Z. LC-PUFA from photosynthetic microalgae: Occurrence, biosynthesis, and prospects in biotechnology. *Appl. Microbiol. Biotechnol.* 91, 905–915, 2011.
13. Khozin-goldberg, I., Leu, S., Boussiba, S. Microalgae as a Source for VLC-PUFA Production.
14. Ratledge, C. Fatty acid biosynthesis in microorganisms being used for Single Cell Oil production. *Biochimie.* 86, 807–815, 2004.
15. Dolch, L., Maréchal, E. Inventory of Fatty Acid Desaturases in the Pennate Diatom Phaeodactylum tricornutum. *Marine Drugs.* 1317–1339 (2015).
16. Vaezi, R., Napier, J.A., Sayanova, O. Activities from Unicellular Microalgae. *Marine Drugs.* 11, 5116–5129, 2013.
17. Yu, S.Y., Li, H., Tong, M., Ouyang, L.L., Zhou, Z.G. Identification of a Δ 6 fatty acid elongase gene for arachidonic acid biosynthesis localized to the endoplasmic reticulum in the green microalga Myrmecia incisa Reisigl. *Gene*, 493, 219–227, 2012.
18. Zhou, X., Robert, S.S., Petrie, J.R., Frampton, D.M.F., Mansour, M.P., Blackburn, S.I., Nichols, P.D., Green, A.G., Singh, S.P. Isolation and characterization of genes from the marine microalga Pavlova salina encoding three front-end desaturases involved in docosahexaenoic acid biosynthesis. *Phytochemistry.* 68, 785–796, 2007.
19. Petrie, J.R., Liu, Q., Mackenzie, A.M., Shrestha, P., Mansour, M.P., Robert, S.S., Frampton, D.F., Blackburn, S.I., Nichols, P.D., Singh, S.P. Isolation and Characterisation of a High-Efficiency Desaturase and Elongases from Microalgae for Transgenic LC-PUFA Production. *Mar Biotechnol.* 12, 430–438, 2010.
20. Tonon, T., Harvey, D., Larson. Identification of a very long chain polyunsaturated fatty acid v 4-desaturase from the microalga Pavlova lutheri 1. *FEBS Lett.* 553, 440–444, 2003.
21. Hamilton, M.L., Powers, S., Napier, J.A., Sayanova, O. Heterotrophic Production of Omega-3 Long-Chain Polyunsaturated Fatty Acids by Trophically. *Mar. Drugs.* 14, 2016.
22. Zhu, B., Tu, C., Shi, H., Yang, G., Pan, K. Overexpression of endogenous delta-6 fatty acid desaturase gene enhances eicosapentaenoic acid accumulation in Phaeodactylum tricornutum. *Process Biochem.* 57, 43-49, 2017.
23. Peng, K.T., Zheng, C.N., Xue, J., Chen, X.Y., Yang, W.D., Liu. J.S., Bai, W., Li, H.Y. Delta 5 Fatty Acid Desaturase Upregulates the Synthesis of Polyunsaturated Fatty Acids in the Marine Diatom Phaeodactylum tricornutum. *J Agric Food Chem.* 62, 8773–8776, 2014.
24. Wang, X., Liu, Y., Wei, W., Zhou, X., Yuan, W. Enrichment of long-chain polyunsaturated fatty acids by coordinated expression of multiple metabolic nodes in the oleaginous microalga Phaeodactylum tricornutum. *J Agric Food Chem*, 65, 7713-7720, 2017.
25. Manandhar-shrestha, K., Hildebrand, M. Characterization and manipulation of a DGAT2 from the diatom Thalassiosira pseudonana: Improved TAG

accumulation without detriment to growth, and implications for chloroplast TAG accumulation. *Algal Res.* 12, 239–248, 2015.

26. Chen, C.Y., Chen, Y.C., Huang, H.C., Huang, C.C., Lee, W.L., Chang, J.S. Engineering strategies for enhancing the production of eicosapentaenoic acid (EPA) from an isolated microalga Nannochloropsis oceanica CY2. *Bioresour. Technol.* 147, 160–167, 2013.
27. Kaye, Y., Grundman, O., Leu, S., Zarka, A., Zorin, B., Didi-Cohen, S., Khozin-Goldberg, I., Boussiba, S. Metabolic engineering toward enhanced LC-PUFA biosynthesis in Nannochloropsis oceanica: Overexpression of endogenous δ12 desaturase driven by stress-inducible promoter leads to enhanced deposition of polyunsaturated fatty acids in TAG. *Algal Res.* 11, 387–398, 2015.
28. Poliner, E., Pulman, J.A., Zienkiewicz, K., Childs, K., Benning, C., Farr, E.M. A toolkit for Nannochloropsis oceanica CCMP1779 enables gene stacking and genetic engineering of the eicosapentaenoic acid pathway for enhanced long-chain polyunsaturated fatty acid production. *Plant Biotechnol J.* 16, 298–309, 2018.
29. Shi, H., Luo, X., Wu, R., Yue, X. Production of eicosapentaenoic acid by application of a delta - 6 desaturase with the highest ALA catalytic activity in algae. *Microb. Cell Fact.* 1–15, 2018.
30. Aasen, I.M., Ertesvåg, H., Heggeset, T.M.B., Liu, B., Brautaset, T., Vadstein, O., Ellingsen, T.E. Thraustochytrids as production organisms for docosahexaenoic acid (DHA), squalene, and carotenoids. *Appl. Microbiol. Biotechnol.* 100, 4309–4321, 2016.
31. Ren, L.J., Huang, H., Xiao, A.H., Lian, M., Jin, L.J., Ji, X.J. Enhanced docosahexaenoic acid production by reinforcing acetyl-CoA and NADPH supply in Schizochytrium sp. HX-308. *Bioprocess Biosyst. Eng.* 32, 837–843, 2009.
32. Wang, F., Bi, Y., Diao, J., Lv, M., Cui, J., Chen, L. Biotechnology for Biofuels Metabolic engineering to enhance biosynthesis of both docosahexaenoic acid and odd - chain fatty acids in Schizochytrium sp. *Biotechnol. Biofuels.* 1–14, 2019.
33. Khozin-goldberg, U.I.I. Identification and Characterization of D 12, D 6, and D 5 Desaturases from the Green Microalga Parietochloris incisa. *Lipids*, 45, 519–530, 2010.
34. Iskandarov, U., Khozin-Goldberg, I., Ofir, R., Cohen, Z. Cloning and Characterization of the D 6 Polyunsaturated Fatty Acid Elongase from the Green Microalga Parietochloris incisa. *Lipids.* 44, 545–554, 2009.
35. Fraser, T., Y, A.K.S., Napier, J.A., Qi, B., Lazarus, C.M. Identification of a cDNA encoding a novel C18- ω9 polyunsaturated fatty acid-species elongating activity from the docosahexaenoic acid. FEBS Lett. 510, 158-165, 2002.
36. Jiang, M., Guo, B., Wan, X., Gong, Y., Zhang, Y., Hu, C. Isolation and Characterization of the Diatom Phaeodactylum Δ5-Elongase Gene for Transgenic LC-PUFA Production in Pichia pastoris. *Mar. Drugs.* 4, 1317–1334, 2014.

37. Tonon, T., Harvey, D., Qing, R., Li, Y., Larson. Identification of a fatty acid v 11-desaturase from the microalga Thalassiosira pseudonana. *FEBS Lett.* 563, 28-34, 2004.
38. Robert, S.S., Petrie, J.R., Zhou, X., Mansour, M.P., Blackburn, S.I., Green, A.G., Singh, S.P., Nichols, P.D. Isolation and Characterisation of a Δ 5-fatty Acid Elongase from the Marine Microalga Pavlova salina. *Mar. Biotechnol.* 11, 410–418, 2009.
39. Hsiao, T.Y., Holmes, B., Blanch, H.W. Identification and Functional Analysis of a Delta-6 Desaturase from the Marine Microalga Glossomastix chrysoplasta. *Mar. Biotechnol.* 9, 154–165, 2007.
40. Xiaolei, M., Jianzhong, Y., Baohua, Z., Kehou, P., Jin, P., Guanpin, Y. Cloning and characterization of a delta-6 desaturase encoding gene from Nannochloropsis oculata. *Chinese Journal of Oceanology and Limnology.* 29, 290–296, 2011.

9

Health Perspective of Nutraceutical Fatty Acids; (Omega-3 and Omega-6 Fatty Acids)

Sneha Sawant Desai and Varsha Kelkar Mane*

Department of Biotechnology, University of Mumbai, Kalina, Santacruz (E), Mumbai, Maharashtra, India

Abstract

Dietary intervention is identified as a key measure in maintenance of human health. PUFAs affect a multitude of physiological processes as evident from epidemiological studies and clinical trials. The chief sources of PUFA have predominantly been seafood, especially fish oil. However, religious as well as individual preferences, accessibility and awareness about the risks associated with the intake of fatty fish has restricted fish oil supplementation in dietary products. A pressing need to find a more acceptable alternative source that would be consistent in composition, sensory properties as well as be economical has brought microalgae to the forefront as a potential source of PUFAs. The absence of cholesterol and lack of fishy odor has further promoted the use of algal dietary supplements. PUFA profiles vary greatly amongst the microalgal species making them an exhaustive resource for common as well as rare unsaturated fatty acids. There is, however, a need to exploit these vital reserves to understand the functional properties of their varied PUFA contents. The present review throws light on this untapped resource more specifically in regards to omega-3 and omega-6 fatty acids and their role as disease preventing functional foods.

Keywords: Essential fatty acids, microalgae, nutraceuticals, antioxidants, cardiac regulation, neuronal health, inflammatory responses, membrane architecture

**Corresponding author*: drvkelkar@mu.ac.in

Alok Kumar Patel and Leonidas Matsakas (eds.) Nutraceutical Fatty Acids from Oleaginous Microalgae: A Human Health Perspective, (227–248) © 2020 Scrivener Publishing LLC

9.1 Introduction

The global nutraceuticals market is remunerative due to an increased demand for functional foods and beverages. The market size is expected to reach $302,306 million by 2022 with a Compound Annual Growth Rate (CAGR) of 7.04% from 2016 to 2022. Functional foods are divided into: omega fatty acid fortified foods, probiotic fortified foods, ionized salt, wheat flour market and other fortified foods. The market of functional foods containing PUFA is particularly at the booming stage [1]. PUFAs are the major components of the phospholipids of the plasma membrane and play a significant role in regulation of membrane fluidity as well as signal transduction. They act as precursors for the synthesis of varying eicosanoids, leukotrienes, prostaglandins and resolvins that further serve as anti-inflammatory, anti-arrhythmic as well as anti-aggregators [2].

9.1.1 Biochemistry of Fatty Acids

Omega fatty acids can be subdivided into: Omega-3 and Omega-6 PUFA. Omega-3 PUFA family includes α-linolenic acid (ALA), eicosapentaenoic acid (EPA) and docosahexaenoic acid (DHA) (Figure 9.1 a-c) whereas linoleic acid (LA), γ-linolenic acid (GLA) and arachidonic acid (AA) belong to the omega-6 family (Figure 9.2 a-d) [3]. The human body produces fatty acids by a process catalyzed by fatty acid synthase complex (FAS). This takes place in four sequential stages that increase the fatty acid chain length by addition of two carbon atoms at a time. The primary product of this fatty acid synthesis is palmitic acid (C16:0). The longer chain fatty acids are formed from palmitic acid by reactions catalyzed by *elongases* and *desaturases*. The *elongases* 1,3,6 and 7 catalyze the elongation of saturated and monounsaturated fatty acids whereas *elongases* 2,4 and 5 catalyze the elongation of PUFA. The *desaturases* introduce a double bond between an existing double bond and the terminal methyl moiety [4]. Mammals however, lack both Δ^{12} and Δ^{15} *desaturases* and hence though capable of producing saturated fatty acids they are unable to synthesize the simplest members of the PUFA family, i.e., linoleic acid (LA 18:2 ω-6 PUFA) and α-linolenic acid (18:3 ω-3 PUFA) [5]. These fatty acids are termed as essential fatty acids since though required for various physiological processes they are not produced by the human body. Dietary supplementation is thus essential to make these fatty acids available for various physiological activities [4]. Fish and fish oil are commonly exploited sources for PUFA; however,

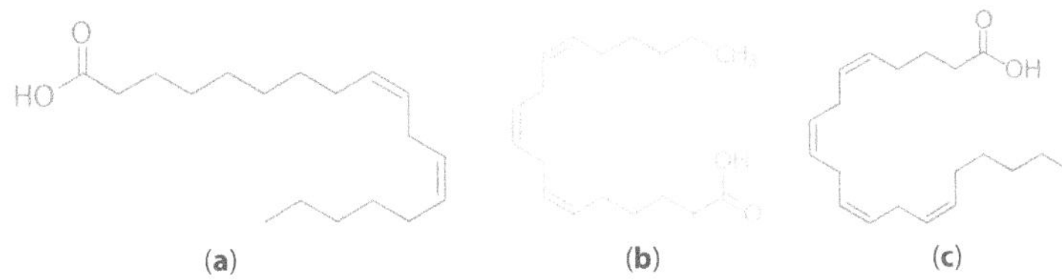

Figure 9.1 Omega-6 PUFA: (a) – Linoleic acid; (b) – γ-Linolenic acid; (c) – Arachidonic acid.

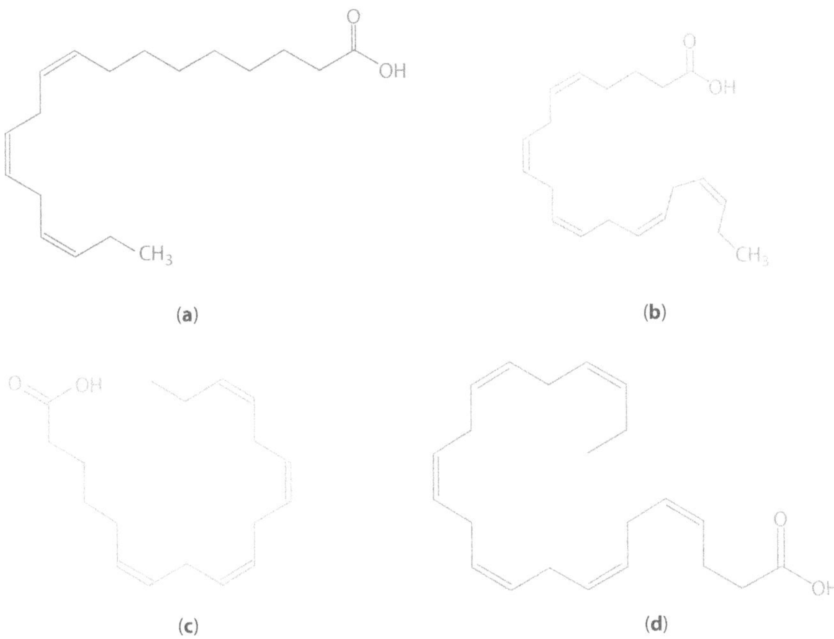

Figure 9.2 Omega-6 PUFA: (a) α-Linolenic acid; (b) Eicosapentaenoic acid; (c) Stearidonic acid; (d) Docosahexaenoic acid.

the presence of toxins, unpleasant odor, poor oxidative stability and go vegan factor has limited their application as food additives [6]. Followed by fish and fish oil, meat, poultry, eggs and dairy products are a few other dietary sources of PUFA [7]. Vegans and vegetarians, however, exclude meat, fish and dairy products from their diet because of cultural or religious reasons or because of their personal preferences thus limiting their intake of the essential PUFAs to plant sources [8]. For such groups of individuals, algae represent a potential source of the essential PUFAs [Table 9.1].

Table 9.1 Commercially significant PUFAs from different microalgal species [6].

PUFA	Microalgal producers	Application
α-linolenic acid	*Ankitrodesmus* spp., *Botryococcus* spp., *Chlamydomonas* spp., *C. moewusii, C. vulgaris, D. bardawil, D. salina, D. tertiolecta, M. pusilla, Muriellopsis* spp., *N. atomus, P. subcapitata, S. acutus, S. obliquus, S. quadricauda, Tetraselmis* spp., *T. suecica*	Cardiovascular disease prevention
Arachidonic acid	*Porphyridium* spp., *N. atomus, P. boryanum*	Nutritional supplement
Docosahexaenoic acid	*Crypthecodiuimu* spp., *Schizochytrium* spp., *Pyramimonas* spp.	Infant formula, nutritional supplement, aquaculture feed
Eicosapentaenoic acid	*Pavlova* spp., *Nannochloropsis* spp., *Monodus* spp., *Phaeodactylum* spp., *C. minutissima, C. vulgaris, Nannochloropsis* spp., *N. atomus, Tetraselmis* spp., *T. suecica*	Nutritional supplements, aquaculture feed
γ-linolenic acid	*Spirulina* spp., *C. homosphaera, Chlorococcum* spp., *D. primolecta*	Nutritional supplements
Hexadecatetraenoic acid	*Ankitrodesmus* spp., *Chlamydomonas* spp., *C. moewusii, D. bardawil, D. tertiolecta, T. suecica, T. cylindrical*	

(*Continued*)

Table 9.1 Commercially significant PUFAs from different microalgal species [6]. (*Continued*)

PUFA	Microalgal producers	Application
Linolenic acid	*B. braunii, C. moewusii, C. protothecoides, C. vulgaris, Chlorococcum* spp., *D. bardawil, D. primolecta, D. tertiolecta, N. atomus, N. oleoabundans, P. subcapitata, S. obliquus, T. Suecica*	Anti-inflammatory, acne reduction, moisture retention
Stearidonic acid	*M. pusilla, T. suecica*	Increase in tissue EPA concentration

9.1.2 Overview of Fatty Acid Synthesis

Primary producers of PUFAs are the photosynthetic organisms particularly algae for EPA and DHA [9]. Nutritional value of microalgae is primarily related to the content of essential fatty acids [10]. The PUFA content of algae represent 10-70% of the total fatty acid content. C18 PUFA is produced in high amounts in chlorophytes whereas rhodophytes and glaucophytes are rich in ARA (C20:4 ω-6) and EPA (C20:5 ω-3) [11]. Like plants, in autotrophic microalgae the first step in fatty acid synthesis is catalyzed by acetyl-CoA carboxylase which converts acetyl-CoA to malonyl-CoA (Figure 9.3). Fatty acid synthase complex further catalyzes the elongation reactions from acyl group to form palmitic (C16:0) and stearic acid (C18:0). Stearic acid is desaturated by Δ^9–desaturase to form oleic acid which is further desaturated by Δ^{12}–desaturase to form linoleic acid (C18:2 ω-6). Linoleic acid is subsequently desaturated as well as elongated in a series of reactions to form γ-linolenic acid, arachidonic acid, α-linolenic acid, eicosapentaenoic acid and docosahexanaeoic acid (Figure 9.2) [12]. In the final step of the desaturase-elongase pathway microalgae utilize Δ4 desaturase to introduce the final double bond. This reaction involves molecular oxygen and thus the pathway is also termed as "aerobic pathway". Many marine microalgae synthesize DHA and EPA by utilizing a recently discovered anaerobic polyketide synthase (PKS) pathway which involves successive elongation using malonyl-CoA. The double bond introduced alongwith malonyl-CoA in most of the cycles is

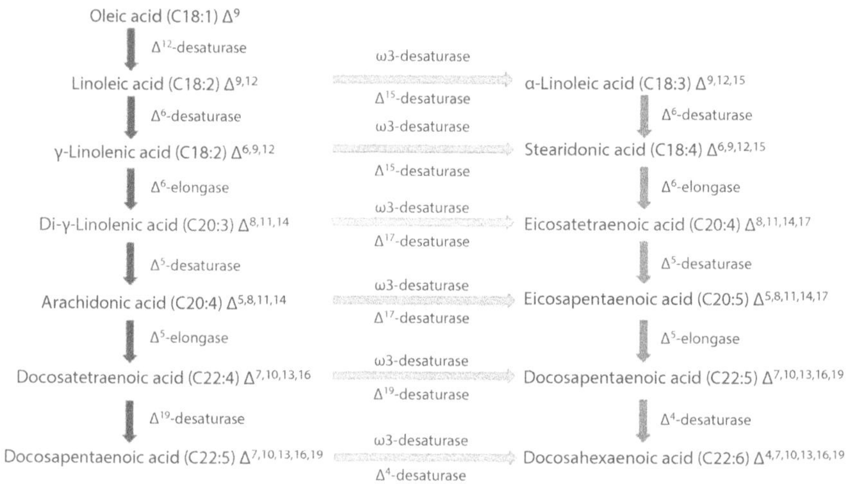

Figure 9.3 Biosynthesis of ω-3 and ω-6 polyunsaturated fatty acids by the action of desaturase - elongase pathway in algae [3].

removed without the participation of molecular oxygen, hence the pathway is termed as "anaerobic pathway" [13]. In the anaerobic pathway, PUFA synthase a huge enzyme complex known to possess multi-catalytic domains such as acyltransferase, malonyl-CoA transferase, ketoacyl synthase, ketoacyl reductase, dehydratase, enoyl reductase, chain length factor and acyl carrier protein domains carry out the elongation of PUFA [14].

9.1.3 Strategies for PUFA Accumulation in Microalgae

Culture conditions such as light, temperature, pH, presence and absence of nutrients, particularly nitrogen and phosphorus, have been known to alter the metabolic pathways and hence the biochemical profile of microalgae like *Chlorella* sp, *Nannochloropsis* sp, *Scenedesmus* sp. etc. [6]. Nitrogen being the growth limiting factor for microalgae is one of the first nutrients to be used from the medium during algal cultivation and hence a number of cultivation strategies have been employed for manipulating lipid content of microalgae by varying the nitrogen sources as well as their concentrations [15–21]. An increase of 10-20% in the lipid content of diatoms, green microalgae and other species has been reported under nitrogen stress conditions [22]. Biphasic strategy involving microalgal cultivation in the presence of sufficient nutrients followed by nutrient starvation is one of the widely employed strategies for lipid accumulation [23]. Microalgae are the richest sources of PUFA particularly EPA and DHA which are synthesized under autotrophic, heterotrophic as well as mixotrophic conditions. Different strategies have been employed to increase EPA and DHA production: a. altering the cultural conditions for maximizing lipid production. b. increasing lipid production through strain improvement. c. genetic engineering for improved EPA and DHA production [6]. Nitrogen depletion and temperature variations are reported to increase linolenic acid and PUFA content in *Haematococcus pluvialis* [24]. A 2.5-fold increase in the EPA content of *Nannochloropsis gaditana* has been observed as a response to high nitrogen concentrations [25]. At high light intensity and in the absence of nitrogen, elevated levels of PUFA were observed in *H. pluvialis* [26]. Under heterotrophic growth conditions, a 39% increase in EPA content was observed in *Nannochloropsis sp.* and *Phaeodactylum tricornutum* whereas 30-40% increase in DPA was observed in *Schizochytrium limacinum* and *Thraustochytrium* species [27]. Induced mutagenesis has been a commonly used strain improvement technique. An increase of more than 30% in the DHA and EPA content has been reported in a UV induced mutant strain of *Pavlova lutheri* [28]. On the other hand, 19.84% increase in the EPA content was noted in *P. tricornutum* under UV stress [29]. Strategies involving a high aeration rate, moderate light intensity and phosphate limitation are reported to stimulate ARA synthesis in *Porphyridium purpureum* [30, 31]. The Δ12 desaturase that catalyzes the synthesis of linoleic acid from oleic acid has been cloned and overexpressed in *Nannochloropsis oceanica*. The genetically engineered strain displayed a 50-75% increase in the ARA content [32]. Overexpression of Δ6 desaturase from *Mortierella alpina* and Δ15 desaturase from *Gibberella fujsikuroi* increased the synthesis of ω-3 PUFAs, particularly ALA and stearidonic acid (SDA) in *Synechocystis sp* [33]. Upregulation of genes encoding the malonyl-CoA acyl carrier protein transacylase (MCAT) led to an 8% increase in EPA content whereas the overexpression of diacylglycerol acyltransferases (DGAT) increased the TAG associated PUFA content by 184% in *N. ocenica* [34, 35]. Also, a 40% increase has been reported in the EPA content of *P. tricornutum* due to an overexpression of glyceraldehyde-3- phosphate acyltransferase [36]. *Nannochloropsis* has been widely exploited for the commercial production of EPA and DHA [37]. Companies have established technologies for obtaining photosynthetic EPA from *Nannochloropsis*. A2 EPA pure is a high EPA (>65%) containing oil commercialized by Aurora Algae [38].

9.2 Health Benefits of PUFA

9.2.1 Omega-6 Fatty Acids

9.2.1.1 Linoleic Acid (LA)

Approximately 20% of fatty acids in human plasma phospholipid comprises of linoleic acid. Cardiolipin, a signature phospholipid of mitochondria which participates in the electron transport chain is made up of four types of fatty acid residues; 60-70% of cardiolipin is linoleic acid (Figure 9.4) [39, 40]. Cardiolipin interacts with mitochondrial membrane proteins, enzymes and metabolite carriers. Amongst these, cardiolipin interacts with high affinity with electron transport chain complexes involved in oxidative phosphorylation as well as ADP/ATP carriers [40]. Linoleic acid has numerous metabolic fates that encompass: (a) β-oxidation, (b) elongation to form C26:2 n-6 (17, 20 – hexacosadienoic acid), (c) desaturation and elongation to ARA (precursor of prostaglandins, leukotrienes, thromboxanes, endocannabinoids) and DHA, (d) synthesis of the oxidative stress metabolites 4-hydroxynenal and malonaldehye, and (e) 9 and 13- hydroxy- and oxo-derivatives through the action of 12/15 lipooxygenase [41]. Linoleic acid is reported to upregulate osteoblast differentiation by acting as an agonist for peroxisome proliferator-activated receptor-α/δ [42]. It is also known to enhance monocyte chemotaxis and adhesion to human aortic endothelial cells [43].

9.2.1.2 γ-Linolenic Acid (GLA)

GLA is an important precursor of prostaglandins. It is reported to play a significant role in the treatment of arthritis, heart disease, obesity,

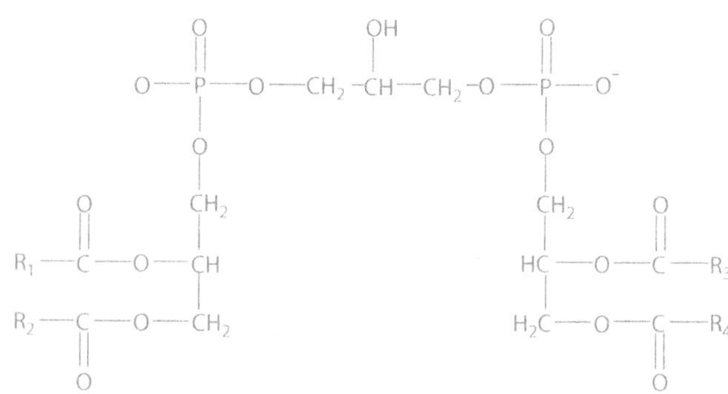

Figure 9.4 Structure of cardiolipin.

alcoholism, depression, schizophrenia, Parkinson's disease, multiple sclerosis and zinc deficiency [6]. Due to its anti-inflammatory potential, GLA has been studied in rheumatoid arthritis, atopic dermatitis, acne as well as psoriasis. It has been found to reduce the expression of COX-2 and prointerleukin-1, thus downregulating the inflammatory responses [44]. GLA reduces IKB phosphorylation as well as degradation, thereby blocking the transmigration of NF-KB to the nucleus as evident from the reduction in nuclear p65 protein expression. This results in reduction of the NF-KB activation in GLA treated macrophages eventually leading to wound healing [45]. GLA has also a beneficial effect on acne vulgaris in terms of reduced lesion number and severity, corresponding to a reduction in inflammation and interleukin-8 as was demonstrated by histologic analyses [46]. Improvement in atopic dermatitis has been related to ingestion of GLA-rich capsules [47]. Augmented intake of GLA increases the content of its elongation product dihomo-γ-linolenic acid (DGLA) in PBMCs and neutrophils. DGLA may be further metabolized to ARA by Δ^5-desaturase [48]. DGLA gives rise to inflammation-suppressing eicosanoids such as PGE1 and 15-hydroxy-eicosatrienoic acid by acting as substrate for COX and 15-LOX. DGLA and its metabolites are also potent inhibitors of platelet aggregation and inflammation [49].

9.2.1.3 *Arachidonic Acid (ARA)*

ARA is known to impart flexibility, selective permeability and fluidity to the cell membrane. It also regulates the function of membrane proteins involved in cell signaling and plays a significant role in maintaining the cellular and organellar integrity. Free ARA affects neuronal excitability and synaptic transmission. It acts on voltage-gated sodium, calcium, potassium, chloride and proton channels thus playing a critical role in the regulation of electric activity of brain, muscles and heart [50]. It prevents ischemia-induced heart arrhythmia by regulating the activity of sodium ion channels [51]. The potassium ion channels in gastric, pulmonary artery, vascular smooth muscle cells and cardiac atrial cells are directly activated by ARA, thus significantly modulating the cortical neuronal excitability. ARA is reported to be an activator of membrane associated, magnesium dependent, neutral sphingomyelinases [50]. It also regulates enzymatic activities and hence is implicated in apoptotic and necrotic events as well as events during embryogenesis, thus impacting the overall physiological and pharmacological health of newborns [50, 52]. Lower levels of PUFA, particularly ARA, have been related to autism [53]. The Food and Agriculture Organization and World Health Organization have endorsed that infant formula should be supplemented with ARA due to its positive impact on the newborn nervous system [54]. ARA supplementation in elders improved cognitive functions by stimulating the proliferation of neural stem/progenitor cells or newborn neurons and hippocampus neurogenesis [55, 56]. It alters neuromuscular signaling and enhances neurotransmitter firing from the nerve cells [57]. By eliciting cell surface membrane lipid peroxidation, ARA has also been reported to kill tumor cells *in vitro*. *In vitro* inhibition of growth of human cervical carcinoma cells and methyl cholanthrene-induced sarcoma cells by ARA has been reported [50]. Metabolites of ARA, namely prostaglandins, have been reported to play essential role in skeletal muscle growth and development by modulating proliferation, differentiation, migration, fusion and survival of myoblasts [58]. Another metabolite of ARA, Lipoxin A4, was found to activate termination of neutrophil infiltration and enhance macrophage uptake of apoptotic cells in animal models [59–65]. It terminates leukotriene C4-induced bronchoconstriction in asthmatic subjects, reduces eczema severity and inhibits the activity of innate lymphoid cells type 2 [66, 67]. It inhibits tumor necrosis factor, IL-6 and ROS generation thus displaying anti-diabetic potential [68, 69]. Prostaglandins (PGI2 and PGE2) and leukotrienes (B4 and D4) are reported to promote wound healing by modulating the angiogenesis factors as well as endothelial functions [70]. ARA derived endocannabinoid mediated signaling controls pathophysiological processes such as apetite, pain and mood [71, 72]. Anandamide, an endocannabinoid, regulates human sperm motility and improves renal functions as well as chronic inflammatory disorders of the gastrointestinal tract by modulating gut homeostasis, gastrointestinal motility, visceral sensation and inflammation [50, 73–76]. ARA metabolites have also been known to play a critical role in the immune resistance towards allergens and parasites [77].

9.2.2 Omega-3 Fatty Acids

9.2.2.1 Alpha-Linolenic Acid (ALA)

Inflammation plays a significant role in atherosclerosis and is an important risk factor in cardiovascular disease and stroke [78]. ALA rich diet is reported to reduce proinflammatory cytokines and hence inflammation [79]. It has been demonstrated that dietary ALA induces an anti-inflammatory response by inhibiting IL-6, IL-1β and TNF-α production in peripheral blood mononuclear cells cultured from hypercholesterolemic patients with high ALA intake [80]. Following cerebral ischemia, neuronal necrosis stimulated by glutamate excitotoxicity due to over-activation of N-methyl-D-aspartate (NMDA) receptors takes place within the core. This is a major mechanism responsible for neuronal cell death. Studies suggest that ALA acts as a potent neuroprotective agent against focal and global ischemia in animal models. It is reported to rectify the potassium channel leading to hyperpolarization, thus increasing the magnesium block of the calcium channel associated with NMDA receptors, facilitating the glutamate mediated excitotoxic neuronal cell death [79]. It is also known to enhance the levels of brain derived neurotrophic factor that is involved in neuronal maintenance, learning and memory, neuronal survival and neurogenesis [79, 81]. In a study assessing the dietary intake of ALA amongst 75,000 women, ALA intake was associated with decreasing the risk of cardiac deaths, which may be due to the anti-arrhythmicity [80]. Improvement in symptoms of children with attention deficit hyperactivity disorder has been reported [82].

9.2.2.2 Stearidonic Acid (SDA)

SDA is an efficient precursor of eicosapentaenoic acid. It reduces the tumor necrosis factor, a cytokine involved in carcinogenesis in whole blood [83]. It acts as a protective agent in tumorigenesis. Studies have shown reduction in cyclooxygenase transcription and translation via downregulation of nuclear kappa light chain enhancer of activated B cells and peroxisome proliferator activated receptor-γ. Additionally, anti-cancer activity of SDA has been observed on breast cancer cells. SDA induces inhibition of inflammatory mediators viz., leukotriene synthesis thus playing a significant role in palliating rheumatoid asthma and mild asthma [84].

9.2.2.3 Docosahexanoic Acid (DHA)

DHA is a major fatty acid both in the brain and the retina. A major DHA accretion spurt is observed in the uterus during the last gestational period signifying the importance of DHA in brain development [85]. Sufficient dietary supplementation of DHA is required during brain development due to rapid neurogenesis [86]. DHA plays a significant role in coordinating hepatic glucose, and amino acid and fatty acid metabolism, thus ensuring sufficient protein utilization particularly in the early development [87]. DHA provides neuroprotection against chronic and acute inflammation [87, 88]. In one of the studies on human CHME3 microglial cells, DHA and EPA supplementation stimulated microbial phagocytosis, reduction in secretion of inflammatory cellular markers and enhanced neutrophil synthesis [89]. DHA is converted to oxylipins such as resolvin and neuroprotectin D1 by 15- LOX. These metabolites function as lipid mediators affecting neuroinflammation. Resolvin is also reported to activate nuclear factor kappa light chain enhancer [88]. In an earlier report with primary astrocytes, DHA was found to induce the expression of brain-derived nuclear factor (BDNF), a critical neurotropic factor essential for regulation of synaptic transmission and an important biomarker for evaluating neurologic disorders [88, 90]. DHA is also involved in cell signaling and is a structurally significant fatty acid in the grey matter of the brain as well as retinal tissues in humans and other animals [91]. The cell phospholipid content is enhanced by 2-fold or greater in response to DHA supplementation. Dietary DHA is rapidly incorporated into sn-2 position of phosphotidylethanolamine, which is subsequently incorporated into rod photoreceptors in the retina or grey matter in the brain [92]. DHA being the principle fatty acid in the grey matter of the brain, significantly affects major depressive and bipolar disorders, Alzheimer's disease, Parkinson's disease and amyotrophic lateral sclerosis [93]. Neuroprotectin D1 is reported to induce neuronal survival via secretase and peroxisome proliferator-activated receptor-γ in Alzheimer's disease models [94]. In case of seizures or brain injury, neuroprotectin-D1 is also reported to elicit signals for sustaining synaptic and circuit integrity that are anti-inflammatory, thus inducing cells survival [95]. DHA administration reduces the inflammatory processes that destroy the motor neurons in the brain stem, spinal cord and cortex. It is also reported to increase the total glutathione levels in microglial cells and in turn enhances their antioxidative capacities [96]. Mental and visual development is known to exhibit a positive correlation to the DHA dietary content. It also suppresses the development of stroke-related behavioral changes and as a result increases the lifespan of these patients. Decreased systolic blood pressures as well as hippocampal acetylcholine levels are also correlated with dietary DHA. DHA supplementation is reported to alter the HDL/LDL cholesterol levels along with a decrease in total cholesterol/HDL levels, suggesting a reduced risk of coronary artery disorders. It also improves learning ability and prevents senile dementia [91]. It is also reported to play a crucial role in lipid metabolism and cell membrane function, as well as eye development. It functions as a precursor for autocoid signaling molecule and as an activator of gene transcription factors such as peroxisome proliferator activated receptor subsequently altering metabolic processes, cholesterol homeostasis and inflammation [97, 98]. Additionally, it interacts with Toll-like and G-protein coupled receptors. It is positively correlated to dopamine and serotonin production as well as activities of their respective receptors [92]. Low levels of plasma DHA are reported in individuals with retinitis pigmentosa, thus leading to speculations that DHA deficiency could lead to central retinal cones defects [98]. It is reported to induce antiplatelet aggregation, TAG lowering effect and an anti-arrhythmic effect [99]. 4-Hydroxy DHA hinders endothelial cell proliferation as well as emergent angiogenesis via peroxisome proliferator-activated receptor-γ [100]. Overall, DHA has a positive impact on hypertension, arthritis, depression, atherosclerosis, diabetes mellitus, myocardial infarction, thrombosis, heart disease and cancers, thus reflecting its significance in the diet [91].

9.2.2.4 Eicosapentaenoic Acid (EPA)

A study found low incidence of coronary artery disease in Greenland Eskimos due to the higher levels of EPA and lower levels of ARA in their plasma as compared to that in the plasma of Eskimos in Denmark and corresponding Danes [101]. EPA is decreasing platelet adhesion as well as reactivity with a corresponding increase in the bleeding time. Antithrombotic effect of EPA was determined due to its ability to inhibit the synthesis of thromboxane A2 that causes platelet aggregation and vasoconstriction [102]. Blood vessels convert EPA to Δ^{17}- Prostacyclin (PGI3), an anti-aggregator that inhibits platelet aggregation [100]. EPA also functions as a substrate for anti-atherogenic eicosanoids [103]. EPA and DHA inhibit atherosclerotic plaque formation by decreasing the macrophages in the atherosclerotic plaque as well as reducing the production of platelet-derived growth factor [102]. EPA enhances the bioavailability of nitric oxide after inhibiting superoxide production by neutrophils, thus causing vasodilatory effect subsequently influencing endothelial function [104]. An earlier report suggested that supplements containing at least 60% EPA of the total omega-3 content with a dose rate of 200-2200 mg/day had positive effects on depressive symptoms. EPA has thus been therapeutically used for the treatment of depression [105]. Controlled oxidative breakdown of EPA or DHA with lipooxygenases gives rise to resolvins, docosatrienes and protectins. Resolvins derived from EPA are of E-series and those derived from DHA are of D-series. Resolvin E1 decreases inflammation by curbing the stimulation of transcription factor NF-KB [106]. EPA also inhibits inflammatory responses by stimulating cell signaling mediated by anti-inflammatory eicosanoids. Additionally, it also inhibits delta-5 desaturase essential for ARA synthesis [102]. The immunomodulatory effect of EPA includes modification of lipid mediator responses such as decrease in arachidonate derived cytokines, enhanced synthesis of resolvins via cyclooxygenase and lipooxygenase; increase in phagocytosis, stalling T-cell proliferation and MHC class I and II expression [107].

EPA is known to stimulate breast cancer cell apoptosis through activation of a neutral sphingomyelinase-mediated pathway [108]. EPA also alters carbohydrate and lipid metabolism, effectively decreasing adiposity and providing anti-obesity effects [109]. EPA enhances the synthesis of peroxisome proliferator-activated receptor-α thus stimulating fatty acid oxidation and reducing triglycerides. It is also known to upregulate peroxisome proliferator-activated receptor-γ thus improving insulin sensitivity [110].

9.3 Conclusion

PUFA and their derivatives play a significant role in almost all physiological and pathological processes. They function as potent cellular modulators, signaling agents. They are involved in development, maintenance as well as functioning of the nervous system and are essential for cellular activities (Figure 9.5). There is, however, still a need to establish more conclusive relationships between PUFAs and insulin metabolism, obesity, liver function, diabetic nephropathy, asthma and mental health. The potential of algae as sources of the essential PUFAs is still untapped globally. The incorporation of algae in food products may offer great potential to prevent chronic disease progression attributed to the presence of long-chain PUFAs. As such, in order to cope with the projected increase in global population and a corresponding decrease in food availability there is a necessity to exploit these organisms for the welfare of humans.

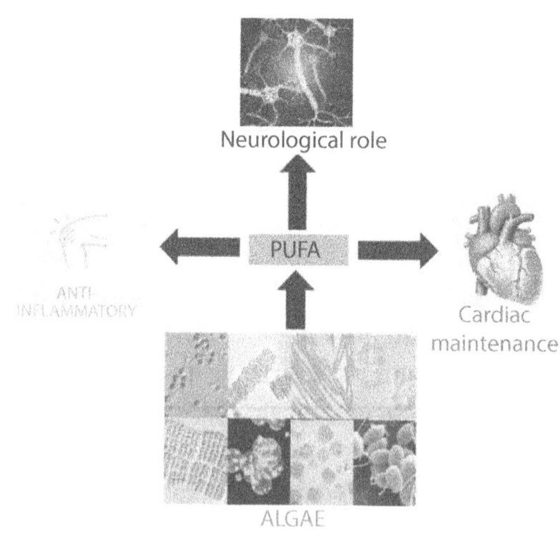

Figure 9.5 Primary roles of PUFA: Anti-inflammatory, neurological development and cardiac maintenance.

References

1. Prasad, E. Nutraceuticals Market by Type (Functional Food (Probiotics Fortified Foods, Omega Fatty Acid Fortified Food, Branded Ionized Salts, Branded Wheat Flour Market, and Other Functional Food), Functional Beverages (Fruits and Vegetable Juices & Drinks, Dairy & Dairy Alternative Drinks, Non-Carbonated Drinks (Bottled water, Tea, and Coffee, and Others (Herbal tea, Sports drinks, Energy drinks), Dietary Supplements (Proteins & Peptides, Vitamins & Minerals, Herbals (Ayurveda extracts, plant extracts, algal extracts, phytochemicals), and Others(Fatty acids, Fiber) and Personal Care) - Global Opportunity Analysis and Industry Forecast, 2014-2022.
2. Ratledge, C. Microbial production of polyunsaturated fatty acids as nutraceuticals. In: McNeil, B., Archer, D., Giavasis, I., Harvey, L. (Eds) Microbial production of food ingredients, enzymes and nutraceuticals. Woodhead publishing, Cambridge, UK, p. 531-558, 2013.
3. Das, L., Bhaumik, E., Raychaudhari, U., Chakraborty, R. Role of nutraceuticals in human health. *J. Food. Sci. Technol.*, 49, 173, 2012.
4. Burdge, G.C., Calder, P.C. Introduction to fatty acids and lipids, in: *Intravenous Lipid Emulsions*, B. Koletzko (Ed.), 1-16, Karger, Munich, 2015.
5. Dunbar, B.S., Bosire, R.V., Deckelbaum, R.J. Omega 3 and omega 6 fatty acids in human and animal health: An African perspective. *Mol. Cell. Endocrinol.*, 398, 69, 2014.
6. Sathasivam, R., Radhakrishnan, R., Hashem, A., Abd_Allah, E.F. Microalgae metabolites: A rich source for food and medicine. *Saudi. J. Biol. Sci.*, 26, 709, 2019.
7. Fayat-Moore, F., Baghurst, K., Meyer, B.J. Four models including fish, seafood, red meat and enriched foods to achieve Australian dietary recommendations for *n*-3 LCPUFA for All Life stages. *Nutrients*, 7, 8602, 2015.
8. Burdge, G.C., Tan, S-Y., Henry, C.J., Long-chain *n*-3 PUFA in vegetarian women: a metabolic perspective. *J. Nutr. Sci.*, 6, e58, 2017.
9. Li-Beisson, Y., Thelen, J.J., Fedosejevs, E., Harwood, J.L. The lipid biochemistry of eukaryotic algae. *Prog. Lipid. Res.*, 74, 31, 2019.
10. Kumar, B.R., Deviram, G., Mathimani, T., Duc, P.A. Microalgae as rich source of polyunsaturated fatty acids. *Biocatal. Agric. Biotechnol.*, 17, 583, 2019.
11. Mimouni, V., Couzinet-Mossion, A., Ulmann, L., Wielgosz-Collin, G. Lipids from microalgae. In: Levine, I.A. and Fleurence, J (Ed.). *Microalgae in health and disease prevention*. 2018, Academic Press, UK, pp. 109–131.
12. Robertson, R., Guiheneuf, F., Schmid, M., Stengel, D.B., Fitzgerald, G., Ross, P., Stanton, C. Algae-derived polyunsaturated fatty acids: implications for human health. In: Catala, A. (Ed.). *Polyunsaturated fatty acids : sources, antioxidant properties and health benefits*. 2013, Nova publishers, USA, pp.

13. Sun, X-M., Ren, L-J., Zhao, Q-Y., Ji, X-J., Huang, H. Enhancement of lipid accumulation in microalgae by metabolic engineering. *Biochim. Biophys. Acta.,* 1864, 552, 2019.
14. Hayashi, S., Satoh, Y., Ujihara, T., Takata, Y., Dairi, T. Enhanced production of polyunsaturated fatty acids by enzyme engineering of tandem acyl carrier proteins. *Sci. Rep.,* 6, 1, 2016.
15. Li, T., Xu, J., Gao, B., Xiang, W., Li, A., Zhang, C. Morphology, growth, biochemical composition and photosynthetic performance of *Chlorella vulgaris* (Trebouxiophyceae) under low and high nitrogen supplies. *Algal Res.,* 16, 481, 2016.
16. Li, X.Y., Zhao, F.J., Yu, D.D. Effect of nitrogen limitation on cell growth, lipid accumulation and gene expression in *Chlorella sorokiniana. Braz. Arch. Biol. Technol.,* 58, 462, 2015.
17. Campos, H., Boeing, W.J., Dungan, B.N., Schaub, T. (2014) Cultivating the marine microalga *Nannochloropsis salina* under various nitrogen sources: Effect on biovolume yields, lipid content and composition, and invasive organisms. *Biomass Bioenerg.,* 66, 301, 2014.
18. Ren, H.Y., Liu, B.F., Ma, C., Zhao, L., Ren, N.Q. (2013) A new lipid-rich microalga *Scenedesmus sp.* strain R-16 isolated using Nile red staining: effects of carbon and nitrogen sources and initial pH on the biomass and lipid production. *Biotechnol. Biofuels,* 6, 143, 2013.
19. Ngangkham, M., Ratha, S.K., Prasanna, R., Saxena, A.K., Dhar, D.W., Sarika, C., Prasad, R.B.N. Biochemical modulation of growth, lipid quality and productivity in mixotrophic cultures of *Chlorella sorokiniana. SpringerPlus.,*1, 33, 2012.
20. Ordog, V., Stirk, W.A., Balint, P., Staden, J.V., Lovasz, C. Changes in lipid, protein and pigment concentrations in nitrogen-stressed *Chlorella minutissima* cultures. *J. Appl. Phycol.,* 24, 907, 2012.
21. Nigam, S., Rai, M.P., Sharma, R. Effect of nitrogen on growth and lipid content of *Chlorella pyrenoidosa. Am. J. Biochem. Biotechnol.,* 7, 124, 2011.
22. Yen, H-W., Hu, I-C., Chen, C-Y., Ho, S-H., Lee, D-J., Chang, J-S. Microalgae based biorefinery from biofuels to natural products. *Bioresour. Technol.,* 135, 166, 2013.
23. Mathimani, T., Uma, L., Prabaharan, D. Formulation of low cost seawater medium for high cell density and high lipid content of *Chlorella vulgaris* BDUG 91771 using central composite design in biodiesel perspective. *J. Clean. Prod.,* 198, 575, 2018.
24. Lei, A., Chen, H., Shen, G., Hu, Z., Chen, L., Wang, J. Expression of fatty acid synthesis genes and fatty acid accumulation in *Haematococcus pluvialis* under different stressors. *Biotechnol. Biofuels,* 5, 1, 2012.
25. Camacho-Rodríguez, J., Cerón-García, M., González-López, C., Fernández-Sevilla, J., Contreras-Gómez, A., Molina-Grima, E. A low-cost culture medium for the production of *Nannochloropsis gaditana* biomass optimized for aquaculture. *Bioresour. Technol.,* 144, 57, 2013.

26. Damiani, M.C., Popovich, C.A., Constenla, D., Leonardi, P.I. Lipid analysis in *Haematococcus pluvialis* to assess its potential use as a biodiesel feedstock. *Bioresour. Technol.,* 101, 3801, 2010.
27. Adarme-Vega, T.C., Lim, D.K., Timmins, M., Vernen, F., Li, Y., Schenk, P.M. Microalgal biofactories: a promising approach towards sustainable omega-3 fatty acid production. *Microb. Cell Factor,* 11, 96, 2012.
28. Chauton, M.S., Reitan, K.I., Norsker, N.H., Tveterås, R., Kleivdal, H.T. A technoeconomic analysis of industrial production of marine microalgae as a source of EPA and DHA-rich raw material for aquafeed: research challenges and possibilities. *Aquaculture,* 436, 95, 2015.
29. Liang, Y., Beardall, J., Herald, P. Effect of UV radiation on growth, chlorophyll fluorescence and fatty acid composition of *Phaeodactylum tricornutum and chaetoceros muelleri* (bacillariophyceae). *Phycologia,* 45, 605, 2006.
30. Su, G., Jiao, K., Chang, J., Li, Z., Guo, X., Sun, Y., Zeng, X., Lu, Y., Lin. L. Enhancing total fatty acids and arachidonic acid production by the red microalgae *Porphyridium purpureum. Bioresour. Bioprocess.,* 3, 1, 2016.
31. Su, G., Jiao, K., Li, Z., Guo, X., Chang, J., Ndikubwimana, T., Sun, Y., Zeng, X., Lu, Y., Lin. L. Phosphate limitation promotes unsaturated fatty acids and arachidonic acid biosynthesis by microalgae *Porphyridium purpureum. Bioprocess. Biosyst. Eng.,* 39, 1129, 2016.
32. Kaye, Y., Grundman, O., Leu, S., Zarka, A., Zorin, B., Didi-Cohen, S., Khozin-Goldberg, I., Boussiba, S. Metabolic engineering toward enhanced LC-PUFA biosynthesis in Nannochloropsis oceanica: overexpression of endogenous Δ12 desaturase driven by stress-inducible promoter leads to enhanced deposition of polyunsaturated fatty acids in TAG. *Algal. Res.,* 11, 387, 2015.
33. Chen, G., Qu, S., Wang, Q., Bian, F., Peng, Z., Zhang, Y., Ge, H., Yu, J., Xuan, N., Bi, Y. Transgenic expression of delta-6 and delta-15 fatty acid desaturases enhances omega-3 polyunsaturated fatty acid accumulation in *Synechocystis sp.* PCC6803. *Biotechnol. Biofuels,* 7, 1, 2014.
34. Chen, J.W., Liu, W.J., Hu, D.X., Wang, X., Balamurugan, S., Alimujiang, A., Yang, W.D., Liu, J.S., Li, H.Y. Identification of a malonyl-CoA acyl carrier protein transacylase and its regulatory role in fatty acid biosynthesis in oleaginous microalga *Nannochloropsis oceanica. Biotechnol. Appl. Biochem.,* 64, 620, 2017.
35. Xin, Y., Lu, Y., Lee, Y.Y., Wei, L., Jia, J., Wang, Q., Wang, D., Bai, F., Hu, H., Hu, Q., Liu, J., Li, Y., Xu, J. Producing designer oils in industrial microalgae by rational modulation of co-evolving type-2 diacylglycerol acyltransferases. *Mol. Plant.,* 10, 1523, 2017.
36. Niu, Y.F., Wang, X., Hu, D.X., Balamurugan, S., Li, D.W., Yang, W.D., Liu, J.S., Li, H.Y. Molecular characterization of a glycerol-3-phosphate acyltransferase reveals key features essential for triacylglycerol production in *Phaeodactylum tricornutum. Biotechnol. Biofuels,* 9, 1, 2016.
37. Peltomaa, E., Johnson, M.D., Taipale, S.J. Marine cryptophytes are great sources of EPA and DHA. *Mar. Drugs,* 16, 1, 2017.

38. Winwood, R.J. Recent developments in the commercial production of DHA and EPA rich oils from microalgae. *OCL*, 20, 1, 2013.
39. Stanley, W.C., Khairallah, R.J., Dabkowski, E.R. Update on lipids and mitochondrial function: impact of dietary n-3 polyunsaturated fatty acids. *Curr. Opin. Clin. Nutr. Metab. Care*, 15, 122, 2012.
40. Paradies, G., Paradies, V., Benedicts, V.D., Ruggiero, F.M., Petrosillo, G. Functional role of cardiolipin in mitochondrial bioenergetics. *Biochim. Biophys. Acta.*, 1837, 408, 2014.
41. Glick, N.R., Fischer, M.H. The role of essential fatty acids in human health. *Evid. Based. Complement. Alternat. Med.*, 18, 268, 2013.
42. Still, K., Grabowski, P., Mackie, I., Perry, M., Bishop, N. The peroxisome proliferator- activated receptor alpha/delta agonists linoleic acid and bezafibrate upregulate osteoblast differentiation and induce periosteal bone formation in vivo. *Calcif. Tissue. Int.*, 83, 285, 2008.
43. Matesanz, N., Jewhurst, V., Trimble, E.R., McGinty, A., Owens, D., Tomkin, G.H., Powell, L.A. Linoleic acid increases monocyte chemotaxis and adhesion to human aortic endothelial cells through protein kinase C- and cyclooxygenase-2-dependent mechanisms. *J. Nutr. Biochem.*, 23, 685, 2012.
44. Silva, J.R., Burger, B., Kuhl, C.M.C., Candreva, T., dos Anjos, M.B.P., Rodrigues, H.G. Wound healing and omega-6 fatty acids: from wound healing to repair. *Mediators. Inflamm.*, 2018, 1, 2018.
45. Chang, C.S., Sun, H.L., Lii, C.K., Chen, H.W., Chen, P.Y., Liu, K.L. Gamma-linolenic acid inhibits inflammatory responses by regulating NF-κB and AP-1 activation in lipopolysaccharide-induced RAW 264.7 macrophages. *Inflamm.*, 33, 46, 2010.
46. Jung, J.Y., Kwon, H.H., Hong, J.S., Yoon, J.Y., Park, M.S., Jang, M.Y., Suh, D.H. Effect of dietary supplementation with omega-3 fatty acid and gamma-linolenic acid on acne vulgaris: a randomised, double-blind, controlled trial. *Acta. Derm. Venereol.*, 94,521, 2014.
47. Simon, D., Eng, P.A., Borelli, S., Kagi, R., Zimmermann, C., Zahner, C., Drewe, J., Hess, L., Ferrari, G., Lautenschaler, S., Wuthrich, B., Schmid-Grendelmeier, P. Gamma-linolenic acid levels correlate with clinical efficacy of evening primrose oil in patients with atopic dermatitis. *Adv. Ther.*, 31, 180, 2014.
48. Innes, J.K., Calder, P.C. Omega-6 fatty acids and inflammation. *Prostaglandins. Leukot. Essent. Fatty. Acids*, 132, 41, 2018.
49. Sergeant, S., Rahbar, E., Chilton, F.H. Gamma-linolenic acid, dihommo-gamma linolenic, eicosanoids and inflammatory processes. *Eur. J. Pharmacol.*, 785, 77, 2016.
50. Tallima, H., Ridi, R.E. Arachidonic acid: Physiological roles and potential health benefits – A review. *J. Adv. Res.*, 11, 33, 2018.
51. Kang, J.X., Leaf A. Prevention of fatal cardiac arrhythmias by polyunsaturated fatty acids. *Am. J. Clin. Nutr.*, 71, 202S, 2000.
52. Elinder, F., Linn, S.I. Actions and mechanisms of polyunsaturated fatty acids on voltage gated ion channels. *Front. Physiol.*, 8, 43 , 2017.

53. Meguid, N.A., Atta, H.M., Gouda, A.S., Khalil, RO. Role of polyunsaturated fatty acids in the management of Egyptian children with autism. *Clin. Biochem.*, 41, 1044, 2008.
54. WHO and FAO joint consultation. Fats and oils in human nutrition. *Nutr. Rev.*, 53, 202, 1995.
55. Tokuda, H., Kontani, M., Kawashima, H., Kiso, Y., Shibata, H., Osumi, N. Differential effect of arachidonic acid and docosahexaenoic acid on age-related decreases in hippocampal neurogenesis. *Neurosci. Res.*, 88, 58, 2014.
56. Tokuda H, Kontani M, Kawashima H, Akimoto K, Kusumoto A, Kiso Y, Koga Y, Shibata H. Arachidonic acid-enriched triacylglycerol improves cognitive function in elderly with low serum levels of arachidonic acid. *J. Oleo. Sci.*, 63, 219, 2014.
57. Rapoport, S.I. Arachidonic acid and the brain. *J. Nutr.*, 138, 2515, 2008.
58. Korotkova, M., Lundberg, I.E. The skeletal muscle arachidonic acid cascade in health and inflammatory disease. *Nat. Rev. Rheumatol.*, 10, 295, 2014.
59. Esser-von Bieren, J. Immune-regulation and functions of eicosanoid lipid mediators. *Biol. Chem.*, 398, 1177, 2017.
60. Maderna, P., Godson, C. Lipoxins: resolutionary road. *Br. J. Pharmacol.*, 158, 947, 2009.
61. Serhan, C.N., Chiang, N. Resolution phase lipid mediators of inflammation: agonists of resolution. *Curr. Opin. Pharmacol.*, 13, 632, 2013.
62. Serhan, C.N. Pro-resolving lipid mediators are leads for resolution physiology. *Nature*, 510, 92, 2014.
63. Serhan, C.N., Chiang, N., Dalli, J. The resolution code of acute inflammation: novel pro-resolving lipid mediators in resolution. *Semin. Immunol.*, 27, 200, 2015.
64. Börgeson, E., McGillicuddy, F.C., Harford, K.A., Corrigan, N., Higgins, D.F., Maderna, P., Roche, H.M., Godson, G. Lipoxin A4 attenuates adipose inflammation. *FASEB. J.*, 26, 4287, 2012.
65. Reis MB, Pereira PAT, Caetano GF, Leite MN, Galvão AF, Paula-Silva FWG, Frade, M.A.C., Faccioli, L.H. Lipoxin A4 encapsulated in PLGA microparticles accelerates wound healing of skin ulcers. *PLoS. ONE.*, 12, e0182381, 2017.
66. Barnig, C., Levy, B.D. Lipoxin A4: a new direction in asthma therapy? *Expert. Rev. Clin. Immunol.*, 9, 491, 2013.
67. Barnig, C., Cernadas, M., Dutile, S., Liu, X., Perrella, M.A., Kazani, S., Wechsler, M.E., Israel, E., Levy, B.D. Lipoxin A4 regulates natural killer cell and type 2 innate lymphoid cell activation in asthma. *Sci. Transl. Med.*, 5, 174, 2013.
68. Das, U.N. Arachidonic acid and lipoxin A4 as possible endogenous anti-diabetic molecules. *Prostaglandins. Leukot. Essent. Fatty. Acids*, 88, 201, 2013.
69. Das, U.N. Is there a role for bioactive lipids in the pathobiology of diabetes mellitus? *Front. Endocrinol.*, 8, 182, 2017.

70. Pozzi, A., Zent, R. Regulation of endothelial cell functions by basement membrane- and arachidonic acid-derived products. *Wiley. Interdiscip. Rev. Syst. Biol. Med.*, 1, 254, 2009.
71. Piomelli, D. The endocannabinoid system: a drug discovery perspective. *Curr. Opin. Investig. Drugs*, 6, 672, 2005.
72. More surprises lying ahead. The endocannabinoids keep us guessing. *Neuropharmacology*, 76, 228, 2014.
73. Amoako, A.A., Marczylo, T.H., Marczylo, E.L., Elson, J., Willets, J.M., Taylor, A.H., Konje, J.C. Anandamide modulates human sperm motility: implications for men with asthenozoospermia and oligoasthenoteratozoospermia. *Hum. Reprod.*, 28, 2058, 2013.
74. Alhouayek, M., Muccioli, G.G. The endocannabinoid system in inflammatory bowel diseases: from pathophysiology to therapeutic opportunity. *Trends. Mol. Med.*, 18, 615, 2012.
75. Izzo, A.A., Muccioli, G.G., Ruggieri, M.R., Schicho, R. Endocannabinoids and the digestive tract and bladder in health and disease. *Handb. Exp. Pharmacol.*, 231, 423, 2015.
76. Moradi, H., Oveisi, F., Khanifar, E., Moreno-Sanz, G., Vaziri, N.D., Piomelli, D. Increased renal 2-arachidonoylglycerol level is associated with improved renal function in a mouse model of acute kidney injury. *Cannabis. Cannabinoid. Res.*, 1, 218, 2016.
77. Tallima, H., El-Ridi, R. Arachidonic acid: physiological roles and potential health benefits–A review. *J. Adv. Res.*, 11, 33, 2018.
78. Van der Valk, F.M., Van Wijk, D.F., Stroes, E.S.G. Novel anti-inflammatory strategies in atherosclerosis. *Current Opinion in Lipidology*, vol. 23, no. 6, pp. 532–539, 2012.
79. Blondeau, N., Lipsky, R.H., Bourourou, M., Duncan, M.W., Gorelick, P. B., Marini, A.B. Alpha-linolenic acid: an omega-3fatty acid with neuroprotective – ready for use in the stroke clinic? *Biomed. Res. Int.*, 2015, 1, 2015.
80. Stark, A.H., Crawford, M.A., Reifen, R. Update on alpha-linolenic acid. *Nutr. Rev.*, 66, 326, 2008.
81. Heurteaux, C., Laigle, C., Blondeau, N., Jarretou, G., Lazdunski, M. Alpha-linolenic acid and riluzole treatment confer cerebral protection and improve survival after focal brain ischemia. *Neuroscience*, 137, 241, 2006.
82. Spector, A.A., Kim, H-Y. Emergence of omega-3 fatty acids in biomedical research. *Prostag. Leukot. Essent. Fatty. Acids*, 140, 47, 2019.
83. Lenihan-Geels, G., Bishop, K.S., Ferguson, L.R. Alternative sources of omega-3 fats: can we find a sustainable substitute for fish. *Nutrients*, 5, 1301, 2013.
84. Krol, B., Keiltyka-Dadasiewicz, A. Contemporary evidence on stearidonic acid health – promoting effects. *Agro. Food. Industry Hi-Tech.*, 26, 43, 2015.
85. Yavin, E., Glozman, S., Green, P. Docosahexaenoic acid accumulation in the prenatal brain: prooxidant and antioxidant features. *J. Mol. Neurosci.*, 16, 229, 2001.

86. Weiser, M.J., Butt, C.M., Mohajeri, M.H. Docosahexaenoic Acid and Cognition throughout the Lifespan. *Nutrients,* 8, 99, 2016.
87. Trepanier, M.O., Hopperton, K.E., Orr, S.K., Bazinet, R.P. N-3 polyunsaturated fatty acids in animal models with neuroinflammation: An update. *Eur. J. Pharmacol.,* 785, 187, 2016.
88. Sun, G.Y., Simonyi, A., Fritsche, K.L., Chuang, D.Y., Hannink, M., Gu, Z., Greenlief, C.M., Yao, J.K., Lee, J.C., Beversdorf, D.Q. Docosahexanoic acid: an essential nutrient and a nutraceutical for brain health and diseases. *Prostaglandins. Leukot. Essent. Fatty. Acids,* 136, 3, 2018.
89. Hjorth, E., Zhu, M., Toro, V.C., Vedin, I., Palmblad, J., Cederholm, T., Freund-Levi, Y., Faxen-Irving, G., Wahlund, L.O., Basun, H., Eriksdotter, M., Schultzberg, M. Omega-3 fatty acids enhance phagocytosis of Alzheimer's disease-related amyloid-beta42 by human microglia and decrease inflammatory markers. *J. Alzheimers. Dis.,* 35, 697, 2013.
90. Jiang, L.H., Shi, Y., Wang, L.S., Yang, Z.R. The influence of orally administered docosahexaenoic acid on cognitive ability in aged mice. *J. Nutr. Biochem.,* 20, 735, 2009.
91. Horrocks, L.A., Yeo, Y.K. Health benefits of docosahexanoic acid (DHA). *Pharmacol. Res.,* 40, 211, 2015.
92. Rogers, L.K., Valentine, C.J., Keim, S.A. DHA supplementation: Current implications in pregnancy and childhood. *Pharmacol. Res.,* 70, 13, 2013.
93. Zarate, R., Jaber-Vazdekis, N., Tejera, N., Perez, J.A., Rodriguez, C. Significance of long chain polyunsaturated fatty acids in human health. *Clin. Transl. Med.,* 6, 1, 2017.
94. Zhao, Y., Calon, F., Julien, C., Winkler, J.W., Petasis, N.A., Lukiw, W.J., Bazan, N.G. Docosahexaenoic acid-derived neuroprotectin D1 induces neuronal survival via secretase- and PPARg-mediated mechanisms in Alzheimer's disease models. *PLoS. One.,* 6, 1, 2011.
95. Bazan NG, Musto AE, Knott EJ. Endogenous signaling by omega-3 docosahexaenoic acid-derived mediators sustains homeostatic synaptic and circuitry integrity. *Mol. Neurobiol.,* 44, 216, 2011.
96. Petit, L.K., Varsanyi, C., Tadros, J., Vassiliou, C. Modulating the inflammatory properties of activated microglia with docosahexaenoic acid and aspirin. *Lipids. Health. Dis.,* 12, 1, 2013.
97. Mun, J.G., Legette, L.L., Ikonte, C., Mitmesser, S.H. Choline and DHA in maternal and infant nutrition: synergestic implications in brain and eye health. *Nutrients,* 11, 1125, 2019.
98. Lauritzen, L., Brambilla, P., Mazzocchi, A., Harslof, L.B., Ciappolino, V., Agostoni, C. DHA effects in brain development and function. *Nutrients,* 8, 1, 2016.
99. Shahidi, F., Ambigaipalan, P. Omega-3 polyunsaturated fatty acids and their health benefits. *Ann. Rev. Food. Sci. Technol.,* 9, 345, 2018.
100. Sapieha, P., Stahl, A., Chen, J., Seaward, M.R., Willett, K.L., Krah, N.M., Dennison, R.J., Connor, K.M., Aderman, C.M., Liclican, E., Carughi, A.,

Perelman, D., Kanaoka, Y., Sangiovanni, J.P., Gronert, K., Smith, L.E. 5-Lipooxygenase metabolite 4-HAD is a mediator of the antiangiogenic effect of ω-3 polyunsaturated fatty acids. *Sci. Transl. Med.*, 3, 1, 2011.

101. Spector, A.A., Kim, H-Y. Emergence of omega-3 fatty acids in biomedical research. *Prostag. Leukot. Essent. Fatty Acids,* 140, 47, 2019.
102. Trebaticka, J., Dukat, A., Durackova, Z., Muchova, J. Cardiovascular diseases, depression disorders and potential effects of omega-3 fatty acids. *Physiol. Res.,* 66, 363, 2017.
103. Calder, P.C., Yaqoob, P. Omega-3 polyunsaturated fatty acids and human health outcomes. *Biofactors,* 36, 266, 2009.
104. Din, J.N., Archer, R.M., Harding, S.A., Sarma, J., Lyall, K., Flapan, A.D., Newby, D.E. Effect of ω-3 fatty acid supplementation on endothelial function, endogenous fibrinolysis and platelet activation in male cigarette smokers. *Heart,* 99, 168, 2013.
105. Puri, B.K., Counsell, S.J., Hamilton, G., Richardson, A.J., Horrobin, D.F. Eicosapentaenoic acid in treatment-resistant depression associated with symptom remission, structural brain changes and reduced neuronal phospholipid turnover. *Int. J. Clin. Pract.,* 55, 560, 2001.
106. Farooqui, A.A., Horrocks, L.A., Farooqui, T. Modulation of inflammation in brain: a matter of fat. *J. Neurochem.,* 101, 577, 2007.
107. Calder P.C. The relationship between the fatty acid composition of immune cells and their function. *Prostaglandins. Leukot. Essent. Fatty Acids,* 79, 101, 2008.
108. Wu, M., Harvey, K., Ruzmeto, N., Welch, Z., Sech, L., Jackson, K., Stillwell, W., Zaloga, G., Siddiqui, R. Omega-3 polyunsaturated fatty acids attenuate breast cancer growth through activation of neutral sphingomyelinase mediated pathway. *Int. J. Cancer,* 117, 340, 2005.
109. Siriwardhana, N., Kalupahana, N.S., Moustaid-Moussa, N. Health benefits of n-3 polyunsaturated fatty acids : eicosapentaenoic acid and docosahexaenoic acid. In: Kim, S-K. (Ed.) *Advances in Food and Nutrition Research,* 65, 211, 2012.
110. Calder, P.C. Mechanisms of action of (n-3) fatty acids. *J. Nutr.,* 142, 592S, 2012.

10

Extraction and Purification of PUFA from Microbial Biomass

Amit Kumar Sharma[1*], Venkateswarlu Chintala[2], Praveen Ghodke[3], Parteek Prasher[4] and Alok Patel[5]

[1]Biofuel Research Laboratory, Centre for Alternate Energy Research and Department of Chemistry, School of Engineering, University of Petroleum and Energy Studies (UPES), Dehradun, Uttarakhand, India
[2]Biofuel Research Laboratory, Centre for Alternate Energy Research and Department of Chemical Engineering, School of Engineering, University of Petroleum and Energy Studies (UPES), Dehradun, Uttarakhand, India
[3]Department of Chemical Engineering, National Institute of Technology Calicut, Kerala, India
[4]Department of Chemistry, School of Engineering, University of Petroleum and Energy Studies (UPES), Dehradun, Uttarakhand, India
[5]Biochemical Process Engineering, Division of Chemical Engineering, Department of Civil, Environmental, and Natural Resources Engineering, Luleå University of Technology, Luleå, Sweden

Abstract

Polyunsaturated fatty acids (PUFAs) are essential for human beings to maintain bio functions that would prevent diseases such as depression, neurological diseases, inflammation, autoimmune diseases and cardiovascular disease. Human beings are unable to synthesize essential PUFAs in the body and therefore, they are taken up from different food sources. The major sources of PUFAs are fish, eggs, milk, fruits and seed oil. However, microbes such as microalgae, bacteria and fungi have good capability for the production of PUFAs under controlled conditions. The establishment of a cost-effective, reliable and eco-friendly technique for extraction and purification of PUFAs from microbes is a crucial point. This chapter summarizes and describes the various physico-chemical methods applied to extract and purify PUFAs from microbes.

**Corresponding author*: amit.orgchmistry@gmail.com

Alok Kumar Patel and Leonidas Matsakas (eds.) Nutraceutical Fatty Acids from Oleaginous Microalgae: A Human Health Perspective, (249–280) © 2020 Scrivener Publishing LLC

Keywords: Microalgae, polyunsaturated fatty acids, cell disruption, extraction, purification

10.1 Introduction

Microalgae are one of the most primitive forms of plants (thallophytes) which lack roots, stems and leaves. Microalgae are sunlight driven cell factories that convert CO_2 (carbon dioxide) to special chemicals like carbohydrates, proteins, lipids, vitamins and pigments through cellular activities. Microalgae are microscopic photosynthetic organisms and depending on species, their size varies from micrometers to millimeter [1]. There are about 200,000-800,000 algal species, of which around only 50,000 species have been defined [2]. Like plants, the microalgae species can also perform photosynthesis process and convert light energy (i.e., solar energy) into chemical energy [2]. Microalgae are aquatic species and do not require any agricultural land. They are capable to grow either in fresh water or wastewater in presence of sunlight. However, the growth rate of microalgae can be controlled by species-specific nutrient medium and environmental conditions (light intensity, salinity, pH, temperature, etc.) [3]. Microalgae have many advantages over terrestrial plants, i.e., needs much lower land area to grow, easy cultivation, avoiding food versus fuel dilemma and ability to sequestrate carbon from atmosphere [4–6]. Depending on presence/absence of Organelles such as chloroplast, mitochondria and nuclei, microalgae are classified into two categories, i.e., prokaryotes and eukaryotes [2]. However, based on their taxonomy, microalgae are classified into ten groups, i.e., Chlorophyceae (green algae), Bacillariophyceae (diatoms), Xanthophyceae (yellow-green algae), Chrysophyceae (golden algae), Rhodophyceae (red algae), Phaeophyceae (brown algae), Dinophyceae (dinoflagellates), Cyanobacteria (blue-green algae), Prasinophyceae and Eustigmatophyceae [2]. In addition, microalgae can also be divided into three categories depending on utilization of carbon sources: autotrophic, heterotrophic and mixotrophic. Autotrophic microalgae utilize inorganic carbon sources (e.g., carbon dioxide or bicarbonates) to form chemical energy through photosynthesis process. On the other hand, heterotrophic microalgae are non-photosynthetic and need an external source of organic compounds (e.g., sugars) as energy source. In addition, some microalgae species are able to utilize both organic and inorganic carbon sources; these are known as mixotrophs [7].

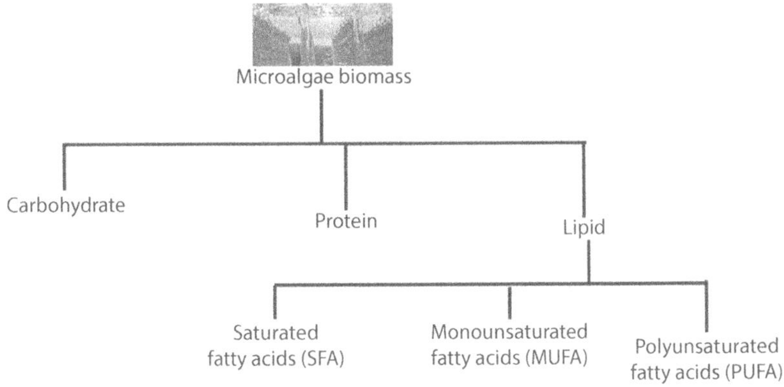

Figure 10.1 Biochemical compositions of microalgae.

10.2 Biochemical Composition of Microalgae

All the microalgae comprise carbohydrates, proteins, and lipids, the percentage of which varies with the type of algae. Depending on minimal nutritional requirements, Chisti (2007) estimate an approximate molecular formula $CO_{0.48}H_{1.83}N_{0.11}P_{0.01}$ for the microalgal biomass [1]. Biochemical compositions of microalgae are shown in Figure 10.1.

10.2.1 Carbohydrates

Microalgae is rich in carbohydrate content (up to 50% of its dry weight or more) due to its higher photo conversion efficiency [1, 3]. Generally, microalgae carbohydrates include glucose, starch, cellulose and different types of polysaccharides. Among them, microalgae carbohydrates such as glucose, starch and cellulose can be utilized for the production of bio-ethanol or bio-methane/hydrogen gases, while polysaccharides have biological functions as storage, protection and structural molecules [1–3]. Carbohydrates are extracted from microalgae by the chemical hydrolysis methods. Before extraction, algal carbohydrates are converted into fermentable sugar using some chemical pretreatments (e.g., hydrogen sulfide and sodium hydroxide) [8]. The composition of the extracted carbohydrate is varied from one species to other species. For example, green microalgae produce starch as energy stock, containing both amylose and amylopectin. The green *alga Tetraselmis suecica* stores 11% and 47% (dry wt., basis) as starch in nutrient replete. Microalgae stored carbohydrates mainly in the form of starch and cellulose [2]. Microalgae like *Chlorella vulgaris*, *Chlamydomonas reinhardtii*, *S. obliquus*, and *Tetraselmis* are some reported microalgae which have scored more than 40% carbohydrates under controlled conditions [2].

10.2.2 Proteins

Proteins have significant commodity value as animal feed. A number of amino acids (building blocks of proteins) are dietary essentials for human beings as they are unable to synthesize them. In the past few decades, microalgae biomass was directly used in the health food market. According to literature, it is reported that more than 75% of the annual microalgae biomass production was used to produce protein in the form of powders, tablets and capsules [3, 9]. A filamentous blue-green alga (Cyanobacterium), *Spirulina* is cultivated in open ponds or photobioreactor under some controlled conditions and used as a dietary supplement as well as food which is available in tablets, flakes and powder form worldwide [10].

10.2.3 Lipids

Lipids are the main constituent of plants/animals cells along with carbohydrates and proteins. Lipids are the biological molecules including fatty acids, glycerolipids, glycerophospholipids, sphingolipids, sterol lipids, prenol lipids, saccharolipids and polyketides that are insoluble in water and soluble in organic solvents such as chloroform, hexane and diethyl ether, etc. [3, 11]. Like carbohydrates, lipids also work both as energy reserves and structural components

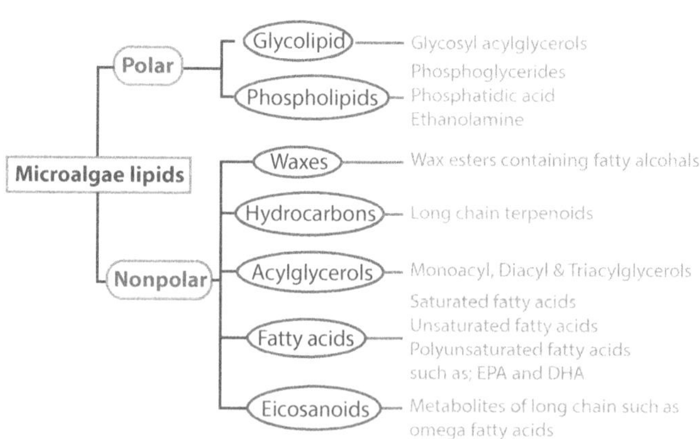

Figure 10.2 Classification of microalgae lipid.

(membranes) of the cell. Generally, the lipids can be categorized into two main classes: neutral (e.g., triacylglycerols, sterols, wax esters and hydrocarbons) and polar lipids (e.g., phospholipids and glycolipids). The classification of microalgae lipids are shown in Figure 10.2. Among them, the fatty acid and its derivatives, i.e., triglycerides, are the significant energy reserves and are used for biodiesel production [5, 11]. Triglycerides are the esters of glycerol with three fatty acids [12]. Furthermore, the fatty acids are the carboxylic acids with the long chain of hydrocarbons and can be classified into saturated, monounsaturated (MUFA) and polyunsaturated fatty acids (PUFA) depending on the presence of double bonds between carbon-carbon atoms of hydrocarbon chain [4]. According to literature, some microalgae can accumulate more than 60% lipid of dry weight under adverse culture conditions [4, 12]. Based on culture conditions (autotrophic, hetrotophic and mixotrophic conditions), the lipid composition of microalgae varies from one species to another species [2, 13–15]. In addition, temperature, pH, nutrient media, light intensity and duration of light dark cycle are also some identified parameters that directly affect saturated/unsaturated fatty acid ratio in microalgae lipids/oils [1, 16].

Generally, microalgae lipids comprise 12-22 carbon chain length with a mixture of saturated and polyunsaturated fatty acids and mainly used for biodiesel fuel production [9, 17]. However, some microalgae species have also shown great potential as a promising source of polyunsaturated fatty acids (PUFAs) [13, 18, 19]. PUFAs are essential for humans and have a beneficial role in regulating depression, neurological diseases, inflammation, autoimmune diseases and cardiovascular disease [9, 20].

10.3 Microalgae as a Source of Polyunsaturated Fatty Acids

Polyunsaturated fatty acids (PUFAs) are those fatty acids in which the constituent hydrocarbon chain possesses two or more double bond in their molecular structure (Figure 10.3). Depending upon position of double bond, PUFAs are categories in two families – omega-3 (ω-3) and omega-6 (ω-6). In ω-3 polyunsaturated fatty acids (omega-3 PUFAs), the 1st double bond present between the 3rd and 4th carbon atom counting from the methyl end of the fatty acid. This family contains Eicosapentaenoic acid (EPA, 20:5, ω-3) Docosahexaenoic Acid (DHA, 22:6, ω-3), α-linolenic acid (ALA, 18:3, ω-3) and docosapentaenoic acid (DPA, 22:5, ω-3) while γ-linolenic acid (GLA, 18:3, ω-6), arachidonic acid (ARA, 20:6, ω-6), are ω-6 PUFAs molecules in which double bond is positioned at the 6th carbon counting from the methyl end of the fatty acid.

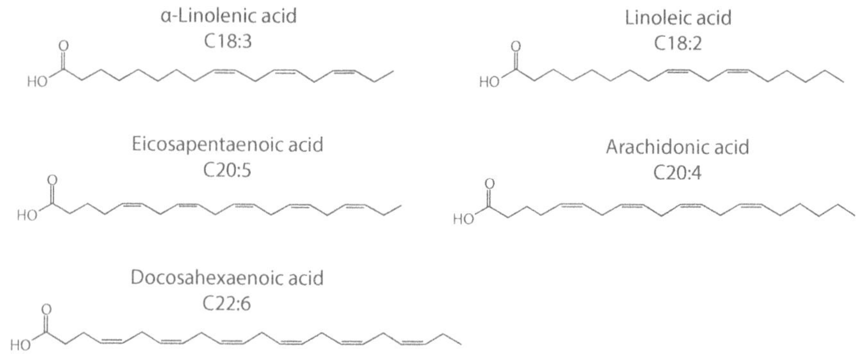

Figure 10.3 Chemical structures of polyunsaturated fatty acids.

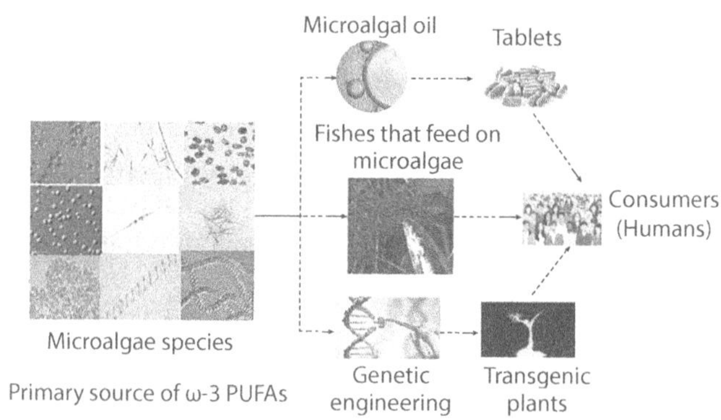

Figure 10.4 Microalgae as a source of polyunsaturated fatty acids [21].

The main sources of PUFAs are fish meat, eggs, and milk, plants (fruits and seeds) microorganisms such as bacteria, fungi and microalgae. Among them, microalgae are considered a more sustainable and promising candidate for the production of omega-3 (n-3) fatty acids due to their higher growth rate, not affecting food chain, sequestering carbon dioxide and higher production of PUFAs content under controlled conditions as presented in Figure 10.4 [3, 19, 21].

10.4 Composition of PUFAs in Microbial Biomass

The fatty acids produced from microalgae with carbon chain length (C_{14} and C_{20}) are generally used for biodiesel production while long chain fatty acids (having carbon chain length more than C_{20}) are commonly polyunsaturated fatty acids (PUFAs). Depending upon microalgae species, the lipid range varies 10-60% of dry weight [5, 11, 12]. Lipid content of microalgae biomass is highly affected under stress conditions. For example, microalgae grown under nitrogen and phosphorous deficient conditions achieved higher lipid content [5, 11]. As reported in literature, different microbes species have different composition of PUFAs and can be improved by regulating growth conditions, e.g., nutrient supply (carbon to nitrogen ratio), pH, light intensity, temperature and attack of contaminants [9, 17, 22]. In contrast to this, some microalgae like Chlorophyceae, Bacillariophyceae, Cryptophyceae, Chrysophyseae and Florideophyceae are rich in EPA whereas as DHA is predominant source in dinoflagellates microalgae. *Nannochlorosis* sp., *Chlorella minutisima*, *Heteromastix rotumda*, *Monodus subterraneus*, *P. tricornutum* are some microalgae species which have achieved 10-40% EPA PUFAs contents under controlled culture conditions while *Schizochytrium* and *Crypthecodinium cohnii* CCMP 316 are found to be rich in DHA [23, 24]. *Schizochytrium* microalgae species attained upto 50% DHA (w/w) of fatty acids by regulating glucose, nitrogen and sodium concentration [25]. *Nannochloropsis*, *Phaeodactylum*, *Schizochytrium*, and *Thraustochytrium*. Species showed enrichment of DHA and EPA (30-40% of fatty acids) under heterotrophic culture conditions [13]. When *Porphyridium purpureum* (red alga) culture under controlled condition of high salinity, optimum pH and temperature and nutrient limitation, it achieved up to 40% ARA of total fatty acids [13, 26]. The composition of PUFAs in various microalgae species are shown in Table 10.1.

10.5 Methods of Lipid Extraction from Microbial Biomass

Lipid extraction from microalgae is a most crucial and challenging step. The extracted lipid is further refined and separated to enrich PUFAs lipids. There is no single method for PUFAs extraction which can be applied to all species of microalgae because different microalgae have different cell structure and size. Currently, the methods used for extraction and purification of PUFAs from fish oil are applied to microalgae biomass [28]. The methods used for extraction should be fast, scable, effective and should not modify the extracted lipid. Presently, there are several methods (e.g., chemical, mechanical, and biological, etc.) which are applied to extract the lipid from biomass. However, these methods are not able to extract whole lipid from microalgae due to rigidity and chemical composition

Table 10.1 PUFAs composition of some microbes.

S.no.	Microalgae species	Polyunsaturated fatty acids (% of total lipid content of dry biomass)	References
1.	*Nannochloropsis gaditana*	EPA 27.4%	[27]
2.	*Pavlova sp.*	EPA 23.5%–25.0%, DHA 8.4%–9.2%	[28]
3.	*Nitzschia closterium*	EPA 2.6%–44.6%, DHA 01%–2.4%	[28]
4.	*Nannochloropsis oculata*	EPA 13.0%–40.0%, DHA 0.0%–0.6%	[15]
5.	*Schizochytrium sp.*	DHA 10%–40%	[15]
6.	*Mortierella elongata*	ARA 58%, EPA 13%	[29]
7.	*Aurantiochytrium sp.*	DHA 33%–55%	[15]
8.	*Cunninghamella echinulate*	GLA 22%	[30]
9.	*Cochlodinium heteroblatum*	EPA 20%, DHA 24%	[30]
10.	*Prorocentram micans*	EPA 3%, DHA 32%	[30]
11.	*Coscinodiscus sp.*	EPA 26%, DHA 4.6%	[31]
12.	*P. tricornutum*	EPA 9.1%–39.0%, DHA 1.1%–5.3%	[28]

of cell wall. Generally, microalgae cell are encapsulated by inflexible cell wall composed of carbohydrate and protein and therefore, interruption of cell wall is essential before extraction of lipid from microalgae [32].

10.5.1 Microalgae Cell Disruption Methods

The efficiency of lipid extraction from microalgae biomass increases with degree of cell disruption [32, 33]. During cell disruption, the microalgae cells are disintegrated which forces the efficient release of intracellular lipids into

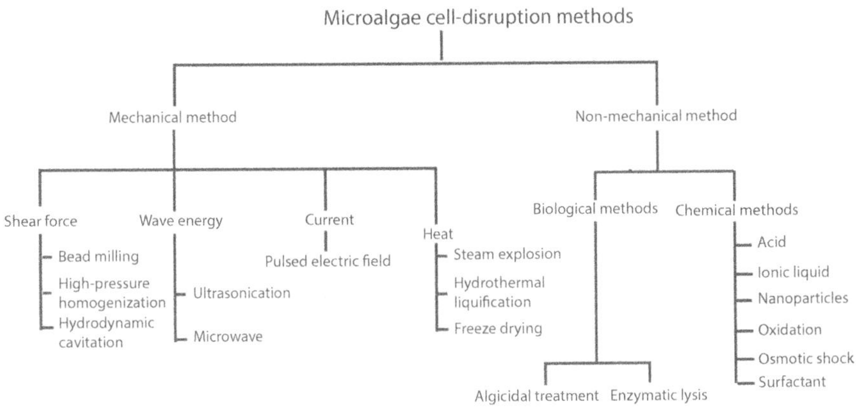

Figure 10.5 Microalgae cell disruption methods.

surrounding medium from the cellular structures, thus supporting the lipid extraction process [32]. Generally cell disruption methods are categorized into two main classes – mechanical and non-mechanical (Figure 10.5).

10.5.1.1 Mechanical Cell Disruption Methods

Mechanicals cell disruption methods generally include bead mill, high-pressure homogenization, lyophilization, microwave, ultrasonication and autoclaving.

Bead milling is getting interest as a most promising cell disruption method for industrial implementation due to its high disruption efficiency, commercially available equipment, high biomass loading rate, noble temperature control and lower operating cost [34]. In this method, a vertical or horizontal cylindrical compartment is filled with microalgae biomass along with a huge number of fine high-velocity steel or glass beads to the desired level and allowed to spin at particular velocity. The beads present inside the compartment collide with each other at higher velocity and the suspended cells are disrupted by compaction or shear forces with energy transfer from the beads to the cells [34]. The process efficiency is highly affected by size of beads, temperature time, number of cycles and spinning speed. Byreddy *et al.* (2016) employed bead mil cell disruption method on *Schizochytrium* microalgae to enhance ω-3 PUFAs [29]. Similarly, Postma *et al.* (2014) used bead mills for disrupting *Chlorella sp.* by a semi continuous mode. The drawback of this method is that it is the most energy consuming method and also leads to extraction of some unwanted chemicals which increases the purification cost.

A high-speed homogenizer (HSH) is an instrument that comprises a stator–rotor assembly and stirrer biomass at high rpm. Generally, it is made up of stainless steel with a variety in designs of stators and rotors. It is assumed that here microalgae cell disruption are carried out by hydrodynamic cavitation mechanism [32, 34]. Again, higher energy consumption is the reason that makes it unfavorable for large scale.

High-pressure homogenization (HPH) is a comparatively simple and most promising rupture technique due to feasibility of scalable and employability to highly concentrated microalgae paste (having 20-25%, w/w). In high-pressure homogenizers (HPHs), the cell disruption takes place by shear forces of the accelerated fluid jet on the stationary valve surface as well as hydrodynamic cavitation from the pressure drop induced shear stress [34, 35]. Some researchers obtained very good results with *Haematococcus, Nannochloropsis oculata, chlorella sps,* microalgae using high-pressure homogenizers cell disruption method [34, 36, 37]. High maintenance cost and poor efficiency with filamentous microbes are the limitations of this process.

Ultrasonication and microwave cell disruption techniques are commonly used to enhance lipid recovery from algal biomass [19, 33, 38]. In ultrasonication method, the energy of high-frequency acoustic waves creates intensive microbubbles in liquid medium which grow and collapse violently to produce cavitation phenomenon. These collapsing cavitation bubbles attained up to 5000K temperature and several hundred atmospheric pressure which results into vigorous and efficient cell disruption forcing the release of intracellular material into surrounding medium [32, 33]. Gerde *et al.* applied ultrasonication technique to disrupt the microalgae cells of *Schizochytrium limacinum* and *Chlamydomonas reinhardtii* for enhancement of lipid extraction efficiency [39]. In addition, Neto *et al.* (2013) utrasonicate three microalgae cells (*Chlorella minutissima, Thalassiosira pseudonana and Thalassiosira fluviatilis*) to improve lipid extraction using n-hexane solvent and got positive results [40]. On the other hand, Abirami *et al.* (2016) studied the different pretreatment methods to rupture the cell wall of *Nannochloropsis gaditana* microalgae and found that ultrasonication assisted solvent extraction method was more efficient for ω-3 fatty acids. Sharma *et al.* (2016) used ultrasonication, microwave, freeze drying and autoclave pretreatment methods to increase lipid extraction efficiency of *Chlorella vulgaris* [11]. The results revealed that microwave assisted solvent extraction showed maximum efficiency for this strain.

Microwaves are the electro-magnetic waves ranging from 1 gigahertz (GHz) to about 300 GHz. When microwaves is applied to microalgae cells suspension, these waves interact selectively with the dielectric or polar molecules which results in rising of temperature due to frictional forces by inter- and intramolecular movements [34]. In addition, the presence of free water in microalgae suspension plays a big role in improving microwave-assisted extraction efficiency because the exposer of water with microwaves reaches boiling point fast and expands inside the cells resulting in an increase of pressure. This combination of heat and pressure developed inside cells due to microwaves causes damage of cell wall/membrane subsequent and release of intermolecular materials in surrounding medium.

Lee *et al.* used microwaves frequency of 2,450 MHz for 5 min in a volume of 100 mL for *Botryococcus* sp., *C. vulgaris*, and *Scenedesmus* sp. and found recovered more lipid content from these microalgae strain. They also concluded that microwave is the more easy, simple and effective method to extract lipid [41]. Similarly, Abomohra *et al.* (2016) studied the impact of ultrasonication, microwave, freeze drying, hot water, bead beating, lypholyzation and liquid nitrogen on lipid recovery from *Scenedesmus obliquus* and reported microwave-assisted extraction more efficient than ultrasonication [42]. However, this technique is limited to use of polar solvents and is not suitable for volatile target compound [34]. Furthermore, both ultrasonic and microwave cell disruption techniques are energy-consuming methods and sometimes increase the process cost. So these methods are more suitable for extraction of PUFAs along with other valuable products leading to a biorefinery concept.

Pulsed electric field (PEF) is another method used for cell disruption. In this technique, an external electric field is induced across the cell membrane/wall to generate a critical electrical potential which leads to pore formation in the membrane/wall (electroporation phenomenon) resulting in cell disruption [33, 34]. Currently, this method has been applied successfully to enhance extraction of various valuable compounds from microbes, e.g., *Chlorella vulgaris, Nannochloropsis sp., Artrhospira platensis, Auxenochlorella protothecoides* [34, 43]. The limitation of this technique is that the solution to be treated should be ion-free, i.e., electrically nonconductive. To improve the PEF cell disruption efficiency of marine microalgae, these algae would be prewashed and deionized properly. Thus PEF cell disruption efficiency is directly depends upon medium composition.

The cell disruption of microalgae biomass by autoclaving is the example of heat treatment technique which is carried out at 121°C and 15 lbs. pressure [41]. Sharma *et al.* (2016) confirmed that the autoclave pre-treatment *Chlorella vulgaris* results in more lipid content than non-treated algae [11].

In addition, Lee *et al.* also conducted a study on application of various pretreatment methods for microalgae biomass to increase lipid extraction and found more lipid content with autoclaving microalgae biomass than non-treated algae [41].

10.5.1.2 Non-Mechanical Cell Disruption Methods

Non-mechanical methods include cell disruption with chemicals, enzymes or osmotic shock methods. Among them, enzymatic lysis is a green and energy efficient with lower operating conditions [33, 34, 43]. This is a biochemical process and carried out at room temperature. Generally, cellulase, protease, papain, and lysozyme are most studied enzymes that have been used for cell lysis of different types of microbes [34]. During cell lysis, enzymes promote hydrolysis of cell wall/membrane resulting the release of lipid into a suitable solvent medium [33, 34, 44]. The cell disruption yield and cost in enzymatic processes is mainly affected by the parameter such as types of enzyme [34]. In addition, temperature, pH, salt concentration and lipid yields are some other parameters which affect the process. This is a gentle and simple process which can be scaled up easily. This process gives better yield than mechanical methods such as ultrasonic or microwave. The drawback of this process is that it is a time-consuming, enzyme-specific and less economical process as there are limited enzymes available in the market for algal cell treatments [19, 34, 44].

Osmotic shock is another technique where cell disruption take place due to the sudden change in concentration of water across the cell membrane leading to release of intracellular component [34]. Freezing drying, acid/base treatments are also applied for algae cell rupture to enhance lipid recovery [33, 43].

10.5.2 Lipid Extraction Methods

According to literature, microalgae can accumulate up to 70% lipid of dry weight depending upon culture conditions [26, 28, 32, 45]. Like plants, the extraction of lipid/oil from microalgae is not easy due to its complex cell wall structure and therefore, the extraction is always carried out with different cell disruption methods to get maximum recovery [33, 42]. Lipid extraction methods should be fast, scalable, effective, non-toxic, inexpensive and should not damage the extracted lipid.

Generally, screw press expeller, hydraulic pressure and solvent extraction are the most common methods used to extract oil from plant seeds.

Currently, microalgae extraction methods include chemical, mechanical and biological extractions.

10.5.2.1 Mechanical Extraction Method

This is the simple and most efficient method used for lipid extraction from both edible (e.g., mustard, soybean) and non-edible seeds (e.g., Jatropha, Karanja) at industrial scale [32, 46]. However, this method gives poor performance with microalgae due to complex and rigid cell wall structure of microalgae which limit easy extraction of lipids [16, 32]).

10.5.2.2 Solvent Extraction Methods

This is the most common method used for extraction of lipid from plants and microbes. The principle underlying the organic solvent extraction process for microalgae lipid is based upon the basic chemistry concept of "like dissolving like". Generally, lipid is extracted from microalgae biomass using polar (egg, methanol, acetone, ethyl acetate, and ethanol, etc.) and non-polar solvents (e.g., chloroform, diethyl ether, hexane, toluene, etc.). The proposed mechanism of solvent extraction system is described here - when microalgae cell comes into contact with organic non-polar solvents (chloroform, hexane, etc.), it penetrates into cytoplasm through the cell membrane and forms solvent–lipids complex

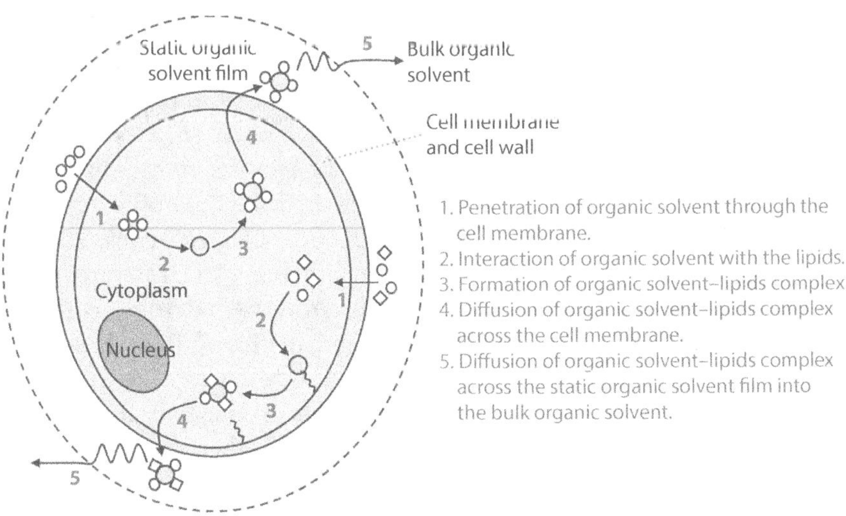

Figure 10.6 Schiamatic diagram of proposed solvent extraction mechanism [32].

with the neutral lipids using van der Waals interactions. Due to concentration gradient, the diffusion of this organic solvent–lipids complex takes place across the cell membrane into the bulk organic solvent surrounding cell wall. This leads to extraction of neutral lipids out of cells in non-polar organic solvents. The mechanism of lipid extraction with a mixture of polar and non-polar solvents is shown in Figure 10.6. The solvents used for lipid extraction should have poor affinity with non-lipid components such as carbohydrate and protein, non-toxic, volatile in nature and cost effective.

Generally, a mixture of non-polar and polar (e.g., chloroform and methanol solvents) is used to extract lipids from plant, microbes and animal tissues at laboratory scale [5, 7, 11, 41, 47]. In 1957, Folch *et al.* developed an extraction method using non-polar (chloroform) and polar (methanol) solvent mixture in 2:1 ratio (volume/volume) from animal tissues and named it the Folch method [48]. However, the drawback of this method was the presence of mineral salts in the crude extract and organic solvent system that affects the extraction efficiency. During the step washing, most of the acid lipids are washed out in the absence of mineral salts. Later, Bligh and Dyer *et al.* (1959) proposed a ternary mixture of chloroform, methanol and water (1:2:0.8, by volume) to extract lipid/oil from fresh tissues [49]. In this organic solvent system, solvent mixture parted into two layers. The lower layer is chloroform with a slight amount of alcohol which contains pure lipids (both polar and non-polar) while upper layer includes mixture of methanol and water solvent system containing non-lipids (carbohydrate and proteins). This is a simple, fast and gentle method carried out at room temperature. Another benefit of this method is that it can also be performed with wet algae. Sharma *et al.* (2016) applied a combination chloroform and methanol solvent mixture along with ultrasonic/microwave treatment to extract oil from *chlorella vulgaris* [11] and observed that microwave-assisted extraction system was most reliable for the tested microalgae. Lee *et al.* (2010) compared the lipid/oil extraction from three microalgae species (*Botryococcus sp.*, *Chlorella vulgaris*, and *Scenedesmus sp*) using chloroform and methanol (1:1) mixture [41]. Similarly, Gupta *et al.* extracted ω-3 PUFAs from the microbes (thraustochytrids) using a blend of chloroform and methanol [50]. However, chloroform solvent is not suitable for food industry due to its toxicity and therefore, it should be replaced by other solvents. In this line, chloroform and methanol were replaced by n-hexane, di ethyl ether, di chloromethane and ethanol or butanol, respectively which were less toxic, but rate and recovery of the lipid extraction was not satisfactory [28, 32]. Sakuragi *et al.* extracted up to 96.7% lipid from the wet biomass of *Euglena gracilis* using liquefied dimethyl ether [16]. Yang *et al.* tried a ternary solvent system of methanol-dimethyl sulfoxide, diethyl ether and hexane (1:1:1. v/v) to extract lipids from *Scenedesmus sp.* The lipid obtained through this solvent system was found more than 80% neutral lipids [51]. Grima *et al.* compared the extraction efficiency of PUFAs from algae with seven blends of chloroform, methanol, hexane, ethanol, butanol and water. They found maximum lipid recovery with chloroform: methanol: water solvent system followed by ethanol: water and hexane: water solvent system [52].

A laboratory scale solvent extraction process is generally carried out at batch scale and limits the extraction efficiency when solvent gets saturated with lipid present inside microbes cell [20, 33, 42, 53]. In addition, during the large scale the batch extraction process requires large amount of solvent and more heat to separate solvent and lipid by distillation techniques which makes the process more expensive.

A continuation organic solvent extraction process i.e, Soxhlet extraction plays a major role in overcoming these problems where solvents evaporated and condensed continuously replenishing the plants/micrbes biomass with fresh solvents to extract lipid [13, 41]. The lipd extraction by solvent extraction and soxhelt method are shown in Figure 10.7. Ramluckan *et al.* studied the effect of 13 different solvents (Acetone, Benzene, Chloroform, Cyclohexane, Diethyl ether, Dichloromethane, Ethanol, Hexane, Isopropanol, Isooctane, Methanol Petroleum ether and Toluene) along with their binary/ternary mixtures e.g., chloroform: methanol, hexane: ethanol and chloroform: ethanol on lipid extraction efficiency from *Chlorella sp.* using soxhlet extraction and achieved good results with chloroform :ethanol solvent system [54]. Moreover, some modified solvent extraction processes (e.g., microwave, ultrasonic and enzymatic assisted

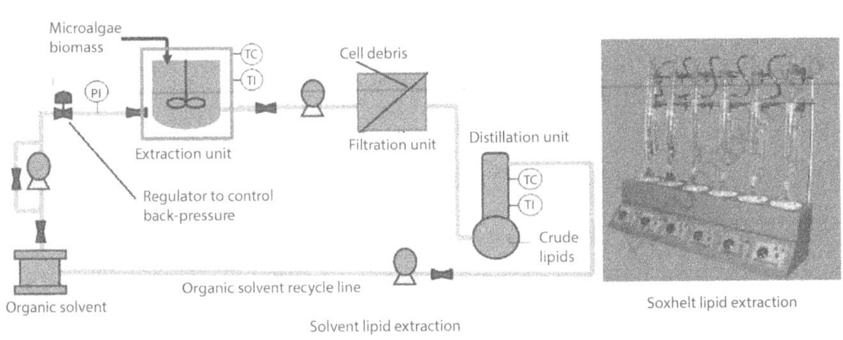

Figure 10.7 Solvent lipid extraction processes [32].

solvent extraction process) were also reported in literature claiming better extraction efficiency [11, 16, 41]. Microwave and ultrasonic assisted solvent extraction can be employed both wet or dry microalgae biomass and reduce extraction time in comparison to conventional solvent extraction process [4]. Teo and Idris explored the effect of a microwave-assisted solvent extraction process with wet Nannochloropsis sp. biomass using different solvent combinations and observed that the microwave-assisted solvent extraction process results into 12.8% lipid extract than conventional solvent extraction (8.47%) using same solvent system [55]. Patel *et al.* employ the ultrasonic and microwave-assisted solvent extraction with wet oleaginous yeast biomass and found that microwave-assisted solvent extraction has greater efficiency than ultrasonic-assisted solvent extraction and conventional extraction [56]. The key limitation of the microwave/ultrasonic extraction methods is higher energy consumption. Enzymatic-assisted solvent extraction process is generally carried out at low temperature and is emerging as an environmentally friendly solvent extraction process [19]. Sati *et al.* compared the effect of enzymatic-assisted extraction process with *Nannochloropsis* sp. and *Scenedesmus* sp. microalgae and observed improved lipid extraction with enzymatic-assisted solvent extraction [55]. Maffei *et al.* carried out enzymatic-assisted solvent extraction using cellulase and mannanase as enzyme with n hexane: propanol (3:2 v/v) from Nannochloropsis sp. and found that the treatment with combined enzymes resulted into more lipids (73%, w/w) than individual enzyme treatments (< 60% lipids, w/w). Zhang *et al.* investigated the effect of different enzyme treatments with cellulase, xylanase and pectinase using chloroform:methanol solvent combination system from microalga Scenedesmus sp. [57]. The results revealed that enzymatic-assisted extraction is more significant than conventional solvent extraction. Generally, enzyme-assisted solvent extraction system is a more green and energy-efficient method than mechanical assisted extractions; however, due to high cost, enzyme specificity with microalgae species and long extraction time are some of the drawbacks which limit its large-scale implications.

10.5.2.3 Green Solvents Extraction Methods

Green solvents are getting more interest due to the hazardous nature of organic solvents. Even a slight amount of these organic solvents (e.g., hexane, chloroform, etc.) in lipid/oil have negative impacts on human health as well as environments. Some examples of various green solvents used to extract microalgae lipid are shown in Figure 10.8. To replace hexane, Tanzi *et al.* carried out lipid extraction using terpenes and obtained good

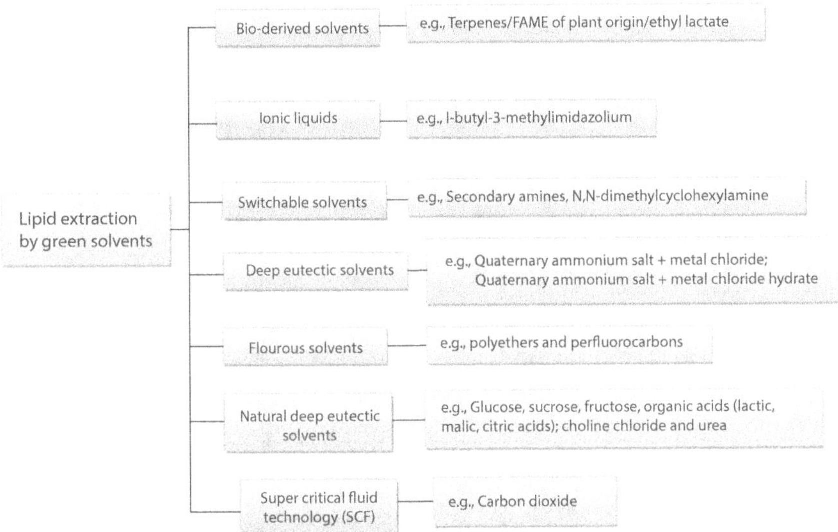

Figure 10.8 Green solvents used for lipid extraction.

results [55]. Terpenes are considered green solvents and obtained from renewable biomass, i.e., pine trees. Similarly, Mahmood *et al.* tried four green solvents named (ethyl acetate, ethyl lactate, cyclopentyl methyl ether, and 2-methyl tetrahydrofuran) to extract lipid from *Chlorella vulgaris* and *Nannochloropsis sp.* The results revealed that green solvents such as 2-methyl tetrahydrofuran and ethyl lactate gave better performance against the benchmark organic solvent, i.e., hexane [58]. In addition, some researchers also examined sugar solvents (e.g., glucose, fructose), organic acids (lactic acid), urea and secondary amines to replace the benchmark organic solvents for extraction of microbial lipids [59].

10.5.2.4 Supercritical Extraction Method

Supercritical fluid extraction (SFE) is a green, safe and newly developed method which can be applied for both extraction and purification of PUFAs from microbes under controlled conditions of temperature and pressure [19, 42]. In this method, Supercritical fluid (e.g., carbon dioxide) is used as an extraction solvent in place of traditional organic solvents, i.e., hexane/chloroform. When microalgae is exposed to the supercritical fluid, i.e., carbon dioxide under control conditions of temperature and pressure, the lipid present in the microalgae cells gets desorbed in the carbon dioxide fluid stream and recovered by condensation (Figure 10.9). Tai *et al.*

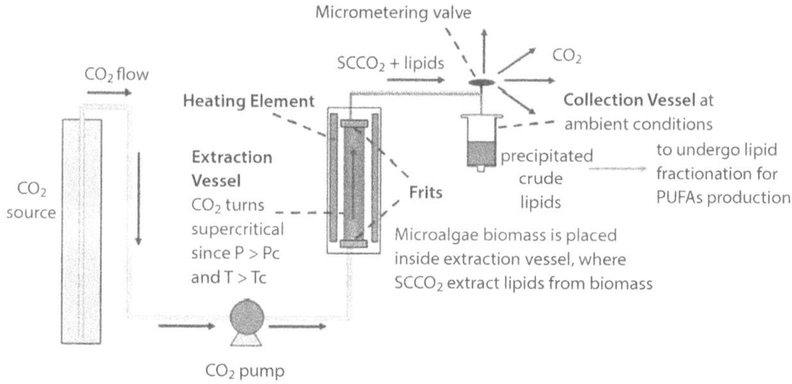

Figure 10.9 Super critical lipid extraction technique from microalgae, modified from Halim *et al.* [32].

reported better extraction efficiency with supercritical lipid extraction method using Scenedesmus sp. Biomass when compared with organic solvent-extraction methods such as Bligh and Dyer's, Folch's, and Soxhlet method [55]. Patil *et al.* observed 28-30% more lipid yield from microwave disrupted *Nannochloropsis salina* using $SCCO_2$ extraction with addition of hexane/ethanol solvents as compared to traditional extraction process [55]. When SC-CO_2 method is employed to *Schizochytrium sp.*, which has a high content of long-chain PUFAs, the extraction of the DHA proportion was increased from 29.3% to 32.2% [28]. In addition, many pilot studies also confirmed the suitability of SC-CO_2 method with polar solvents to extract PUFAs lipids from various microalgae biomass [55, 60].

10.6 Purification and Enrichment of PUFAs

Polyunsaturated fatty acids (PUFAs) such as EPA and DHA are generally used in the health and pharmaceutical industry. The lipid extracted from microbes (e.g., microalgae) is the mixture of saturated and unsaturated fatty acids. Therefore, it is necessary to purify and enrich PUFAs extracted from microalgae biomass. Depending upon carbon chain length and degree of unsaturation, different methods are applied for purification and enrichment of PUFAs. These methods include winterization, urea complexation, molecular distillation, chromatographic separation and enzymatic purification [15, 19, 28]. Different methods for extraction and purification of PUFAs from microalgae are shown in Table 10.2.

Table 10.2 Extraction and purification methods for enrichment of PUFAs from microbes.

S. no.	Microalgae species	Lipid extraction method	Chemical used for extraction	Quantification and purification method	Polyunsaturated fatty acids (total lipid content of dry biomass)	References
1.	*Diacronema vlkianum*	Soxhelt extract method with acid digestion of biomass	Petroleum ether	TLC	ALA 14 mg/100 g dry biomass EPA 3212 mg/100 g dry biomass DHA 836 mg/100 g dry biomass GLA 112 mg/100 g dry biomass	[61]
2.	*Chlorella vulgaris*	Soxhelt extract method with acid digestion of biomass	Petroleum ether	TLC	ALA 661 mg/100 g dry biomass EPA 19 mg/100 g dry biomass DHA 16 mg/100 g	[61]
3.	*Tetraselmis* sp.	Solvent extraction	Chloroform-methanol (2:1 (v/v)	–	EPA 10,410 mg/100 g dry biomass	[62]

(Continued)

Table 10.2 Extraction and purification methods for enrichment of PUFAs from microbes. (*Continued*)

S. no.	Microalgae species	Lipid extraction method	Chemical used for extraction	Quantification and purification method	Polyunsaturated fatty acids (total lipid content of dry biomass)	References
4.	*Nannochloropsis gaditana*	Carbon dioxide supercritical fluid (CO_2-SF)	–	–	EPA 1150 mg/ 100g dry biomass	[27]
5.	*Desmodesmus* sp.	Folch method	Chloroform-methanol	Argentated (silver ion) column chromatography	ALA 92% purify	[63]
6.	*P. tricornutum*	Simultaneous extraction and transesterification	Hexane, ethanol and acetyl chloride	Silica gel column chromatography	EPA 320 mg/100 g dry biomass	[64]
7.	*M. subterraneus*	Simultaneous extraction and transesterification	Hexane, ethanol and acetyl chloride	Silica gel column chromatography	EPA 14,300 mg/ 100 g dry biomass	[64]
8.	*Crypthecodinium cohnii*	Saponification and transmetylation	Ethanol	Winterization and urea complexation	DHA, recovered 19% of total fatty acid with more than 95% purity	[23]

(*Continued*)

Table 10.2 Extraction and purification methods for enrichment of PUFAs from microbes. (*Continued*)

S. no.	Microalgae species	Lipid extraction method	Chemical used for extraction	Quantification and purification method	Polyunsaturated fatty acids (total lipid content of dry biomass)	References
9.	*Spirulina maxima*	Acid digestion and soxhelt extraction	Petroleum ether	TLC	ALA 40 mg/100 g dry biomass GLA 452 mg/100 g dry biomass	[61]
10.	*Nannochloropsis gaditana*	Solvent extraction method	Dichloromethane-ethanol (1:1)	–	EPA 3.7 g/100 g DW dry biomass	[65]
11.	*P. tricornutum*	Supercritical CO_2 using Deep Eutectic Solvents with Microwaves as pre-treatment	Dimethyl carbonate	–	30% in total TFA	[66]
12.	*Nannochloropsis* sp.	Supercritical fluid-carbon dioxide (SF-CO2) extraction	–	–	EPA 5.69 mg/g DHA 0.12 mg/g	[67]

10.6.1 Low-Temperature Crystallization Enrichment

This is the easiest convenient operation and low-cost method to concentrate PUFAs by solidifying the saturated and unsaturated fatty acids lipids through a process of winterization. This method involves the solidification of the saturated and monounsaturated fatty acids by chilling at different temperatures. After removal of major concentration of saturated fatty acids (SFAs), PUFAs are concentrated in filtrate through low-temperature crystallization. Vázquez *et al.* carried out low-temperature crystallization to separate and enrich soybean oil using hexane and acetone in 1:9 ratio (v/v) solvents. Similarly, López-Martínez and his co-worker obtained 39.9%-58.8% concentrated γ-linolenic acid by the low-temperature crystallization method [28]. However, the extensive use of organic solvents, residue problems, and the low separation efficiency are the drawbacks of this method.

10.6.2 Urea Complexation

It is a simple and most efficient technique method for fractionating polyunsaturated fatty acids from plant seeds and microbes. The separation of PUFAs by urea complexation method depends on formation of inclusion complex of urea and FA (fatty acids). During this process, when free fatty acids are mixed with alcoholic solution (methanol or ethanol), urea makes a complex with saturated fatty acids and monounsaturated fatty acids at a certain temperature, with PUFAs remaining in solvents. Now the PUFAs are separated through the process of filtration and nitrogen purging. The formation of urea complex with fatty acids is directly related to the rate of saturation. In addition, the recovery of total PUFAs is highly influenced by the factors such as ratio of grease and urea, the crystallization time and Temperatures. Gu *et al.* performed some experiments to concentrate and purify linolenic acid by urea complexation method and observed upto 91.5% α-linoleic acid content [68]. Similarly, Abdullah and Salimon obtained 92.81% LA from a mixture of FFA extracted from Jatropha oil by urea complex fractionation at the following optimized conditions: urea/FFA ratio (w/w) = 5, crystallization temperature = -10°C and crystallization time = 24 h. This process is carried out in simple equipment at mild temperatures (room temperature) and requires low energy consumption. Furthermore, it is more efficient than low crystallization or selective solvent extraction methods with lower process cost. When urea complexation is employed in combination with chromatographic separation, it allows separation of fatty acids with a high degree of purity [30, 60].

10.6.3 Distillation Method

In this method fatty acids are separated by fractional distillation as different fatty acids have different boiling points. To avoid feasibility of oxidation, the distillation was performed under reduced pressure and lower temperature along with minimal residence time. This may be attained by molecular distillation or short-path distillation [69]. This is a very old technique and high energy consuming process. Furthermore, the fractionation of PUFA-rich oils (e.g., marine oil esters) is not easy because isolation of these components becomes less effective as molecular weight increases. Distillation is commonly used in combination with other purification techniques.

10.6.4 Enzymatic Purification

In this method, the concentration/purification of PUFAs is carried out by enzymatic hydrolysis or esterification reaction using different lipases. During this process, the separation mainly depends on bending of FA (fatty acid) because of the presence of cis-olefinic double bonds. Due to bending of FA, the terminal methyl group comes close to the ester bond and causes a steric hindrance effect on lipases to catalyze hydrolyze [70]. For example, there are 5 and 6 double bonds in the composition of EPA and DHA, respectively, resulting in high steric hindrance and hence, lipases are unable to reach the ester linkage between these FA and glycerol. On the other hand, there are not such barriers to lipases in the case of saturated and mono-unsaturated FA and they could be easily hydrolyzed. Chakraborty *et al.* concentrates ω-3 PUFA in sardine oil using an extracellular lipase which is derived from *Bacillus circulans*, isolated from marine macroalgae, *Turbinaraia conoides*. This lipase enriched sardine oil with 37.7% EPA and 5.1% ALA in TG fraction by hydrolysis reaction in 3 h. When urea complexation was also applied in addition to the above process, it produced up to 51.3% EPA [71]. Rupani *et al.* (2012) tried some commercial lipases of *Candida rugosa*, *Pseudomonas cepacea*, *Pseudomonas fluorescens*, and *Rhizomucor miehei* to hydrolyze the flax seed oil having ALA as a major FA. The results revealed that *Candida rugosa* was able to enrich up to 72% ALA and further use of urea complexation results into 80% ALA as major FA [72]. Jacob *et al.* carried out experiments to improve EPA and DHA in the mixture of FA derived from *Chlorella marina*, *Chaetocerous calcitran*, *Isochrysis galbana*, and *Tetraselmis gracillus* using lipase source *Candida cylinderacea* [14]. The study showed that *Candida cylinderacea* lipases may have very good potential for commercial production of PUFAs in pharmaceutical industries under standardized condition.

10.6.5 Chromatographic Separation

Depending on carbon chain length and degree of unsaturation, FFA can be separated using suitable adsorbents. Different chromatographic methods such as high-performance liquid chromatography (HPLC), reverse-phase HPLC and Silica gel thin-layer chromatography was applied to separate PUFAs from mixture of FFA [13, 19].

Garima *et al.* extracted purified EPA from microalgae *Phaeodactylum tricornutum* using the following steps: extraction of fatty acids from wet biomass using direct saponification, enrichment of PUFAs by urea complexation methods and isolation of EPA by preparative HPLC [73]. Some researchers tried Argentation HPLC method to separate PUFAs with 3-6 double bonds. For example, Moffat *et al.* used Ag-HPLC to separate PUFA triglycerides from sardine oil [30].

Robles Medina *et al.* separated PUFAs from cod liver oil with high degree of purity by Beckman reverse-phase C 18 analytical column [74]. In this line, Shantha *et al.* introduced a simple, reliable and cost-effective thin-layer chromatographic (TLC) method to concentrate longer chain PUFAs extracted from food and marine lipids [75]. Furthermore, methyl esters of tetraene, pentaene and hexaene from some natural sources (menhaden oil, sea scallop lipids, whale lipids, etc.) were improved up to 85% concentraton using TLC plates [75]. Silicic acid column chromatography is also applied to separate and purify PUFAs. It is a simple, rapid scalable technique. Cohen and Cohen concentrated 93% pure EPA from fatty acid methyl esters extracted from *Porphyridium cruentum* using this technique [76].

Nowadays supercritical fluid chromatography (SFC) is a more attractive chromatographic technique for separation and purification of PUFAs and can be used as an alternative to gas chromatography (GC) or high-pressure liquid chromatography (HPLC) due to its ability to process volatile compounds of high boiling point, accomplishing high column efficiency and high separation selectivity [19, 28]. Letisse *et al.* concentrated EPA and DHA content in the mixture of FAMEs extracted from fish cannery waste under supercritical conditions [77]. In another study, Pettinello *et al.* studied the SFC process to enrich EPA-EE using carbon dioxide as the supercritical solvent on batch as well as pilot scale. To perform experiments, they have developed different types of silica adsorption columns for fractionation of the initial mixture containing 68% EPA-EE. Using this chromatographic technique, they concentrated EPA-EE up to 95% and 93% at batch and pilot scale, respectively [78].

10.6.6 Supercritical Fluid Fractionation (SFF)

The supercritical fluid technology is a green method to extract and purify PUFA concentrates from fish oil, sea weeds and microalgae [19, 28, 78]. Fleck *et al.* developed an automated countercurrent column to separate the fraction of FAEE by utilizing SC-CO2 [79]. In addition, Perretti *et al.* performed experiments to modify the FAEE concentration of a mixture by using SC-CO_2 fractionation process and optimized the conditions in terms of pressure, temperature, and flow rate of this selective solvent [80].

10.7 Concluding Remarks

Microalgae have shown very good potential for production of PUFAs. Advances in media optimization and ease of bioreactor scale up study explore new venue for large biomass production that leads to a higher level of omega-3 fatty acids production. However, the extraction and purification of PUFAs from microalgae is a challenging step and under development stage. This chapter reviews the various physico-chemical methods applied for extraction and purification of PUFAs from microbes. The directions to develop methods for extracting and purifying PUFAs from microbes are proposed. The following conclusions are withdrawn from this chapter:

- Solvent extraction process has been generally used to extract lipid from microalgae. Organic solvents, such as dichloromethane, n-hexane, chloroform, methanol and petroleum ether are commonly used to extract lipids. However, these solvents are toxic for the food industry and therefore, they should be replaced by green solvents.
- The supercritical fluid extraction is a green and most promising technique to improve and purify lipid extraction. However, the operation cost of this method is very high and the research should be focused to reduce cost of this process.
- The purification of PUFAs are carried out using various techniques like low-temperature crystallization, urea complex formation, enzymatic purification, supercritical fluid fractionation and chromatographic extraction.

- A combination of suitable extraction and purification process should be employed to get the higher yield and extraction efficiency of PUFAs from microalgae. The PUFAs composition in microalgae can also be improved by bioengineering technology.
- This chapter concludes that efforts should be focused on reducing the production losses and energy cost associated the extraction and purification processes. In addition, a large-scale, viable, energy-efficient and eco-friendly process should be developed.

References

1. Chisti, Y., Biodiesel from microalgae. *Biotechnology advances,* 2007. **25**(3): p. 294-306.
2. Richmond, A., *Handbook of microalgal culture: biotechnology and applied phycology*. Vol. 577. 2004: Wiley Online Library.
3. Chew, K.W., *et al.*, Microalgae biorefinery: high value products perspectives. *Bioresource technology*, 2017. **229**: p. 53-62.
4. Deshmukh, S., R. Kumar, and K. Bala, Microalgae biodiesel: A review on oil extraction, fatty acid composition, properties and effect on engine performance and emissions. *Fuel Processing Technology*, 2019. **191**: p. 232-247.
5. Sharma, A.K., P.K. Sahoo, and S. Singal, Influence of different nitrogen and organic carbon sources on microalgae growth and lipid production. *IOSR Journal of Pharmacy and Biological Sciences*, 2015. **10**(1): p. 48-53.
6. Arora, N., *et al.*, Bioremediation of domestic and industrial wastewaters integrated with enhanced biodiesel production using novel oleaginous microalgae. *Environmental Science and Pollution Research*, 2016. **23**(20): p. 20997-21007.
7. Sharma, A.K., *et al.*, Impact of various media and organic carbon sources on biofuel production potential from Chlorella spp. *3 Biotech,* 2016. **6**(2): p. 116.
8. Nayak, M., A. Karemore, and R. Sen, Performance evaluation of microalgae for concomitant wastewater bioremediation, CO2 biofixation and lipid biosynthesis for biodiesel application. *Algal Research*, 2016. **16**: p. 216-223.
9. Kay, R.A. and L.L. Barton, Microalgae as food and supplement. *Critical reviews in food science & nutrition,* 1991. **30**(6): p. 555-573.
10. Becker, E., Micro-algae as a source of protein. *Biotechnology advances,* 2007. **25**(2): p. 207-210.
11. Sharma, A.K., *et al.*, Exploration of upstream and downstream process for microwave assisted sustainable biodiesel production from microalgae Chlorella vulgaris. *Bioresource technology*, 2016. **216**: p. 793-800.

12. Yao, L., *et al.*, Microalgae lipid characterization. *Journal of agricultural and food chemistry*, 2015. **63**(6): p. 1773-1787.
13. Kumar, B.R., *et al.*, Microalgae as rich source of polyunsaturated fatty acids. *Biocatalysis and agricultural biotechnology*, 2019.
14. Jacob, J.P. and S. Mathew, Effect of Lipases from Candida cylinderacea on Enrichment of PUFA in Marine Microalgae. *Journal of food processing and preservation*, 2017. **41**(1): p. e12928.
15. Béligon, V., *et al.*, Microbial lipids as potential source to food supplements. *Current Opinion in Food Science*, 2016. **7**: p. 35-42.
16. Xue, Z., *et al.*, Edible oil production from microalgae: A review. *European journal of lipid science and technology*, 2018. **120**(6): p. 1700428.
17. Yu, W.-L., *et al.*, Modifications of the metabolic pathways of lipid and triacylglycerol production in microalgae. *Microbial cell factories*, 2011. **10**(1): p. 91.
18. Abirami, S., S. Murugesan, and S. Sivaswamy, Effect of various pretreatment methods prior to extraction of omega 3 fatty acids from Nannochloropsis gaditana. *Int J App Res*, 2016. **2**(10): p. 81-85.
19. Bellou, S., *et al.*, Microbial oils as food additives: recent approaches for improving microbial oil production and its polyunsaturated fatty acid content. *Curr Opin Biotechnol*, 2016. **37**: p. 24-35.
20. Ryckebosch, E., *et al.*, Microalgae as an alternative source of omega-3 long chain polyunsaturated fatty acids. *Lipid Technology*, 2012. **24**(6): p. 128-130.
21. Verma, M.L., *et al.*, *Microbial production of omega-3 polyunsaturated fatty acids*, in *Biotechnological Production of Bioactive Compounds*. 2020, Elsevier. p. 293-326.
22. Singhasuwan, S., *et al.*, Carbon-to-nitrogen ratio affects the biomass composition and the fatty acid profile of heterotrophically grown Chlorella sp. TISTR 8990 for biodiesel production. *Journal of biotechnology*, 2015. **216**: p. 169-177.
23. Mendes, A., T. Lopes da Silva, and A. Reis, DHA concentration and purification from the marine heterotrophic microalga Crypthecodinium cohnii CCMP 316 by winterization and urea complexation. *Food technology and biotechnology*, 2007. **45**(1): p. 38-44.
24. Mühlroth, A., *et al.*, Pathways of lipid metabolism in marine algae, co-expression network, bottlenecks and candidate genes for enhanced production of EPA and DHA in species of Chromista. *Marine drugs*, 2013. **11**(11): p. 4662-4697.
25. Ward, O.P. and A. Singh, Omega-3/6 fatty acids: alternative sources of production. *Process Biochemistry*, 2005. **40**(12): p. 3627-3652.
26. Shanab, S.M., R.M. Hafez, and A.S. Fouad, A review on algae and plants as potential source of arachidonic acid. *Journal of advanced research*, 2018. **11**: p. 3-13.

27. Molino, A., M. Martino, V. Larocca, G. Di Sanzo, A. Spagnoletta, T. Marino, D. Karatza, A. Iovine, S. Mehariya, D. Musmarra, Eicosapentaenoic Acid Extraction from Nannochloropsis gaditana using Carbon Dioxide at Supercritical Conditions. *Marine drugs,* 2019. **17**(2): p. 132.
28. Li, X., *et al.*, Extraction and purification of eicosapentaenoic acid and docosahexaenoic acid from microalgae: A critical review. *Algal Research,* 2019. **43**: p. 101619.
29. Byreddy, A.R., C.J. Barrow, and M. Puri, Bead milling for lipid recovery from thraustochytrid cells and selective hydrolysis of Schizochytrium DT3 oil using lipase. *Bioresource technology,* 2016. **200**: p. 464-469.
30. Medina, A.R., *et al.*, Downstream processing of algal polyunsaturated fatty acids. *Biotechnology advances,* 1998. **16**(3): p. 517-580.
31. Renaud, S.M. and J.T. Luong-Van. Seasonal variation in the chemical composition of tropical Australian marine macroalgae. in *Eighteenth International Seaweed Symposium.* 2006. Springer.
32. Halim, R., M.K. Danquah, and P.A. Webley, Extraction of oil from microalgae for biodiesel production: A review. *Biotechnol Adv,* 2012. **30**(3): p. 709-732.
33. Lee, S.Y., *et al.*, Cell disruption and lipid extraction for microalgal biorefineries: A review. *Bioresour Technol,* 2017. **244**(Pt 2): p. 1317-1328.
34. Gunerken, E., *et al.*, Cell disruption for microalgae biorefineries. *Biotechnol Adv,* 2015. **33**(2): p. 243-60.
35. Passos, F., *et al.*, Algal biomass: physical pretreatments, in *Pretreatment of Biomass.* 2015, Elsevier. p. 195-226.
36. Shene, C., *et al.*, High pressure homogenization of Nannochloropsis oculata for the extraction of intracellular components: Effect of process conditions and culture age. *European Journal of Lipid Science and Technology,* 2016. **118**(4): p. 631-639.
37. Choi, W.Y. and H.Y. Lee, Effective production of bioenergy from marine Chlorella sp. by high-pressure homogenization. *Biotechnology & Biotechnological Equipment,* 2016. **30**(1): p. 81-89.
38. Ryckebosch, E., *et al.*, Microalgae as an alternative source of omega-3 long chain polyunsaturated fatty acids. *Lipid Technology,* 2012. **24**(6): p. 128-130.
39. Gerde, J.A., *et al.*, Evaluation of microalgae cell disruption by ultrasonic treatment. *Bioresource technology,* 2012. **125**: p. 175-181.
40. Neto, A.M.P., *et al.*, Improvement in microalgae lipid extraction using a sonication-assisted method. *Renewable Energy,* 2013. **55**: p. 525-531.
41. Lee, J.-Y., *et al.*, Comparison of several methods for effective lipid extraction from microalgae. *Bioresource technology,* 2010. **101**(1): p. S75-S77.
42. Abomohra, A.E.-F., W. Jin, and M. El-Sheekh, Enhancement of lipid extraction for improved biodiesel recovery from the biodiesel promising microalga Scenedesmus obliquus. *Energy Conversion and Management,* 2016. **108**: p. 23-29.

43. Zhang, R., *et al.*, Emerging techniques for cell disruption and extraction of valuable bio-molecules of microalgae Nannochloropsis sp. *Bioprocess and biosystems engineering*, 2019. **42**(2): p. 173-186.
44. Show, K.Y., *et al.*, Microalgal drying and cell disruption--recent advances. *Bioresour Technol*, 2015. **184**: p. 258-266.
45. Liu, B. and C. Benning, Lipid metabolism in microalgae distinguishes itself. *Current opinion in biotechnology*, 2013. **24**(2): p. 300-309.
46. Atabani, A., *et al.*, Non-edible vegetable oils: a critical evaluation of oil extraction, fatty acid compositions, biodiesel production, characteristics, engine performance and emissions production. *Renewable and sustainable energy reviews*, 2013. **18**: p. 211-245.
47. Vishwakarma, R., D.W. Dhar, and S. Saxena, Influence of nutrient formulations on growth, lipid yield, carbon partitioning and biodiesel quality potential of Botryococcus sp. and Chlorella sp. *Environmental Science and Pollution Research*, 2019. **26**(8): p. 7589-7600.
48. Folch, J., M. Lees, and G.S. Stanley, A simple method for the isolation and purification of total lipides from animal tissues. *Journal of biological chemistry*, 1957. **226**(1): p. 497-509.
49. Bligh, E. and W. Dyer, A rapid method of total lipid extraction and purification. Cd, *z. J. Biochrm. Physiol*, 1959. **37**: p. 91.
50. Gupta, A., *et al.*, Exploring omega-3 fatty acids, enzymes and biodiesel producing thraustochytrids from Australian and Indian marine biodiversity. *Biotechnology journal*, 2016. **11**(3): p. 345-355.
51. Yang, F., *et al.*, Optimization of medium using response surface methodology for lipid production by Scenedesmus sp. *Marine drugs*, 2014. **12**(3): p. 1245-1257.
52. Grima, E.M., *et al.*, Recovery of microalgal biomass and metabolites: process options and economics. *Biotechnology advances*, 2003. **20**(7-8): p. 491-515.
53. dos Santos, R.R., *et al.*, Comparison between several methods of total lipid extraction from Chlorella vulgaris biomass. *Ultrasonics sonochemistry*, 2015. **22**: p. 95-99.
54. Ramluckan, K., K.G. Moodley, and F. Bux, An evaluation of the efficacy of using selected solvents for the extraction of lipids from algal biomass by the soxhlet extraction method. *Fuel*, 2014. **116**: p. 103-108.
55. Sati, H., *et al.*, Microalgal lipid extraction strategies for biodiesel production: A review. *Algal research*, 2019. **38**: p. 101413.
56. Patel, C., *et al.*, Comparative compression ignition engine performance, combustion, and emission characteristics, and trace metals in particulates from Waste cooking oil, Jatropha and Karanja oil derived biodiesels. *Fuel*, 2019. **236**: p. 1366-1376.
57. Zhang, Z., *et al.*, Effects of fatty acid methyl esters proportion on combustion and emission characteristics of a biodiesel fueled marine diesel engine. *Energy Conversion and Management*, 2018. **159**: p. 244-253.

58. Mahmood, W.M.A.W., C. Theodoropoulos, and M. Gonzalez-Miquel, Enhanced microalgal lipid extraction using bio-based solvents for sustainable biofuel production. *Green Chemistry*, 2017. **19**(23): p. 5723-5733.
59. Kumar, S.J., *et al.*, Sustainable green solvents and techniques for lipid extraction from microalgae: A review. *Algal Research*, 2017. **21**: p. 138-147.
60. Patil, D., Recent trends in production of polyunsaturated fatty acids (PUFA) concentrates. *J Food Res Technol*, 2014. **2**: p. 15-23.
61. Batista, A.P., L. Gouveia, N.M. Bandarra, J.M. Franco, A. Raymundo, Comparison of microalgal biomass profiles as novel functional ingredient for food products. *Algal Research*, 2013. **2**(2): p. 164-173.
62. Makri, A., S. Bellou, M. Birkou, K. Papatrehas, N.P. Dolapsakis, D. Bokas, S. Lipid synthesized by micro-algae grown in laboratory-and industrial-scale bioreactors. *Engineering in Life Sciences*, 2011. **11**(1): p. 52-58.
63. Nagappan, S. and S. Kumar Verma, Co-production of biodiesel and alpha-linolenic acid (omega-3 fatty acid) from microalgae, Desmodesmus sp. MCC34. *Energy Sources, Part A: Recovery, Utilization, and Environmental Effects*, 2018. **40**(24): p. 2933-2940.
64. Belarbi, E.-H., E. Molina, and Y. Chisti, *RETRACTED: A process for high yield and scaleable recovery of high purity eicosapentaenoic acid esters from microalgae and fish oil*. 2000, Elsevier.
65. Ryckebosch, E., S.P.C. Bermúdez, R. Termote-Verhalle, C. Bruneel, K. Muylaert, R. Parra-Saldivar, I. Foubert, Influence of extraction solvent system on the extractability of lipid components from the biomass of Nannochloropsis gaditana. *Journal of Applied Phycology*, 2014. **26**(3): p. 1501-1510.
66. Tommasi, E., G. Cravotto, P. Galletti, G. Grillo, M. Mazzotti, G. Sacchetti, C. Samorì, S. Tabasso, M. Tacchini, E. Tagliavini, Enhanced and selective lipid extraction from the microalga P. tricornutum by dimethyl carbonate and supercritical CO2 using deep eutectic solvents and microwaves as pretreatment. *ACS Sustainable Chemistry & Engineering*, 2017. **5**(9): p. 8316-8322.
67. Leone, G.P., R. Balducchi, S. Mehariya, M. Martino, V. Larocca, G. Di Sanzo, A. Iovine, P. Casella, T. Marino, D. Karatza, Selective Extraction of ω-3 fatty acids from Nannochloropsis sp. using supercritical CO2 extraction. *Molecules*, 2019. **24**(13): p. 2406.
68. Gu, H.-B., *et al.*, Concentration of α-linoleic acid of perilla oil by gradient cooling urea inclusion. *Agricultural Sciences in China*, 2009. **8**(6): p. 685-690.
69. *<Recent trends in production of polyunsaturated fatty acids.pdf>*.
70. Bottino, N.R., G.A. Vandenburg, and R. Reiser, Resistance of certain long-chain polyunsaturated fatty acids of marine oils to pancreatic lipase hydrolysis. *Lipids*, 1967. **2**(6): p. 489-493.
71. Chakraborty, K., *et al.*, Preparation of eicosapentaenoic acid concentrates from sardine oil by Bacillus circulans lipase. *Food chemistry*, 2010. **120**(2): p. 433-442.

72. Rupani, B., *et al.*, Lipase-mediated hydrolysis of flax seed oil for selective enrichment of α-linolenic acid. *European journal of lipid science and technology*, 2012. **114**(11): p. 1246-1253.
73. Grima, E.M., *et al.*, Gram-scale purification of eicosapentaenoic acid (EPA, 20: 5n-3) from wetPhaeodactylum tricornutum UTEX 640 biomass. *Journal of applied phycology*, 1996. **8**(4-5): p. 359-367.
74. Medina, A.R., *et al.*, Obtención de concentrados de ácidos grasos poliinsaturados por el método de los compuestos de inclusión de urea. *Grasas y aceites*, 1995. **42**: p. 174-182.
75. Shantha, N. and R. Ackman, Silica gel thin-layer chromatographic method for concentration of longer-chain polyunsaturated fatty acids from food and marine lipids. *Canadian Institute of Food Science and Technology Journal*, 1991. **24**(3-4): p. 156-160.
76. Cohen, Z. and S. Cohen, Preparation of eicosapentaenoic acid (EPA) concentrate fromPorphyridium cruentum. *Journal of the American Oil Chemists Society*, 1991. **68**(1): p. 16-19.
77. Létisse, M. and L. Comeau, Enrichment of eicosapentaenoic acid and docosahexaenoic acid from sardine by-products by supercritical fluid fractionation. *Journal of separation science*, 2008. **31**(8): p. 1374-1380.
78. Pettinello, G., *et al.*, Production of EPA enriched mixtures by supercritical fluid chromatography: from the laboratory scale to the pilot plant. *Journal of Supercritical Fluids*, 2000. **19**(1): p. 51-60.
79. Fleck, U., C. Tiegs, and G. Brunner, Fractionation of fatty acid ethyl esters by supercritical CO2: high separation efficiency using an automated countercurrent column. *Journal of supercritical fluids*, 1998. **14**(1): p. 67-74.
80. Perretti, G., *et al.*, Supercritical carbon dioxide fractionation of fish oil fatty acid ethyl esters. *Journal of supercritical fluids*, 2007. **40**(3): p. 349-353.

11

Market Perspective of EPA and DHA Production from Microalgae

Jyoti Sharma[1], Pampi Sarmah[2] and Narsi R Bishnoi[1*]

[1]*Department of Environmental Science & Technology, Guru Jambheshwar University of Science & Technology, Hisar, Haryana, India*
[2]*Department of Ecology and Environmental Science, Assam University Silchar*

Abstract
Omega-3 fatty acids have been considered as very vital supplements for human beings as the human body cannot synthesize these polyunsaturated fatty acids. Marine fishes and plant oils are the major sources of these Poly Unsaturated Fatty Acids and have been used for millennia. However, due to increasing demand, and depletion in fish resources, there is an urgent need to find alternative sources of omega-3 fatty acids. Nowadays microalgae serve as a significant feedstock for the production of Eicosapentaenoic acid (20:5, n-3, EPA) and Docosahexaenoic acid (22:6, n-3, DHA). EPA and DHA have anti-inflammatory properties. Modern society has embraced these PUFAs as a dietary supplement because of their health benefits. The present review discusses the sources of EPA and DHA, their benefits, and microalgae as the best feedstock. The market scenario of EPA and DHA has also been illustrated.

Keywords: Microalgae, omega 3 fatty acids, EPA, DHA, fatty acid composition, benefits of microalgae, factors affecting microalgae, market trend

11.1 Introduction

Due to deficit of enzyme desaturase, usually the human body cannot synthesize Omega-3 fatty acids which insert double bond at ω3 position [1].

**Corresponding author*: nrbishnoi@gmail.com

Alok Kumar Patel and Leonidas Matsakas (eds.) Nutraceutical Fatty Acids from Oleaginous Microalgae: A Human Health Perspective, (281–298) © 2020 Scrivener Publishing LLC

In human physiology, Omega-3 fatty acids class involves α-linolenic acid (ALA), Eicosapentaenoic acid (EPA), and Docosahexaenoic Acid (DHA). ALA is derived mainly from vegetable oil and nuts, whereas the other two are mainly found in fish and some microorganisms like bacteria and microalgae [2]. These long-chain Polyunsaturated Fatty acids play a significant role in dietetics, beverages and the pharmaceutical industry. But the conversion of ALA into EPA and DHA is quite a slow process in the human body [3]. Therefore, these must be provided in the form of dietary supplements.

Omega-3 fatty acids play a significant role in the treatment of various common disorders such as hyperlipidaemia, hypertension, hypertonia, premenstrual tension, and cancer. The main role of EPA and DHA are controlling the blood lipids and lipoproteins levels and the biosynthesis of eicosanoids [2].

The daily uptake of EPA and DHA ranging from 0.2-0.3 g/day has been recommended by some of the nutritional institutes and government agencies for normal healthy people, while cardiovascular patients generally are recommended to take 1.0-4.0 g/day [1].

Worldwide consumption of PUFA for the years 2013 and 2015 were 123.8 thousand metric tons and 134.7 thousand metric tons, respectively, at a cost of US$2.3 billion and US$2.5 billion. It is anticipated that the costs of Omega-3 PUFAs demand will double by 2020 [4].

EPA and DHA are the major nutritional ingredients in fish oil of salmon feed. The demand of the omega-3 market has increased, but due to undersupply of pelagic fishery to the global market, the world is facing a global fish crisis. Presently, the fish oil supply is about 1 million metric tonnes per year and around 70% of this fish oil is commonly used for aquafeeds (The Marine Ingredients Organisation IFFO, 2013). EPA and DHA production of various marine fishes is illustrated in Table 11.1. So, the imbalance between the shortage of fish oil supply and increased demand from aquaculture industry has compelled researchers and industries to find alternate sources of PUFAs. The prices of fish oil have also been increased. During 2005, prices were unstable, between 300–800 USD/ton, and in 2012-14 prices were 2,000 USD/ton [5]. The prices will increase by 25% over the next five years as anticipated by the Food and Agriculture Organization (FAO). Microalgae can serve as a potential feedstock for EPA and DHA. Microalgae have been proved as an alternate and renewable and sustainable source as it uses photosynthesis process for its growth. A lot of literature is available on biofuel production using microalgae [6–8], while studies related to value-added products from microalgae have received scant attention from researchers [9–11]. Use of microalgae as a feed highly depends on experience and knowledge from the biofuel industry; along with this some other

Table 11.1 EPA and DHA composition of various marine fishes [1].

Species name	Ecosapentaenoic acid (g/kg of fish)	Docosahexanoic acid (g/kg of fish)
Sardine *(Sardinops sagax)*	6.6	19.0
Herring *(Clupea harengus)*	8.5	8.3
Atlantic salmon *(Salmo salar)*	6.2	5.8
Surf smelt *(Hypomesus pretiosus)*	3.6	5.7
Capelin *(Mallotus villosus)*	3.6	4.6
Horse mackerel *(Trachurus trachurus)*	1.6	5.8
Red porgy *(Pagrus pagrus) cultured*	2.3	4.0
Arctic charr *(Salvelinus alpinus*	1.3	2.8

points must be considered like growth condition of microalgae, and biomass processing [11].

The fatty acids contents in microalgae vary from species to species, e.g., *Crypthecodinium cohnii* contains only DHA; *Skeletonema costatum*; *Nannochloropsis* sp. and *Chaetoceros* sp. contains mostly EPA, while some other species *Porphyridium cruentum* contains primarily ARA. The omega-3 fatty acids synthesis can be enhanced by optimising the growth conditions and metabolic engineering approaches. Therefore, omega-3 PUFAs production from algae have commercial importance through microalgal biotechnology with recent developments in genomics and bioinformatics analyses. The present chapter is aimed at the potential of microalgae for EPA and DHA production and their market perspectives. Factors affecting the growth of microalgae, algal oil extraction, and purification method have been discussed.

11.2 Categories of Omega-3 Fatty Acids and Their Health Benefits

Fatty acids are a chain of hydrocarbons with methyl group at one end of the molecule (omega = ω or n) and carboxylic acid at the other end [12]. Mainly two nomenclatures are used for fatty acids classification. Alpha carbon (α-carbon) named as the carbon atom next to the carboxylic group is the consecutive carbon called β-carbon. The double bonds in nomenclature are indicated as the presence of carboxyl group. Omega-3 fatty acids are PUFAs in which the first double may be placed between third and fourth carbon from the ω carbon. The three most important categories of omega-3 fatty acids are ALA acid (18:3, n-3; ALA), EPA (20:5, n-3; EPA) and DHA (22:6, n-3; DHA) (Figure 11.1) [13]. These three omega-3 fatty acids (ALA, EPA and DHA) comprise three, five, or six double bonds in a long carbon chain of 18, 20, or 22 atoms, respectively. Cis configuration in fatty acids is readily available in nature. Docosahexaenoic acid (C22:6 n-3) is a long-chain PUFA which is present in the human body. Primarily it is the foremost structural lipid in the brain and the retina of the eye. It is also a key component of the heart. EPA and DHA have some mediators which provide strong defence mechanism to the body and have an anti-inflammatory property and also nullify the reactive oxygen species (Chauton *et al.*, 2017) [16]. Recently a large number of producers have focused towards PUFA production as it is a more cost-effective process as compared to biofuel production [14]. Different health benefits of EPA and DHA have been detailed below (Figure 11.2).

11.3 Brain Development

The brain is predominantly composed of fat and contains an elevated level of DHA and AA. These FA, particularly DHA, are believed to facilitate fluidity of neuronal membrane and regulates neurotransmitters [15]. The experiment conducted on pregnant rats concludes that Docosahexaenoic acid deficiency can reduce the brain plasticity and impairs brain function during brain maturation and adulthood [16]. It is evident that n-3 PUFA has been proved vital for the development of the foetal brain; therefore, EPA and DHA must be included in the diet of pregnant women to confirm the continuous supply of these FA to the foetus (Food and Agriculture

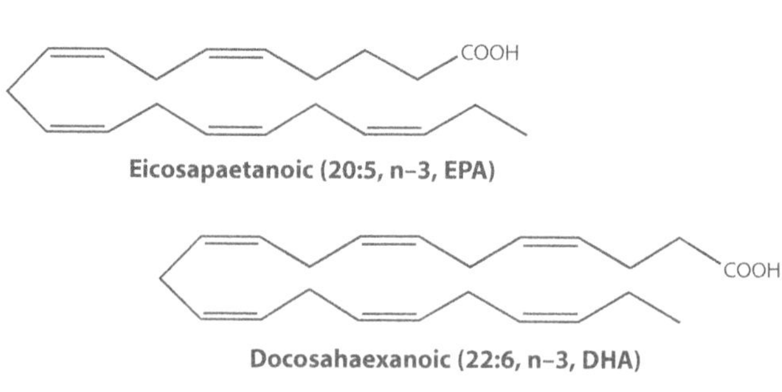

Figure 11.1 Chemical structure of EPA and DHA.

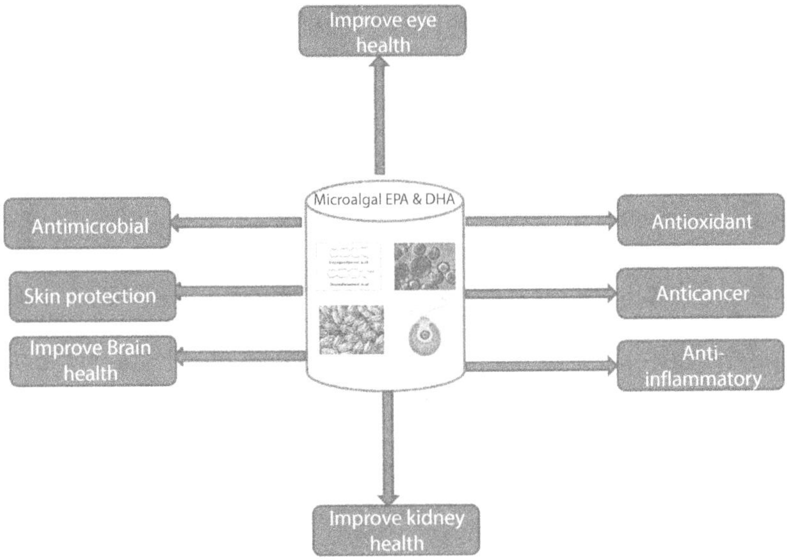

Figure 11.2 Different health benefits of EPA and DHA [modified 15].

Organization of the United Nations, 2010). DHA intake by mother benefits the infant by various means, such as problem-solving skills (at 9 months), improved eye-and-hand coordination (at 2.5 years of age), reduces allergic incidence (at 16 years of age) and asthma and reduces release of inflammatory markers in mother and foetus, which may be linked to preterm labour.

11.4 Cardiovascular Diseases

n-3 polyunsaturated fatty acids play a significant role in preventing cardiovascular diseases (CVD). The intake of n-3 PUFA reduces the overall mortality, mortality due to myocardial infarction (MI) and sudden death due to coronary heart disease (CHD). Observations based on several studies have linked improvement in HDL to LDL ratio, low levels of blood triglycerides and fish-eating habits with a lesser rate of cardiovascular diseases [17, 18]. EPA and DHA are beneficial in sudden cardiac death (SCD), atrial fibrillation and atherosclerosis [19, 20]. An EPA- and DHA-rich diet has the ability to replace n-6 PUFA viz. Arachidonic acid in the membrane of almost every cell. The oxidation of AA may produce prostaglandins, hydroxy fatty acids, thromboxanes, leukotrienes and lipoxins, which when present in a huge quantity may lead to the formation of atheromas and thrombus;

n-6 PUFA enriched diet may induce pro-aggregatory and pro-thrombotic process resulting in CVD.

11.5 Present Sources of Omega-3 PUFAs

Plants (primarily GLVs and nuts) and deep-sea fishes are the richest sources of omega-3 fatty acids. Eicosapentaenoic acid (EPA) alpha-linolenic acid (ALA) and docosahexaenoic acid (DHA) can be easily obtained from these sources [21]. Perillafrutescens seed oil is also considered as a rich source of omega-3 linolenic acid [22]. However, plant sources such as flaxseed butternuts, walnuts, pumpkin seeds, black and red currant seeds, soy, wheat germ, canola oil and green leafy plants like purslane generally contain a high concentration of Alpha-linolenic acid (ALA). Furthermore, with respect to nutritive value the optimum ratio of n-6 to n-3 is observed in camelina, i.e., 1: 14 and hemp seed, i.e., 1: 0.4. Marine-derived EPA and DHA acts as the best source of omega-3 and can be commonly found in oily fishes in high concentration. Fish oil is known to be a better source as compared to plant seed oil due to lower n-6 to n-3 ratio [1]. Among all the fishes, Scombridae, Clupeidae and Salmonidae (oily fish families) contain the maximum percentage of EPA and DHA thus fish oil PUFA is of more significance than plant oil PUFA. The human body may acquire omega-3 in the form of ALA from plant sources; however, the mechanism required for its conversion into PUFA (EPA and DHA) is lacking. Thus, at present, the fish oil could be considered as an ideal source for omega-3 PUFA.

During the last decade, algal oil rich in long-chain polyunsaturated fatty acids (LCPUFAs) have been used as nutraceuticals among which omega-3 LCPUFAs (DHA and EPA) are prominently acknowledged [15]. At present, marine members such as microalgae of families Thraustochytriaceae and Crypthecodiniaceae are commonly used for the production of DHA-enriched algal oil and biomass. Thraustochytridsand Crypthecodiniaceae includes genera Schizochytrium & Ulkenia and Crypthecodiniumisa, respectively. These organisms are widely disseminated in the oceans all over the world. Heterotrophic microalgae can efficiently synthesize a large quantity of DHA from glucose in the absence of light while autotrophic algae use natural light to initiate their metabolic activities to produce DHA. This may limit the fermenter design, size in particular, which in turn may limit the yield. On the other hand, the open pond system is best suited for the growth of autotrophic algae. Practices, such as intensive collection/isolation/screening procedures have been adopted for the production of

DHA and EPA rich oils from the microalgal strains. The most efficient microalgal strain shared properties like high growth rates, growth unaffected by low salinity conditions and the high proportion of DHA/EPA (as a percentage of total lipids at higher temperature fermentation, i.e., 30 °C).

Moreover, an extensive toxicological examination is performed on the viable algal strains to get regulatory approval [23]. Commercially, microalgae *Crypthecodinium cohnii* is used by DSM Nutritional Products to make DHASCOT oil for the infant formula market, which contains 40 to 45% w/w of DHA content and almost no EPA. This algal oil contains 35%-40% DHA and an extremely low level of EPA, i.e., <2%. Moreover, DSM produces algal oil with DHA: EPA ratio of ca 2:1 using different strains of Schizochytrium to achieve the quality of fish oil. The resulting product is known as Life's Omega TM containing as low as 40% total DHA and EPA, i.e., 24% and 12%, respectively. Until now the production of EPA and DHA from algal oil has been restricted to only labs, it is upscaled. Aurora algae have exclusively grown strains of microalgae *Nannochloropsis* in large sea-water open ponds with high solar radiation in geographic locations.

11.6 Why Microalgae?

Microalgae are the primaeval organisms which can be eukaryotic or prokaryotic and withstand harsh environmental conditions [24]. Photosynthetic rate of microalgae is quite higher as compared to other terrestrial organisms. Microalgae have the capability to accumulate lipid content up to 70% but some of the species possess 90% lipid accumulation under optimised conditions [6, 24]. However, lipid accumulation potential and biomass productivity differ from species to species. Some microalgal species like Neochloris, Dunaliella, Nanochloris, *Chlorella*, *Porphyridium* and *Nitzschiahave* have been found to contain 20-50% oil content [25]. In a study carried out by Abou-shanab *et al.* (2013) [26] found that *C. pitchmanii* showed maximum lipid content (51%) while *C. vulgaris* possessed only 26% out of 45 microalgal strains.

Microalgae possess a higher fraction of LCPUFA as compared to traditional sources, whereas the concentration of these n-3 PUFAs differs from species to species. Algal species from Bacillariophyceae (diatoms) and Chrysophyceae dominate in EPA and DHA content [27]. Higher concentration of EPA has been observed in Cryptophyceae, Prasinophyceae, Rhodophyceae, Xanthophyceae, Glaucophyceae and Eustigmatophyceae, whereas DHA was found to be significantly higher in Dinophyceae, Prymnesiophyceae, and Euglenophyceae [28].

Table 11.2 Microalgae metabolites from different microalga species which are being used for marketing purposes [53–56]. (*Continued*)

Algal species	Metabolites
Chroomonas mesostigmatica	EPA, DHA
Guillardia theta	EPA, DHA
Hemiselmis sp	EPA, DHA
Proteomonas sulcata	EPA, DHA
Rhodomonas salina	EPA, DHA
Storeatula major	EPA, DHA
Teleaulax acuta	EPA, DHA
Teleaulax amphioxeia	EPA, DHA
Nannochloropsis spp	EPA, DHA
Skeletonema spp.	EPA, DHA
Isochrysis spp	EPA, DHA
Tetraselmis sp	EPA, DHA
Dunaliella salina	EPA, DHA
Chroothece richteriana	EPA, DHA
Fragilaria sp.	EPA, DHA
Biddulphia aurita	EPA, DHA
Amphora sp.	EPA, DHA
Phaeodactylum tricornutum	EPA, DHA, and Lutein
Nitzschia longissima	EPA, DHA, Fucoxanthin, Lipids
Tetraselmis sp	EPA, DHA, β-carotene, Lutein
Neochloris conjuncta	EPA, DHA and α-linolenic acid
Neochloris terrestris	EPA, DHA and α-linolenic acid
Neochloris texensis	EPA, DHA, and α-linolenic acid

(*Continued*)

Table 11.2 Microalgae metabolites from different microalga species which are being used for marketing purposes [53–56]. (*Continued*)

Algal species	**Metabolites**
Botryococcus braunii	EPA, DHA, Butylated hydroxytoluene (BHT), Linolenic acid, α-linolenic acid
Scenedesmus obliquus	EPA, DHA, Vitamins, α-linolenic acid, Amino acids

The microalgae speciality lies in the fact that they possess a lot of valuable compounds which can be used in various industries, such as the cosmetic, pharmaceutical and food industries (Table 11.2) [6].

11.7 Factors Affecting Growth and Fatty Acid Composition of Microalgae

The yield of polyunsaturated fatty acids depends on various physicochemical parameters such as salinity, nitrate stress, media composition, temperature, pH, light, etc. (Figure 11.3). The high concentration of LCPUFAs is associated with high biomass yield, so optimisation of operational parameters is a necessity to enhance desired lipid content. Nitrate stress has been proved to increase the lipid content of alga [29]. Microalga biomass is going to be the best alternate, the sustainable and economic source for the production of these LCPUFAs. Factors affecting the growth of algae have been discussed below in detail.

Microalgae can tolerate a wide range of pH but it depends on species. Some eukaryotic species can thrive in an acidic environment (pH 5-7), whereas cyanobacteria can easily grow in an alkaline environment (pH 7-9). Rachlin & Grosso [30] studied the effect of pH on growth of *Chlorella vulgaris* and concluded that *Chlorella vulgaris* grow better at optimum pH, i.e., from 7.5-8.5. In acidic pH, growth rate of *Chlorella vulgaris* reduced by 27.3-55% whereas in alkaline conditions the growth rate reduced from 46%-37.2%.

Temperature is also one of the important factors responsible for the growth of microalgae. Microalgae can withstand a wide range of temperature (5-35 °C). But generally, microalgae exhibit its best growth at optimum temperature. Lipid content of *Nannochloropsis oculate* was found to increase two times when cultivated under optimised temperature

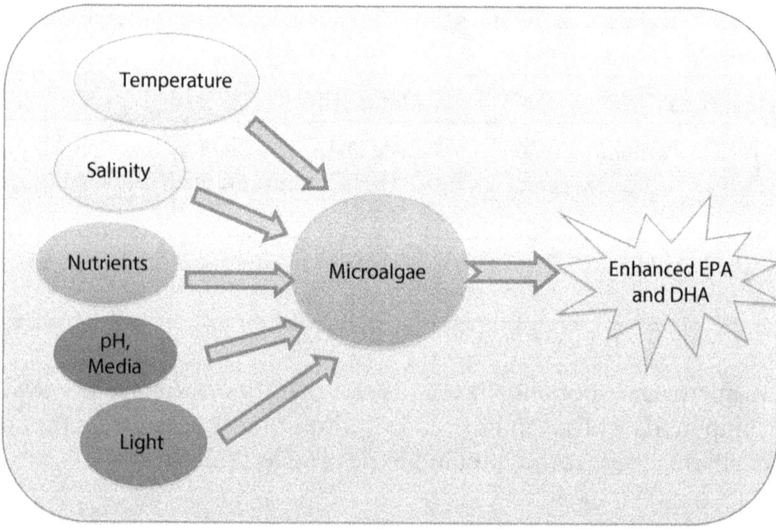

Figure 11.3 Optimisation of different operational parameters for enhanced EPA and DHA.

conditions [31]. *Rhomonas* species exhibited high lipid, carbohydrate, protein and chlorophyll contents at 25-27 °C [32]. Kitaya *et al.*, (2005) [33] carried out a study to find optimum temperature for high lipid accumulation of *Chlorella vulgaris* and *Nannochloropsis oculata*. The optimum temperature recorded for high lipid content was 30 °C and 27 °C, respectively. Salinity can also influence the biomass yield and lipid accumulation potential of microalgae. Microalgae isolated from freshwater sources can grow in fewer saline conditions whereas marine species of algae can grow in the high saline environment [34]. Some microalgal species can accumulate maximum lipid content under salinity stress (High NaCl). Lipid content of *Dunaliella Salina* was found to increase by 67% when cultivated under the increased concentration of NaCl (0.5-1.0 M) [35]. Oleic and palmitic concentration of *Botryococcus braunii* was found to increase by twofold and 1.7-2.25-fold under high salinity concentration of 34mM and 85mM, respectively [36]. Chemical composition of the medium is one of the major aspects affecting the growth of algae. Different media used for the algal growth are BG-11 medium, Allen medium, Fogg's medium, Bold basal medium, Noro medium, TAP medium, AF6 medium, Chu medium and Beneck's medium. The growth of *Chlorococcum* was found to be maximum in BG 11 medium when cultivated in five different media. Whereas maximum lipid content was observed in BBM medium [37].

Fatty acid composition and algal growth are also influenced by varying light intensity and photoperiod. These factors can influence the lipid biosynthetic route. Wahidin *et al.* (2013) [38] studied the effect of different photoperiod and light intensity on the lipid content of *Nannochloropsis* and found that lipid content increased up to 31.3% under 18h:16h light-dark cycle and 100 μmol $m^{-2}s^{-1}$ light intensity. Lipid content of *Isochrysis galbana* was enhanced when cultivated under alternate supply of blue light [39].

Algal growth is also affected by several nutrients like N, P, C Fe, Na, Ca, K, Mg, Cl, and micronutrients such as Mn, Zn, Cu, Mo, B, Co, Se and V.

Optimum nutrient concentration is required to maintain the cellular mechanism of algae-like respiration, photosynthesis, protein synthesis and transportation [40]. Nutrient starvation is one of the most widely accepted schemes for high lipid accumulation [41]. Lipid content was increased upto 16.41% in *C. vulgaris* and 15.31% in *N. oculata* when grown under nitrogen-limited environment Converti *et al.* (2009) [31]. Nitrogen-starved conditions enhanced the lipid content from 23-46% as was observed in *Chaetoceros mulleri* and biomass decreased from 19 $mgL^{-1}d^{-1}$ to 12 $mgL^{-1}d^{-1}$ [42]. Phosphorus starvation also leads to enhanced lipid content. Phosphorus contributes to energy generation, signal pathway and photosynthesis. Lipid content of Scendesmus sp. was found to increase by 53% under phosphorus starvation [43].

Iron supplementation in the medium enhances the lipid accumulation potential of algae by reducing the stationary period of algal growth. In a study carried out by [44], lipid concentration of *Chlorella vulgaris* was found to increase by 3-7-fold under high concentration of iron. Magnesium ions have the same effect on algal fatty acid composition as that of iron. Gorain *et al.* (2013) [45] conducted a study and observed that n lipid content (1.44-fold rise) increased with increase in the concentration of magnesium in the medium.

11.8 Algal Oil Extraction, Purification and Its Refining Techniques

The algal oil extraction for the production of DHA and EPA is a very challenging task due to the shorter life cycle of algae. The PUFAs are usually exposed to potential oxidation and reacted with oxidised radicals which in turn leads to the production of rancid, which is an odourless oil and not suitable for human health [22]. The oxidation process should be eliminated during the extraction process and oil storage areas. To improve colour, clarity, odour and remove particulate material and chemical contaminants, the crude algal oil is kept in a cool place, usually under a nitrogen blanket for refining. The pigments (i.e., carotenoids, chlorophyll), free fatty acids, phosphatides (i.e., lecithin), trace metals, sterols (i.e., cholesterol), and diacyl glycerides (MAGs and DAGs) and waxes, etc., contain various impurities. The chemical refining process is used to remove free fatty acids (FFAs) from the source material and physical process is not suitable for algal oils due to high temperature. The neutralization process is important to remove the contaminants and was carried out by the degumming process, wherein water is added to eliminate the phosphatides, sterols. Removal of the soap from oil using water followed by physical separation of the aqueous phase and drying under vacuum condition. The next step is bleaching which is used to remove the debris from the material by filtration, where absorbent clay or activated carbon is used to remove colour pigments, oxidation products and trace metals. The bleached oil is further de-waxed by winterisation to remove high melting waxes and to improve its clarity. To extract pure oil contents from algae, certain deodoriser has been added to the oil components under vacuum pressure.

11.9 Microalgae as a Boon for Long-Chain Omega-3 PUFAs

Omega-3 has been incorporated into pharmaceutical drugs in high concentration. In upcoming years several companies will hire the Omega-3. Some of the algal species can be considered as a potential source for EPA and DHA. Moreover, fish and krill have consumed algae directly. Different species of algae having the potential to produce the metabolites like EPA, DHA andα-linolenic acid has been elucidated in Table 11.2. The source of algae is immense scope for expanding omega-3 in the global market. Omega-3 (PUFAs) plays a vital role in maintaining good heart health, and protecting against cancer and certain birth defects. It will be growing continuously based on the research outcomes. Based on the regional market analysis, the Omega-3 Polyunsaturated Fatty Acid (PUFA) has been extracted from fish oils, krill oils and squid oils; and algal oils which possess EPA and DHA. The consumption of omega-3 in 2013 was 123.8 thousand metric tons, worth US$2.3 billion; in 2014 this increased to 134.7 thousand metric tons valued at US$2.5 billion and it is anticipated that these figures will double by 2020. From the statistical reports, it was found that there are 42 key global players and 106 major players across North America (29); Europe (35); Asia-Pacific (30) and Rest of World (12).

Microalgae are the promising source of value-added products such as antioxidants, fatty acids, carbohydrates, proteins and carotenoids [46]. The

Omega-3 long-chain polyunsaturated fatty acids (n-3 LC-PUFAs) play an important role in health benefits. This fatty acid is important in controlling and preventing mainly cardiovascular disease, and also hypertension hyperlipidemia, dementia and Alzheimer's disease, depression, cancer, and inflammation [47]. Algae have been used as both food and feeder for centuries. Algae can serve as alternative protein sources to meet global food demand [10]. Several species of *Chlorella*, *Spriogyra* and *Sprulina* have been used as a food source in India, Asia, the US, Israel and Australia (Enzing *et al.*, 2014) [49]. Microalgae have been considered an important component of food supplies in the EU food and feed market [48]. The potential role of microalgae for the production of value-added compounds is being widely discovered. The market scenario for commercial applications of microalgae is still not developed so well; still, research and innovation is actively engaged in exploring newer compounds innovation from algae [49]. The dried algae parts have been used as food and feeder for cattle. The production of value-added compounds from microalgae is receiving meager attention from researchers as compared to other crops. From the year 2000, the production of value-added products from algae have increased fivefold [24]. The habitats of algae make an important contribution towards the distribution of value-added compounds in the algal cells. The conditions of cultivation of algae in open and closed systems also play a vital role in the distribution of different compounds. The n-3 fatty acids [a-linolenic acid (ALA), eicosapentaenoic acid (EPA), docosapentaenoic acid (DPA n-3), and docosahexaenoic acid (DHA)] have become less abundant in American diets, and the average ratio of n-6 to n-3 fatty acids has increased from as little as 1:1 to as much as 30:1. Dietary intakes of n-3 and n-6 fatty acids are critical determinates of the proportions of bioactive 20- and 22-carbon n-6 and n-3 highly unsaturated fatty acids (HUFAs) in tissue phospholipids (2). Based on 1999 data, the ratios of total n-6 to n-3 and LA to ALA have been increased. The ratios of total n-6 to n-3 and LA to ALA have increased from 1909 to 1999. The ratio of total n-6 to n-3 has been increased 42% from 6.7 in 1909-C to 9.6 in 1999 and also the ratio of LA to ALA increased 38% from 7.3 in 1909-C to 10.0 in 1999 [49]. The percentage contribution of energy from each food category (or specific commodity) was calculated by dividing the calories contributed by each food category by the total estimated per capita calorie consumption for each year and then multiplying by 100%. Similarly, the percentage contribution of fatty acids from each food category was calculated by dividing the amount of the fatty acids derived from the food by the total estimated per capita consumption of fatty acids that year and multiplying by 100% [50, 51]. The EPA and DHA are considered as key components of various biological pathways in humans that affect health and disease. Fiester (2005) [52] suggested flaxseed as a source of omega-3 but flaxseed does not provide the same benefits as EPA and DHA. Mammals lack the conversion process of ALA to EPA and DHA. Animal and marine products are considered as main EPA and DHA sources.

References

1. Khan, W.A., Chun-mei, H., Khan, N., Iqbal, A., Lyu, S., Shah, F., Bioengineered Plants Can Be a Useful Source of Omega-3 Fatty Acids. *BioMed. Res. Intl*, 2017.
2. Kalogeropoulos, N., Chiou, A., Gavala, E., Christea, M., Andrikopoulos, N.K., Nutritional evaluation and bioactive microconstituents (carotenoids, tocopherols, sterols and squalene) of raw and roasted chicken fed on DHA-rich microalgae. *FRIN* 43, 2013.
3. Peltomaa, E., Johnson, M.D., Taipale, S.J., Marine Cryptophytes Are Great Sources of EPA and DHA. *Mar. Drugs*, 16(1), 3, 2018.
4. Matthan, N.R., Ooi, E.M., Horn, L. Van, Neuhouser, M.L., Woodman, R., Lichtenstein, A.H., n.d. Plasma Phospholipid Fatty Acid Biomarkers of Dietary Fat Quality and Endogenous Metabolism Predict Coronary Heart Disease Risk: Observational Study. *Am Heart J.*, 3(4), 2014.
5. http://www.fao.org/home/en/
6. Chisti, Y., Biodiesel from microalgae. *Biotechnol. Adv.* 25, 294–306, 2007.
7. Scott, S.A., Davey, M.P., Dennis, J.S., Horst, I., Howe, C.J., Lea-smith, D.J., Smith, A.G., Biodiesel from algae: challenges and prospects. *Curr. Opin. Biotechnol.* 21, 277–286, 2010.
8. Guedes, A.C., Amaro, H.M., Barbosa, C.R., Pereira, R.D., Malcata, F.X., Fatty acid composition of several wild microalgae and cyanobacteria, with a focus on eicosapentaenoic, docosahexaenoic and α-linolenic acids for eventual dietary uses. *FRIN* 44, 2721–2729, 2011.
9. Brennan, L., Owende, P., Biofuels from microalgae — A review of technologies for production, processing, and extractions of biofuels and co-products. *Renew. and Sust. Eng. Rev.*, 14, 557–577, 2010.
10. Burr, G.S., Barrows, F.T., Gaylord, G., Wolters, W.., Apparent digestibility of macro-nutrients and phosphorus in plant-derived ingredients for Atlantic salmon, Salmo salar and Arctic charr, Salvelinus alpinus. *Aquac. Nutr.* 17, 570–577, 2011.
11. Skrede, A., Mydland, L.T., Ahlstrøm, Ø., Reitan, K.I., Evaluation of microalgae as sources of digestible nutrients for monogastric animals. *J ANIM FEED SCI.* 20(1), 131-142, 2011.
12. Sharma, J., Kumar, S. S., Bishnoi, N. R., Pugazhendhi, A., Enhancement of lipid production from algal biomass through various growth parameters. *J. of Mol. Liq.*, 269, 712-720, 2018.

13. Michel, J., Microalgal Nutraceuticals, In *Handbook of Marine Microalgae*, Academic Press, pp. 255-267, 2015.
14. Abdo, S.M., Ali, G., El-baz, F.K., Potential production of omega fatty acids from microalgae. *Int. J. Pharm. Sci. Rev. Res*, 210-215, 2015.
15. Sayeda, M. A., Ali, G. H., & El-Baz, F. K., Potential production of omega fatty acids from microalgae. *Int. J. Pharm. Sci. Rev. Res*, 210-215, 2015.
16. Chauton, M., Chauton, M.S., Inge, K., Henrik, N., Tveterås, R., Kleivdal, H.T., A techno-economic analysis of industrial production of marine microalgae as a source of EPA and DHA-rich raw material for aquafeed: Research challenges and possibilities A techno-economic analysis of industrial production of marine microalgae as a source. *Aqua*, 436, 95–103, 2017.
17. Bhatia, H.S., Agrawal, R., Sharma, S., Huo, Y., Ying, Z., Omega-3 Fatty Acid Deficiency during Brain Maturation Reduces Neuronal and Behavioral Plasticity in Adulthood 6. *PloS one*, 6(12), 2011.
18. Sidhu, K.S., Health benefits and potential risks related to consumption of fish or fish oil. *Regul Toxicol Pharm*, 38, 336–344, 2003.
19. Ruxton, C.H.S. Reed, S.C., Simpson, M.J.A. & Millington, K.M., The Health benefits of Omega-3 polyunsaturated fatty acids: a review of the evidence. *J Hum Nutr Diet*. 20, 275–285, 2004.
20. Lavie, C.J., Milani, R.V., & Ventura, H.O., Obesity and cardiovascular disease: risk factor, paradox, and impact of weight loss. *J. Am. Coll. Cardiol*, 53(21), 1925-1932, 2009.
21. Swanson, D., Block, R., Mousa, S.A., Omega-3 Fatty Acids EPA and DHA: Health Benefits Throughout Life. *Adv. in Nut.*, 1(2) 1–7, 2012.
22. Bishop, K.S., Erdrich, S., Karunasinghe, N., Han, D.Y., Zhu, S., Jesuthasan, A., Ferguson, L.R., Cancer, A., Zealand, N., Zealand, N., Zealand, N.N., Zealand, N., An Investigation into the Association between DNA Damage and Dietary Fatty Acid in Men with Prostate Cancer. *Nutrients*, 7(1), 405–422, 2015.
23. Nabavi, S.F., Bilotto, S., Russo, G.L., Orhan, I.E., Habtemariam, S., Daglia, M., Devi, K.P., Loizzo, M.R., Tundis, R., Nabavi, S.M., Russo, G.L., Omega 3 polyunsaturated fatty acids and cancer: lessons learned from clinical trials. *Canc. and Met. Rev.*, 34(3), 359-380, 2015.
24. Singh, B., Guldhe, A., Rawat, I., Bux, F., Towards a sustainable approach for development of biodiesel from plant and microalgae. *Renew. Sus.. Eng Rev*. 29, 216–245, 2014.
25. Choi, H.G. & Yu, S., Influence of crude glycerol on the biomass and lipid content of microalgae. *Biotechnol. Equip*. 29, 37–41, 2015.
26. Abou-shanab, R.A.I., Ji, M., Kim, H., Paeng, K., Jeon, B., Microalgal species growing on piggery wastewater as a valuable candidate for nutrient removal and biodiesel production. *J. Environ. Manage*. 115, 257–264, 2013.
27. Bigogno, C., Khozin-goldberg, I., Boussiba, S., Lipid and fatty acid composition of the green oleaginous alga Parietochloris incisa, the richest plant source of arachidonic acid. *Phytochem*, 60, 497–503, 2002.

28. Lang, I., Hodac, L., Friedl, T., Feussner, I., 2011. Fatty acid profiles and their distribution patterns in microalgae: a comprehensive analysis of more than 2000 strains from the SAG culture collection. *BMC plant biol,* 11(1), 124, 2011.
29. Sharma, J., Kumar, V., Kumar, S. S., Malyan, S. K., Mathimani, T., Bishnoi, N. R., & Pugazhendhi, A., Microalgal consortia for municipal wastewater treatment–Lipid augmentation and fatty acid profiling for biodiesel production. *J. of Photochem. and Photobiol. B: Biol.*, 111638, 2019.
30. Rachlin, J.W. & Grosso, A., The Effects of pH on the Growth of Chlorella vulgaris and Its Interactions with Cadmium Toxicity. *Arch. Environ. Contam. Toxicol.* 508, 505–508, 1991.
31. Converti, A., Casazza, A.A., Ortiz, E.Y., Perego, P., Borghi, M. Del, Effect of temperature and nitrogen concentration on the growth and lipid content of Nannochloropsis oculata and Chlorella vulgaris for biodiesel production. *Chemical Engineering and Processing: Process Intensification* 48, 1146–1151, 2009.
32. Renaud, S.M., Thinh, L., Lambrinidis, G., Parry, D.L., Effect of temperature on growth, chemical composition and fatty acid composition of tropical Australian microalgae grown in batch cultures. *Aqua.*, 211, 195–214, 2002.
33. Kitaya, Y., Azuma, H., Kiyota, M., Effects of temperature , CO_2 / O_2 concentrations and light intensity on cellular multiplication of microalgae, Euglena gracilis. *Adv Space Res.,* 35, 1584–1588, 2005.
34. Rashid, N., Ur Rehman, M.S., Sadiq, M., Mahmood, T., Han, J.I., Current status, issues and developments in microalgae derived biodiesel production. *Renew. Sustain. Energy Rev.* 40, 760–778, 2014.
35. Takagi, M., Yoshida, T., 2006. Effect of Salt Concentration on Intracellular Accumulation of Lipids and Triacylglyceride in Marine Microalgae Dunaliella Cells. *J. Biosc. Bioeng.*, 101, 223–226, 2006.
36. Rao, A.R., Dayananda, C., Sarada, R., Effect of salinity on growth of green alga Botryococcus braunii and its constituents. *Biores.* 98, 560–564, 2007.
37. Kirrolia, A., Bishnoi, N.R., Singh, R., Effect of shaking, incubation temperature, salinity and media composition on growth traits of green microalgae Chlorococcum sp . *J. Algal. Biom. Utiliz.,* 3, 46–53, 2012.
38. Wahidin, S., Idris, A., Raehanah, S., Shaleh, M., The influence of light intensity and photoperiod on the growth and lipid content of microalgae Nannochloropsis sp. *Bioresour. Technol.*, 129, 7–11, 2013.
39. Yoshioka, M., Yago, T., Yoshie-stark, Y., Arakawa, H., Morinaga, T., Effect of high frequency of intermittent light on the growth and fatty acid profile of Isochrysis galbana. *Aqua.*, 338, 111–117, 2012.
40. Show, P.L., Tang, M.S.Y., Nagarajan, D., Ling, T.C., Ooi, C., n.d. A Holistic Approach to Managing Microalgae for Biofuel Applications. *Int. J. of Mol. Sc.*, 18(1), 215, 2017.
41. Singh, Priyanka, Kim, Y.J., Zhang, D., Yang, D.C., Biological Synthesis of Nanoparticles from Plants and Microorganisms. *Trends Biotechnol.* 34, 588–599, 2016.

42. Gao, Y., Yang, M., Wang, C., Bioresource Technology Nutrient deprivation enhances lipid content in marine microalgae. *Bioresour. Technol.* 147, 484–491, 2013.
43. Xin, L., Hong-ying, H., Jia, Y., Lipid accumulation and nutrient removal properties of a newly isolated freshwater microalga, Scenedesmus sp. LX1, growing. *N. Biotechnol.* 27, 59–63, 2010.
44. Liu, A., Chen, W., Zheng, L., Song, L., Identification of high-lipid producers for biodiesel production from forty-three green algal isolates in China. *Prog. Nat. Sci. Mater. Int.* 21, 269–276, 2011.
45. Gorain, P.C., Bagchi, S.K. &, Mallick, N., Effects of calcium , magnesium and sodium chloride in enhancing lipid accumulation in two green microalgae. *Environ. Technol.* 37–41, 2013.
46. Sarmah, P., Rout, J., Biochemical profile of five species of cyanobacteria isolated from polythene surface in domestic sewage water of Silchar town, Assam (India). *Curr. Trends Biotechnol. Pharm.* 12, 7–15, 2018.
47. McManus, A., Howieson, J., Nicholson, J., Review of literature and resources relating to the health benefit of regular consumption of seafood as part of a healthy diet. Center of Excellence Science Seafood & Health (CESSH), Curtin Health Innovation Research Institute. 1–113, 2009.
48. Draaisma, R.B., Wijffels, R.H., Slegers, P.M., Brentner, L.B., Roy, A., Barbosa, M.J., Food commodities from microalgae. *Curr. Opin. Biotechnol.* 24, 169–177, 2013.
49. Enzing, C., Ploeg, M., Barbosa, M., Sijtsma, L., Micro-algal production systems, Microalgae-based products for the food and feed sector: an outlook for Europe. *JRC Scientific and Policy Reports,* 19–37, 2014.
50. Cherubini, F., The biorefinery concept: using biomass instead of oil for producing energy and chemicals. *Energ Convers Manage.*, 51(7), 1412–1421, 2010.
51. Stevens, J. R., & Stephens, D. W., The adaptive nature of impulsivity. 2010.
52. Fiester, A., Ethical issues in animal cloning. *Perspect. Biol. Med.* 48, 328–343, 2005.
53. Krzemińska, I., Pawlik-Skowrońska, B., Trzcińska, M., Tys, J., Influence of photoperiods on the growth rate and biomass productivity of green microalgae. *Bioprocess Biosyst. Eng.* 37, 735–741, 2014.
54. Orcutt, D. M., & Patterson, G. W., Sterol, fatty acid and elemental composition of diatoms grown in chemically defined media. *Comp. Biochem. Physiol. B, Biochem.*, 50(4), 579–583, 1975.
55. Renaud, S.M., Thinh, L. Van, Parry, D.L., The gross chemical composition and fatty acid composition of 18 species of tropical Australian microalgae for possible use in mariculture. *Aqua.*, 170, 147–159, 1999.
56. Sathasivam, R., Radhakrishnan, R., Hashem, A., Abd_Allah, E.F., Microalgae metabolites: A rich source for food and medicine. *Saudi J. Biol. Sci.* 26, 709–722, 2019.

12

Oleaginous Microalgae – A Potential Tool for Biorefinery-Based Industry

Riti Thapar Kapoor

Assistant Professor, Amity Institute of Biotechnology, Amity University, Noida, Uttar Pradesh, India

Abstract

Rapid depletion in fossil fuel reserves due to continuously increasing population and change in their lifestyle have prompted scientists and policymakers to explore unconventional energy resources. Biofuel as inexhaustible energy is the significant substitute for fossil fuel in a recent scenario to explore green energy resources. Oleaginous microalgae synthesize lipid, sugar and protein, thus attracting interest in biofuel production, nutraceuticals formulation and bioremediation. Oleaginous microalgae show rapid growth, require less land area for large-scale cultivation and convert solar energy into fatty acids which are turned to biodiesel via trans-esterification process. Hence, biofuel production from microalgae has several advantages such as the pruning of the nation's dependence on fossil fuel imports, job opportunities and an increase in income of poor farmers.

Keyword: Biofuel, industrial applications, oleaginous microalgae

12.1 Introduction

Environmental pollution due to anthropogenic activities, urbanization and rapid industrialization has a detrimental effect on the health of human beings and the ecosystem. Energy acquisition is a driving force for the socioeconomic development of the nation [1]. Recently expenditure of energy has been doubled in which fossil fuel consumption is more than 80% [2, 3]. The demand for fuel was 85 million barrels of liquid fuel/day

Email: rkapoor@amity.edu

Alok Kumar Patel and Leonidas Matsakas (eds.) Nutraceutical Fatty Acids from Oleaginous Microalgae: A Human Health Perspective, (299–330) © 2020 Scrivener Publishing LLC

in 2006 and it is predicted that it will increase to 107 million barrels by 2030 [4]. The International Energy Agency [5] reported that there will be a 50% rise in energy demand especially by China and India by 2030 [6]. The burning of fossil fuels leads to generation of greenhouse gases such as CO_2, NO, volatile organic compounds and hydrocarbons which may increase earth's temperature and are responsible for global warming [7]. This has created significant interest in recommending biofuel as one of the sustainable energy sources [8]. Biofuel has emerged as an unconventional fuel to fossil fuel [9, 10] as it is a benign, bio-degradable and carbon-neutral energy resource [11]. It can be directly used in diesel engine without any moderation [12]. Biodiesel is better than common diesel as these are composed of oxygenated esters of long-chain fatty acid [13]. Biodiesel combustion in the diesel engine can withdraw unburned and polycyclic aromatic hydrocarbons [14]. It can reduce particulate matter and CO emission and it does not produce any chemical substances that are harmful to the environment [15, 16]. Developed nations such as the United States, Brazil, Malaysia and European countries are using biodiesel as it does not affect the environment [17, 18]. More than three hundred plants have been recognized as the main source for biodiesel production [19]. Several food crops such as soybean, sunflower, cottonseed, coconut and groundnut have been used as a feedstock for biodiesel but they are not desirable due to food verses fuel conceptualization [20, 21]. Edible oil becomes unfeasible in the long run due to an increase in the gap between demand and supply ratio [22]. High price of edible oil adversely affected the cost of biodiesel; due to this non-edible oil-containing crops such as jatropha, rice bran, linseed and castor were used for biodiesel production and these are regarded as second-generation biodiesel sources [23–25]. These are easily available as feedstock and can be produced on wasteland without the application of fertilizer and are less expensive in comparison to edible oil (Table 12.1). The biodiesel generation from non-edible oil crops requires chemical processing due to the presence of free fatty acids [26]. Due to the high impurities, its purification is needed which may add to the cost of biodiesel production. Biofuel production from food crops grown in agricultural land will affect food security and prices whereas the cultivation of non-edible oil crops will result in competition with food crops. Due to the above-mentioned challenges, microalgae cultivation has received global attention as a biofuel feedstock (Table 12.2).

Microalgae have been considered as raw material for third-generation biodiesel production [29, 30] and fourth-generation focuses on the development of engineered genomes of microalgae which may lead to maximum biofuel production at minimum cost [31]. Microalgae are photosynthetic microbes that can form biomass by water and CO_2 in the presence of

Table 12.1 Utilization of non-edible and edible crops for biofuel production [27].

Crops	Oil yield (weight %)
Non-edible crops	
Jatropha	28
Hemp	33
Castor	48
Edible oil crops	
Sunflower	40
Soybean	18
Palm	36
Rapeseed	41
Corn	44

Table 12.2 Lipid content in microalgae [27, 28].

Microalga	Oil content (% dry weight)
Botryococcus braunii	25-75
Chlorella species	28-32
Crypthecodinium cohnii	20
Cylindrotheca species	16-37
Isochrysis species	25-33
Monallanthus salina	20
Nannochloris species	25-35
Nannochloropsis species	31-68
Neochlorisoleo abundans	35-54
Nitzschia species	45-47
Schizochytrium species	50-77

sunlight [29]. They can be easily cultivated and fulfil the global biofuel requirement [29]. Reports revealed microalgae cultivation exhibits twenty -five times higher yield than palm oil and two hundred fifty times more than soybean oil [32]. Kligerman and Bouwer [33] stated that microalgae play a significant role in wastewater remediation and biodiesel production. Singh *et al.* [34] reported that microalgae can generate thirty times more energy/unit area in comparison to first- and second-generation biofuel. Recent studies reported an increase in biomass, lipid accumulation and carbon capture capacity can be obtained by genetic modification of microalgae [35, 36]. A biofuel is a precious tool in checking global warming [37], augmenting economy of our country and enhancing energy security of the nation [38]. The glycerol can be obtained after the production of biodiesel which can be used in the formulation of cosmetics and lubrication of machines. Thus, the use of microalgae is helpful in cost-effective biofuel production and environment protection [39].

12.2 Industrial Applications of Microalgae

The use of *Nostoc* microalgae by the Chinese for their survival during famine about two thousand years ago has been reported. Aztecs used *Spirulina* (*Arthrospira platensis* and *Arthrospira maxima*) from Texcoco lake of Mexico in AD 1300. For the preparation of "tecuitlat" Spanish chroniclers were used to collect microalgae from the lakes. The Chad population of Africa used to collect *Spirulina* from Kossorom lake and consumed it as a food material. Different species of *Nostoc* such as *commune, flagelliforme* and *punctiforme* are consumed in China, Mongolia and the US as a foodstuff. *Spirogyra* and *Oedogonium* are used as a food component in Burma, Thailand, Vietnam and India. The biotechnological applications of microalgae began in 1940 and gained momentum with the first seminar on algae cultivation, which was held at Stanford University in 1952. *Spirulina* was declared as the best food at the United Nations World Food Conference in 1974 and WHO stated that *Spirulina* is a rich source of iron and protein and can be given to children without any side effect.

12.3 Use of Microalgae as Biofertilizer

Most of the microalgae can fix atmospheric nitrogen and can be utilized as biofertilizer. Cyanobacteria act as a soil conditioner and maintain soil fertility [40]. *Nostoc, Anabaena, Tolypothrix* and *Aulosira* used as inoculants in rice fields as they can fix a significant amount of atmospheric nitrogen. Microalgae enhances the growth of crop plants by production of growth-promoting substances like vitamin B_{12}, 3-IAA, 3-IPA and 3-methyl indole etc. The thermo-chemical decay of algal biomass at elevated temperature without oxygen leads to the formation of biochar which acts as a bio-fertilizer and enhances carbon sequestration.

12.4 Microalgae as a Food Component

Microalgae have been used as a nutritional supplement throughout the world as they are rich sources of carbohydrate, protein and fibre. Microalgae also contain vitamins and minerals potassium, iron, iodine, magnesium and calcium. *Chlorella vulgaris, Haematococcus pluvialis, Dunaliella salina* and *Spirulina maxima* are widely used as a food supplement for human beings. *Spirulina platensis* is an eminent source of protein [41], pigments [42], fatty acids [43], vitamins and phenolic compound [44]. *Dunaliella salina*, halophilic microalgae is a potential source of beta-carotene which can be used as orange coloured dye and vitamin C supplement. *Chlorella* and *Spirulina* are used in production of powder, tablet and capsules due to their high protein content and nutritive value [45]. Microalgae are also incorporated in formulation of pasta, snacks and drinks either as food colourant or nutritional supplement [46].

12.5 Microalgae as a Nutraceutical

The market of nutraceuticals is considered as the most dynamic sector in the food industry, forming 20% of the total food market. The global market for nutraceuticals was around $250 billion in 2014 and it will increase to $385 billion by 2020 due to rise in demand by consumers [47]. Nutraceutical has emerged as a tool for health improvement and disease prevention [48]. Microalgae are rich in polyunsaturated fatty acids [49], carotenoids [50], phycobilins [51], essential amino acids [52] and vitamin [46]. Microalgae have been used in preparation of confectionery products, high protein beverages and dry milk products in which soybean is the main constituent [53]. Microalgae are rich sources of vitamin A, vitamin C, vitamin B complex, vitamin D and E [54]. Vitamin B12 helps in DNA repair, and a high level of folate and vitamin B12 is related to reduction in risk of breast cancer [55]. *Dunaliella tertiolecta, Tetraselmis suecica, Chaetoceros calcitrans* and *Porphyridium cruentum* synthesize vitamin E [56]. Astaxanthin shows a hundred times more antioxidant activity in comparison to vitamin E and protects against diseases such as diabetes, heart diseases, obesity, cancer, neurodegenerative and eye diseases [48]. Diatoms contain fucoxanthin, which shows antidiabetic, anti-photoaging, anticancer and neuroprotective characteristics [57, 58].

Phycobiliproteins are found in *Spirulina, Porphyridium, Galdieria*, cryptophyta and glaucophyta which shows anti-inflammatory, hepatoprotective, neuroprotective and atheroprotective properties [59]. The peptides of marine algae act as antihypertensive, antitumor, antimutagenic and anti-coagulant agents [60]. *Pavlova, Tetraselmis* and *Nannochloropsis* species contain phytosterol which lowers cholesterol and prevents cardiovascular diseases. Phytosterol acts as secondary messenger and helps in neurotransmission [61]. *Chlorella* and *Spirulina* have been used in formulation of antioxidant capsules. Spirulina biomass as an extract or its processing with confectionary food products enhances functionality of digestive tract and helps in maintenance of intestinal health. *Spirulina* and *Chlorella* are used as food colorant, production of health drink and green tea [62]. *Dunaliella salina* is rich in α-carotene, β-carotene, lutein, lycopene and vitamin C [27]. Bishop *et al.* [63] reported that β-carotene obtained from *Dunaliella* inhibits cancer cells. *Scenedesmus* contains astaxanthin antioxidant and vitamin [64]. *Aphanizomenon flos-aquae* has omega-3 alpha-linolenic acid which is required for good health [65].

Microalgae are unexploited sources which can serve as functional compound and provide active ingredients for combating malnutrition Figure 12.1. The proteins of microalgae help in repairing of degraded tissues. Microalgal polyglycans are used in prebiotic supplement and promote gut microflora [66]. Microalgae reduces hypertension, blood glucose level, protects against renal failure, increases growth of intestinal *Lactobacillus* and alleviation of hyperlipidemia [62]. *Chlorella* has β-1,3-glucan, which acts as immune-stimulator and free radical scavenger reduces lipid level in blood [67]. Yamaguchi [68] reported that *Chlorella* protects from atherosclerosis, hypercholesterolemia and tumor.

12.6 Pigments and Carotenoids

Microalgae are an abundant source of chlorophyll and carotenoids. *Chlorella* and *Spirulina* were used as health food in Japan, Taiwan and Mexico [69]. The commercialization of β-carotene obtained from *Dunaliella salina* was done in the 1980s [70] and astaxanthin from *Haematococcus pluvialis* [71] in the 1990s. Carotenoid and pigment act as antioxidant which protects against chronic diseases [72]. Due to the rich protein content, *Arthrospira* is an important component of diet [73]. *Dunaliella* powder as food supplement for human consumption is supplied by Cognis Nutrition and Health company. *Aphanizomenon flos-aquae* promotes good health in human beings [74].

Several other carotenoids have been obtained such as lutein from *Muriellopsis* speciesand *Scenedesmus almeriensis* [75], *Chlorella zofigiensis* yields astaxanthin, canthaxanthin and lutein [76], *Scenedesmus komareckii* gives canthaxanthin [77], zeaxanthin, phytoene and phytofluene from *D. salina* [78], *Botryococcus braunii* produces echinenone [79] and fucoxanthin from *Phaeodactylum tricornutum* [80]. Fucoxanthin is another microalgal pigment which helps in weight loss [81].

12.7 Phycobilins

Phycocyanin, phycoerythrin and allophycocyanin are water-soluble pigments present in *Arthrospira* and *Porphyridium* [82]. These have wide application as food supplements and in cosmetic industries. These are used in flow cytometry and immunology experiments [83] and used as photosensitisers in photodynamic therapy for cancer treatment [84]. *Spirulina* is a rich source of phycocyanin [85]. Phycobilins are used in industry, clinical and immunology laboratories for research work.

12.8 Fatty Acids

Microalgae contain lipids in the range between 1-70% of dry weight but may increase up to 90% under specific conditions [86]. Plants and animals are unable to synthesize polyunsaturated fatty acids (PUFA) due to lack of requisite enzymes. Fish oil contains long-chain polyunsaturated fatty acids but due to unpleasant smell and taste, oxidative unstability and toxins accumulation, fish oil is not widely used [87]. Microalgae are rich in long-chain PUFA such as γ-linolenic acid, arachidonic acid, eicosapentaenoic and docosahexaenoic acid which are used to cure diabetes, respiratory disorders, obesity, skin infections, arthritis, Parkinson's and heart diseases [88]. They also yield different types of sterols such as brassicasterol, sitosterol and stigmasterol which show pharmaceutical applications [89]. The phytosterol market is currently about US$ 300 million and continuously growing at 7-9% per annum. *Porphyridium purpureum, Phaeodactylum tricornutum, Isochrysis galbana, Nannochloropsis* species and *Nitzschia laevis* have significantly produced EPA [90].

Omega-3 fatty acid derives EPA and DHA which helps in lowering of blood cholesterol and development of fetal brain [91]. Omega-6 fatty acid produces arachidonic acid which initiates prostaglandin and leucotriene synthesis and improves function of central nervous system [92]. *Crypthecodinium cohnii, Isochrysis galbana, Phaeodactylum tricornutum* and *Porphyridium cruentum* significantly produce arachidonic acid and *Monoraphidium, Scenedesmus* and *Nannochloropsis* species produce PUFA [93]. Seambiotic and Aurora companies are producing EPA at pilot-scale and rest biomass of microalgae is used in production of biofuel [94]. Transparency Market Research project reveals that the global nutraceuticals market will cross US$279 billion with annual growth rate of 7.3% by the end of 2021. The EPA/DHA market will increase up to 11% with sales US$4 billion by 2022 [95].

12.9 Animal Nutrition

Application of nutraceuticals is not limited to human beings but it plays a significant role in the animal health area. Approximately 30% of microalgae are used for animal feed as reported by Becker [46]. Microalgae are mainly used for cultivation of larvae, juvenile shell, finfish and zooplanktons which can be used as a feed material for juvenile animals [96]. *Chlorella, Tetraselmis, Isochrysis, Phaeodactylum, Nannochloropsis, Skeletonema* and *Thalassiosira* are also used for animal feed. Microalgae

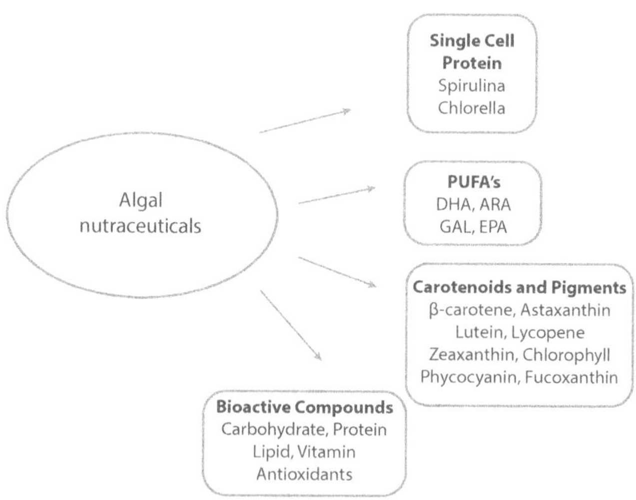

Figure 12.1 Types of microalgae nutraceuticals [74].

improves external appearance, physiology, immunity and fertility of animals [97]. Arthrospira enhances red and yellow pattern in carp whereas *Haslea ostrearia* increases blue-green colouration on gills and labial palps of oyster and enhances its market value by 40% [98].

Dunaliella *salina, Haematococcus pluvialis* and *Spirulina* are used for cultivation of prawn, salmonid fish and ornamental fish. Astaxanthin pigment obtained from *Haematococcus* species is used in aquaculture and poultry industries as a feed supplement for salmon, trout and shrimp.

12.10 Safety Related Issues Related to Microalgal Nutraceuticals

The allergenicity reports due to microalgae consumption are very meagre. However, evaluation of the presence of toxins, heavy metal and pathogenic microbes in the nutraceuticals prepared from microalgae are needed. Some reports reveal development of anaphylaxis and poisoning symptoms after consumption of wild *Spirulina* and it may be due to neurotoxins production by its contamination with *Microcystis* [99]. The high doses of kainic and domoic acids obtained from red algae showed neurotoxic effects on mice and other animal models [100]. Cyanotech, Earthrise Farms and Solazyme companies supply spirulina and Algalin protein in the market and this is generally regarded as safe (GRAS). Hence, there is a need that microalgal nutraceuticals should be developed under controlled cultivation conditions coupled with regular monitoring of quality parameters [101].

12.11 Application in Pharmaceutical Industry

Microalgae are a rich source of novel bioactive compounds which are utilized in the pharmaceutical industry [102]. *Arthrospira, Botryococcus braunii, Chlorella vulgaris, Dunaliella salina, Haematococcus pluvialis* and *Nostoc* were used for the isolation of bioactive compounds [103]. The bioactive compounds are mainly secondary metabolites such as organic acids, amino acids and enzymes, etc. The bioactive components exhibit antimicrobial, antiviral, antifungal, antioxidant and anti-inflammatory activities [104]. *Dunaliella salina* produces β-carotene [86]. *Haematococcus* species are abundant in astaxanthin and this compound exhibits better antioxidant activity in comparison to vitamin E [105]. Researches revealed that astaxanthin protects the human body against different diseases such as cancer, diabetes, neurodegenerative and eye disorders [106]. *Spirulina* contains phycocyanin and vitamin E and it shows antioxidant activity [107]. *Spirulina* is used in the treatment of diabetes, arthritis, hypertension, anaemia, cardiovascular diseases and cancer [108]. Microalgae are abundant source of antiviral compounds such as cyanovirin-N which works against human immunodeficiency virus [109]. The sulfur containing polysaccharides obtained from *Spirulina platensis* have antiviral activity which protects cell from the entry of cytomegalo, Herpes and measles virus [110]. The compounds produced from marine cyanobacteria is effective in killing cancer cells by inducing protein kinase C (PKC) signaling [111]. Cryptophycin is an anticancer compound extracted from *Nostoc* ATCC 53789 [112] whereas scytonemin isolated from *Stigonema* shows anti-inflammatory properties [113].

12.12 Utilization of Microalgae in Cosmetics Production

Microalgae are used in the formulation of cosmetics [27]. Microalgae extracts are used in formulation of sun protection cream, anti-aging cream and hair care products. *Chlorella* and *Arthrospira* have been utilized in the production of skin care products [114]. The cosmetic products made up of *Arthrospira* remove skin aging signs and exert a tightening effect; similarly *Chlorella vulgaris* promotes collagen synthesis which helps in regeneration of tissues and ridge inhibition. Penta-pharm (Switzerland) company commercialized skincare products from *Nannochloropsis oculata* and *D. salina* which shows skin-tightening properties and promotes energy metabolism of skin [114].

12.13 Microalgal Application in Wastewater Treatment

Water pollution due to manufacturing industries has become a serious environmental issue. Dye present in water bodies exhibits toxicity to aqueous environment but does not reduce growth of microalgae [115]. Therefore, algae are a better alternative for bioremediation of coloured wastewater in contrast to bacteria and fungi as they do not need extra carbon sources [116]. Algae generates energy from sunlight and its mass cultivation is not expensive. *Chlorella* and *Oscillitoria* can degrade azo dyes to aromatic amines which further converts into simple compounds or CO_2 [117]. More than thirty azo compounds can be degraded by *Chlorella pyrenoidosa, Chorella vulgaris* and *Oscillateria tenuis* [117]. The biosorption of dye by microalgae can be a productive and cost-effective method for decolourization of industrial effluent; it may be a more feasible alternative than other available material.

12.14 Factors Affecting Lipid Production in Microalgae

Microalgal species can survive under diverse environmental conditions. The lipid production in microalgae is dependent on species and can be affected by different factors such as light intensity, temperature, nutrient availability, salt and metal stress.

12.14.1 Light Intensity

Light is needed for microalgae growth during photosynthesis. The sufficient light intensity supports the lipid production in microalgae as it enhances storage of photoassimilates, which are further converted into chemical energy [118]. *Nannochloropsis* species showed accumulation of highest amount of lipids (47% of dry weight) under high light intensity [119]. Lipid content in *Scenedesmus abundans* was significantly increased with increase in light intensity [120]. *Botryococcus* species showed maximum lipid content (35.9%) at 6000 lux [121]. Sun *et al.* [122] found 33% lipid production in *Neochloris oleoabundans* HK-129 at 14,800 lux intensity. The high light intensity causes photoinhibition, damages microalgal photosystem and reduces lipid accumulation in microalgae [123]. The low light intensity forms polar lipids whereas high light intensity increases triacyglycerol accumulation [124].

12.14.2 Temperature

Microalgae growth and lipid production increase with increase in temperature [125]. *Chlorella vulgaris* accumulated maximum lipids at 25°C whereas decrease in temperature resulted in reduction in lipid content [126]. In *Scenedesmus obliquus*, lipid content was 18-40% dry weight, when temperature was enhanced from 20 to 28°C [123]. Converti *et al.* [126] found lipid content in *Nannochloropsis oculata* was increased from 7.9-14.9% with increase in temperature from 20-25°C. It has been observed that a limited amount of CO_2 reduces metabolism of microalgae and decreases lipid content [127]. Widjaja *et al.* [128] stated growth and lipid production in *Chlorella vulgaris* were enhanced with increased CO_2 concentration. *Chlamydomonas* species showed 65% lipid content and productivity was 169 mg/ liter/day with 4% CO_2 supply [129]. Maximum lipid content (34% wt) was obtained when *Chlorococcum littorale* was cultivated with 5% CO_2 [130]. Ho *et al.* [131] grew *Sclerosaurus armatus* at 2% CO_2 concentration and 22.4% lipid content was obtained. The lipid contents in *Ochromonas danica* [132] and *Nannochloropsis salina* [133] were increased under high temperature.

12.14.3 Nutrient Availability

Nutrient starvation or limitation is a viable approach to increase lipid content with control of cell cycle [134]. Under nutrient stress, triacylglycerols are formed [135]. The microalgal species can improve lipid accumulation and undergo transformation under nutrient stress [136]. Under nutrient deficiency *Dunaliella salina* showed decrease in lipid content from 25 to 9% but carbohydrate content was increased from 16-56% [137]. The limitation in nitrogen, phosphorus and sulfur are recognized as a lipid induces in microalgae species (Table 12.3). Mandal and Mallick [138] cultivated *Scenedesmus obliquus* under phosphorus starvation and found significant enhancement in lipid content from 10 to 29.5%. Sato *et al.* [139] found that sulfur starvation enhanced TAG accumulation in *Chlamydomonas reinhardtii.* Pribyl *et al.* [140] reported 57.25% increase in lipid content when *Chlorella vulgaris* was cultivated under nutrient depletion condition for one week. Rodolfi *et al.* [141] reported that decrease in nitrogen and silicate contents increased lipid content in algae. The nitrogen deficiency enhanced lipid content from 32-60% in *Nannochloroposis* [141]. Significant growth rate with high lipid contents was observed in *Ellipsoidion* when NH_4^+ was used as compared to nitrate and urea as nitrogen source [142].

12.14.4 Salinity Stress

The salinity stress enhances lipid content in microalgae. Pal *et al.* [119] studied salt stress effect on the growth of *Nannochloropsis* and reported maximum 47% fatty acid content with 13 g/liter NaCl. Kirrolia *et al.* [144] reported significant rise in lipid content in *Scenedesmus* species under salinity. Takagi *et al.* [145] observed 70% elevation in intracellular lipid content in *Dunaliella* cells under salt stress. *Chlorella vulgaris* showed 21% increase in lipid content under NaCl stress [146].

Table 12.3 Microalgae-based nutraceuticals and their application [74].

Nutraceuticals	Product name	Microalgae	Applications
Polysaccharides	Carragenan/alginate	*Porphyridium cruentumj, Rhodella* species	Thickeners, flocculant, lubricant
Amino acids	Mycosporine-like amino acids	*Aphanizomenon flos-aquae*	Sunscreen agent
Phycobilins	Phyocyanin, Fhycoerythrin	Cyanobacteria, Rhodophyta, Cryptophyta, Glaucophyta	Natural pigment, antioxidant
Polyhydroxyalkonate	Ester	*Nostoc* species, *Synechocystis*	Biodegradable plastics
Carotenoids	β-carotene	*Dunaliella salina*	Natural food colorant, antioxidant, anti-cancer properties
	Astaxanthin	*Chlorella zofingiensis Haematococcus pluvialis*	Pigment and antioxidant
	Canthaxanthin	*Chlorella* species	Pigmenter

(*Continued*)

Table 12.3 Microalgae-based nutraceuticals and their application [74]. (*Continued*)

Nutraceuticals	**Product name**	**Microalgae**	**Applications**
	Zeaxanthin	*Chlorella ellipsoidea Dunalielle salina* (mutant)	Anti-oxidant, food pigmenter
	Lutein	*Scenedesmus, Muriellopsis* species	Antioxidant
	Phytoene	*Dunalielle*	Antioxidant
	Echinenone	*Botryococcus braunii,*	Antioxidant
	Fucoxanthin	*Phaeodactylum tricornutum*	Antioxidant
Fatty acids	Arachidonic acid (AA)	*Parietochloris incisa*	Nutritional supplement
	Eicosapentaenoic acid (EPA)	*Nannochloropsis* species, *Phaeodactylum tricornutum, Monodus subterraneus*	Nutritional supplement
	Docosahexaenoic acid (DHA)	*Crypthecodinium cohnii, Schizochytrium* species, *Ulkenia* species	Nutritional supplement
	Sterol	Many species	Nutraceutical
	Squalene	*Aurantiochytrium* species	Nutritional supplement
	γ-linolenic acid	*Spirulina* species	Nutritional supplement

12.14.5 Metal Stress

The lipid production in microalgae can be affected by metal ions. Ren *et al.* [147] reported the effect of iron, magnesium and calcium on biomass and lipid production in *Scenedesmus* sp. R-16 and found that lipid content increased to 43, 35 and 47%, respectively. *Chlorella minutissima* showed resistance to copper and cadmium ions and lipid content was significantly increased to 93.9 and 21.1% respectively [148]. Yeesang and Cheirsilp [121] reported twofold rise in lipid content when *Botryococcus* species was cultivated in the presence of iron (0.74 mM). The high lipid content was reported in *C. protothecoides, C. vulgaris* and *C. pyrenoidosa* in the presence of 4 mg/liter copper ion.

12.15 Application of Microalgae in Biofuel Production

Microalgae are small manufacturing units that produce various useful products including eco-friendly biofuel Table 12.5.

12.15.1 Advantages of Using Microalgae for Biofuel Production

1. Microalgae cultivation is not dependent on agriculture land as it can be easily cultivated in degraded land, marine water and industrial effluent, etc. Hence cultivation of microalgae does not compete with agriculturally important food crops.
2. It shows fast growth, i.e., short cropping cycle (only ten days).
3. Microalgae cultivation can be done around the year as it is not dependent on weather conditions.
4. It consumes CO_2, a greenhouse gas responsible for a rise in earth's temperature and climate change (Table 12.4). They can easily sequester carbon dioxide thereby reducing greenhouse gas emission. For the generation of one kilogram of microalgae approximately two kilograms of CO_2 consumption takes place [28].
5. They grow easily in the presence of air, water, mineral salts and light. There is no need of herbicide and pesticides for the cultivation of microalgae.

6. They utilize nitrogen and phosphorus present in wastewater released from municipality and reduce nutrient load in wastewater.
7. The carbohydrate component of microalgae can be used for bioethanol production whereas oil content can be used for biodiesel and residual biomass for methane production.
8. They have significant amount of lipids in their thylakoid membrane. The microalgal biofuel is non-toxic and highly biodegradable. It has potential to completely replace fossil fuel due to the presence of lipid in significant amount.
9. Microalgal biomass left after biodiesel production can be used in preparation of nutraceuticals, therapeutics, eicosapentaenoic acid (EPA), docosahexaenoic acid (DHA), biocontrol agents, fertilizers and animal feed, etc.

Table 12.4 Impact of nutrient stress condition on lipid content of microalgae [143].

Microalgae	Nutrient stress factor	Lipid content (%)
Chlorella vulgaris	Nitrogen starvation	53
Chlorella zofingiensis	Nitrogen and phosphorus starvation	42
Scenedesmus species	Nitrogen starvation	31
Chlorella protothecoides	Phosphorus starvation	32
Chlorella lobophora	Sulfur starvation	50

Table 12.5 Benefits of using microalgal biofuel [157].

Microalgal biofuel	Applications
Indigenous effect	Reduced utilization of fossil fuel, reliability in supply, easy availability, indigenous distribution
Environmental friendly	Reduction in emission of greenhouse gases, biodegradable, no air pollution, better carbon sequestration
Commercial applications	Sustainable option, development of agricultural sector, no import of biodiesel, more industrial investment, increase in the government revenue through taxes

12.16 Biodiesel

Biodiesel is a biodegradable, non-toxic and sustainable fuel [149], which is produced by transesterification of triacylglycerols with an acid or alkali catalyst [150]. Direct transesterification is a fast and cost-saving method in comparison to another method in which lipids are first extracted from algal biomass then transesterification takes place. Biodiesel production from microalgae can be an alternative to fossil fuel and for large-scale production microalgae with high growth rate and lipid content should be identified [151].

Biofuel include biodiesel, bioethanol and biogas which can be produced by fermentation and trans-esterification [152]. Transesterification of microbial oil used oil and animal/plant fat with ethanol/butanol and methanol results in biodiesel production [153]. Triglycerides initiate biodiesel production in microalgae by the following steps: (1) Carbon dioxide forms acetyl Co-A (2) Formation of acyl-CoA from chain elongation and (3). Synthesis of triglycerides from acyl-CoA [154]. The oxygen atoms are present in the biodiesel whereas petro-diesel are hydrocarbons which contains hydrogen and carbon.

12.17 Biogas

The organic waste forms biogas or biomethane by anaerobic digestion. Biogas is composed of 55-75% CH_4 and 25-45% CO_2. The rate of biogas production with microalgae is low due to frequent degradation of micro algal cells by bacteria and less C:N ratio, which forms ammonia and it inhibits biogas production.

12.18 Hydrogen

Hydrogen is an important energy resource because it does not generate greenhouse gases and generates H_2O as a byproduct [155]. Microalgae can directly generate hydrogen in the presence of sunlight and water under anaerobic conditions. Microalgae can produce hydrocarbon which can be converted into diesel, kerosene and gasoline. *Botryococcus braunii* has been reported to generate hydrocarbons with oil production [156].

12.19 Biosyngas

Biosyngas is produced by microalgae gasification in presence of O_2 and it produces CO, H_2, CH_4, H_2O, hydrocarbon and ashes. A high temperature (800-1200°C) is required for gasification and biomass should not contain more than 20% water [157].

12.20 Ethanol

Bioethanol can be used as biofuel but ethanol production by microalgae is in its initial phase of research. Nowadays bioethanol is produced by fermentation of sugars but cellulose and hemicellulose can also be hydrolyzed into sugar [158]. The cellular composition of microalgae is very simple and it does not contain lignin so its processing is easier. Microalgal cells contain an ample amount of polysaccharides which can form sugars [159].

12.21 Cultivation of Microalgae for Biofuel Production

Research on microalgae as a sustainable energy source began in 1970. Microalgae are able to use sunlight, water and atmospheric CO_2 to accumulate biomass in the form of lipids. During the photosynthesis, neutral lipids are accumulated as triacylglycerols (TAGs) in microalgal cells and by transesterification, triacylglycerols can be further converted into various types of fatty acid methyl esters which are an essential component of biodiesel. Microalgae can be considered as raw material for biodiesel production as they have 20-50% oil content but some microalgae have more than 80% oil content [160]. Microalgae can be cultivated by the following methods.

12.21.1 Open Microalgal System

The open pond system is utilized for microalgae cultivation as they use atmospheric CO_2 available in the atmosphere. The location is an important criteria for an open pond system in which microalgae can get sufficient sunlight. The open pond can be natural or artificial and may be of different types such as circular, tanks and raceway ponds, etc. *Chlorella* is cultivated in circular concrete ponds and rotating arms are also constructed in these ponds which ensure proper mixing of the microalgal culture and prevents deposition of biomass.

12.21.2 Closed Microalgal System

Photobioreactors are used as a closed microalgal system but they are designed to allow proper mixing, improve accessibility of light and gas exchange facilities for cell growth. Photobioreactor can solve different problems associated with open cultivation system and researchers have concentrated on designing of photobioreactors for large-scale microalgae cultivation [161]. Tubular photobioreactors are considered as most applicable. Bubble column and airlift photobioreactor are used as they produce higher microalgal biomass [162]. The excess of oxygen adversely affect microalgae growth and due to this auxiliary tank is used to separate generated oxygen [163].

12.21.3 Hybrid Microalgal System

A hybrid system has both an open and a closed system which can be used to produce a significant amount of microalgal biomass [164]. Microalgae are firstly cultivated in closed photobioreactor then shifted to an open system for more biomass production [165]. This system provides promising option for biofuel production by microalgae.

Despite the advancements in microalgae cultivation system, bottlenecks are yet to be addressed efficiently for enhancing biofuel productivity. Reports have revealed that open raceway ponds can be made easily at low cost but they produce less amount of triacylglycerols; significant amount of biomass and oil production can be obtained in a closed photobioreactor but it is an expensive method and may not be feasible. Recent technologies for improving microalgal yield can save expenditure.

12.22 Current Research Status in India

India has a wide coastal region as well as different rivers that cover a large surface area. The algal cultivation in less than 3% of India's total land can make us self-reliant in biofuel production [34]. Recently scientists have developed a car based on B-20 biodiesel under the New Millennium India Technology Leadership project by marine microalgae which was jointly started by collaboration of CSIR, Ministry of Earth Sciences, CSMCRI, IIT-Kharagpur, IICT-Hyderabad, NIOT-Chennai and NIO-Goa. The scientists of Central Salt Marine and Chemical Research Institute (CSMCRI), Bhavnagar prepared biodiesel from mats of microalgae. A team of researchers from Indian Institute of Chemical Technology are trying to reduce the cost of microalgae oil from Rs 500 to Rs 20 per litre. Dr. MGR Algae Biofuel Research Center has launched a biodiesel plant at Sivakasi, Tamil Nadu for production of biofuel from algae.

12.23 Concluding Remarks and Future Prospectives

The microalgae biofuel is safe, cost-effective and can fulfil energy demand in future. To date, genetically unmodified microalgae have been exploited and modification in algal strain is mandatory to promote the feasibility of biofuel. Genetic engineering has been used for the production of bioactive compounds and biofuel from algal cell factories. The coordinative research between different streams such as biochemistry, chemistry and molecular biology of oleaginous microalgae is required for the generation of biofuel. The genetic modification in potential microalgal strain will ensure maximum production of triacylglycerol. Biofuel production from microalgae can be developed as an industry and it may provide jobs. The engineering of microalgae to augment biofuel generation is essential for commercialization of algal fuel. Further focused research is required to make our country energy independent.

12.24 Acknowledgements

The author is thankful to Dr. Chanderdeep Tandon, Director, Amity Institute of Biotechnology, Amity University, Noida, for his valuable suggestions during the preparation of this manuscript.

References

1. Walker, G., Simcock, N., Day, R., Necessary energy uses and a minimum standard of living in the United Kingdom: energy justice or escalating expectations? *Energ Res Soc Sci.* 18, 129-138, 2016.
2. Pfenninger, S., Keirstead, J., Renewables, nuclear or fossil fuels? Scenarios for Great Britain's power system considering costs, emissions and energy security. *Appl Energ.* 152, 83-93, 2015.

3. Bentley, R.W., Introduction. In: *Introduction to peak oil*. Springer International Publishing. pp. 1-8, 2016.
4. Revellame, E., Hernandez, R., French, W., Holmes, W., Alley, E., Biodiesel from activated sludge through in situ transesterification. *J. Chem Technol Biotechnol.* 85(5), 614-620, 2010.
5. International Energy Agency. World energy outlook 2007. http://www.iea.org/textb ase/nppdf /free/2007/weo 2007.pdf.
6. Atabani, A.E., Silitonga, A.S., Badruddin, I.A., Mahlia, T.M.I., Masjuki, H.H., Mekhilef, S. A., Comprehensive review on biodiesel as an alternative energy resource and its characteristics. *Renew Sust Energ Rev.* 16, 2070-2093, 2012.
7. Reuter, M., Buchwitz, M., Hilboll, A., Richter, A., Schneising, O., Hilker, M., Decreasing emissions of NOx relative to CO_2 in East Asia inferred from satellite observations. *Nature Geosci.* 7, 792-795, 2014.
8. Agarwal, A.K., Gupta, J.G., Dhar, A., Potential and challenges for large-scale application of biodiesel in automotive sector. Prog. *Energy Combustion Sci.* 61, 113-149, 2017.
9. Bruce, E.R., Opportunities for renewable bioenergy using microorganisms. *Biotechnol. Bioeng.* 100, 203-212, 2008.
10. Vasudevan, P.T., Briggs, M., 2008. Biodiesel production-current state of the art and challenges. *J. Ind. Microbiol. Biotechnol.* 35, 421-30, 2008.
11. Hossain, A.B.M., AlEissa, M.S., Biodiesel fuel production from palm, sunflower waste cooking oil and fish byproduct waste as renewable energy and environmental recycling process. *British Biotechnology Journal,* 10, 1-9, 2016.
12. Khan, S.A., Rashmi, Hussain, M.Z., Prasad, S., Banerjee, U.C., Prospects of biodiesel production from microalgae in India. *Renew. Sustain. Energy. Rev.* 13, 2361-2372, 2009.
13. Hoekman, S.K., Broch, A., Robbins, C., Ceniceros, E., Review of biodiesel composition, properties, and specifications. *Renew Sust Energ Rev.* 16, 143-169, 2012.
14. Mardhiah, H.H., Ong, H.C., Masjuki, H.H., Lim, S., Lee, H.V., A review on latest developments and future prospects of heterogeneous catalyst in biodiesel production from non-edible oils. *Renew Sustain Energy Rev.* 67, 1225-1236, 2017.
15. Basha, S.A., Gopal, K.R., Jebaraj, S., A review on biodiesel production, combustion, emissions and performance. *Renew Sust Energ Rev.* 13, 1628-1634, 2009.
16. Popovicheva, O., Engling, G., Lin, K.T., Persiantseva, N., Timofeev, M., Kireeva, E., Diesel/biofuel exhaust particles from modern internal combustion engines: microstructure, composition, and hygroscopicity. *Fuel.* 157, 232-239, 2015.
17. Knothe, G., Krahl, J., Van Gerpen, J., *The biodiesel handbook*. Elsevier. 2015.
18. Tsoutsos, T.D., Tournaki, S., Paraíba, O., Kaminaris, S.D., The used cooking oil-to-biodiesel chain in Europe assessment of best practices and environmental performance. *Renew Sust Energ Rev.* 54, 74-83, 2016.

19. Bart, J.C.J., Palmeri, N., Cavallaro, S., *Biodiesel science and technology: from soil to oil.* Woodhead Publishing Limited, 2010.
20. Daud, N.M., Abdullah, S.R.S., Hasan, H.A., Yaakob, Z., Production of biodiesel and its wastewater treatment technologies: a review. *Process Saf Environ Prot.* 94, 487-508, 2015.
21. Churchill, G.V., Srinivasan, C.A., Experimental investigations on combustion and emission characteristics of biodiesel blends in CI engine. *Int Res J Eng Technol.* 4(1), 1523-1529, 2017.
22. Chapagain, B.P., Yehoshua, Y., Wiesman, Z., Desert date (*Balanites aegyptiaca*) as an arid lands sustainable bioresource for biodiesel. *Bioresour Technol.* 100, 1221-1226, 2009.
23. Zhang, H., Zhou, Q., Chang, F., Pan, H., Liu, X.F., Li, H., Production and fuel properties of biodiesel from *Firmiana platanifolia* L.f. as a potential non-food oil source. *Indus Crop Prod.* 76, 768-771, 2015.
24. Kamel, D.A., Farag, H.A., Amin, N.K., Fouad, Y.O., Biodiesel synthesis from non-edible oils by transesterification using the activated carbon as heterogeneous catalyst. *Int J Environ Sci Technol.* 14, 785-794, 2017.
25. Soccol, C.R., Neto, C.J.D., Soccol, V.T., Sydney, E.B., Costa, E.S.F.D., Medeiros, A.B.P., Vandenberghe LPDS., Pilot scale biodiesel production from microbial oil of *Rhodosporidium toruloides* DEBB 5533 using sugarcane juice: performance in diesel engine and preliminary economic study. *Bioresour Technol.* 223, 259–268, 2017.
26. Kumar, G., Srivastava, R., Singh, R., Exploring Biodiesel: Chemistry, Biochemistry, and Microalgal Source, *Int J Green Energy.* 10, 775-796, 2013.
27. Priyadarshani, I., Rath, B., Commercial and industrial applications of micro algae-a review, *J. Algal Biomass Utilization.* 3(4), 89-100, 2012.
28. Chisti, Y., Biodiesel from microalgae. *Biotechnol Adv.* 25(3), 294-306, 2007.
29. Piloto-Rodríguez, R., Sánchez-Borroto, Y., Melo-Espinosa, E.A., Verhelst, S., Assessment of diesel engine performance when fueled with biodiesel from algae and microalgae: an overview. *Renew Sustain Energy Rev.* 69, 833-842, 2017.
30. Wu, S., Song, L., Sommerfeld, M., Hu, Q., Chen, W., Optimization of an effective method for the conversion of crude algal lipids into biodiesel. *Fuel.* 197, 467-473, 2017.
31. Dutta, K., Daverey, A., Jih-Gaw, L.J.G., Evolution retrospective for alternative fuels: first to fourth generation. *Renew Energ.* 69, 114-122, 2014.
32. Ahmad, A.L., Yasin, N.H.M., Derek, C.J.C., Lim, J.K., Microalgae as a sustainable energy source for biodiesel production: a review. *Renew Sust Energ Rev.,* 15, 584-593, 2011.
33. Kligerman, D.C., Bouwer, E.J. Prospects for biodiesel production from algae-based wastewater treatment in Brazil: a review. *Renew Sustain Energy Rev.* 52(C), 1834-1846, 2015.
34. Singh, A.K., Singh, A.K., Singh, S., Choudhary, P., Singh, N.K., An overview on: next generation cost effective biofuel production from algae. *Int. J. Curr. Microbiol. App. Sci.* 7, 4139-4149, 2018.

35. Beacham, T.A., Sweet, J.B., Allen, M., Large scale cultivation of genetically modified microalgae: a new era for environmental risk assessment. *Algal Res.* 25, 90-100, 2017.
36. Shumbulo, E., Ki, D., Microalgae to biofuels: Promising' alternative and renewable energy –a review. *Renew Sustain Energy Rev.* 81, 743-755, 2018.
37. Creutzig, F., Ravindranath, N.H., Berndes, G., Bolwig, S., Bright, R., Cherubini, F., Chum, H., Corbera, E., Delucchi, M., Faaij, A., Fargione, J., Bioenergy and climate change mitigation: an assessment. *Gcb Bioenerg.* 7, 916-944, 2015.
38. Hughes, S.R., Moser, B.R., Gibbons, W.R., Moving toward energy security and sustainability in 2050 by reconfiguring biofuel production. In: *Convergence of food security, energy security and sustainable agriculture.* Springer, Berlin Heidelberg. 15-29, 2014.
39. Arazo, R.O., Luna, M.D.G.D., Capareda, S.C., Assessing biodiesel production from sewage sludge-derived bio-oil. *Biocatal Agric Biotechnol.*, 10, 189-196, 2017.
40. Song, T., Martensson, L., Eriksson, T., Zheng, W., Rasmussen, U. Biodiversity and seasonal variation of the cyanobacterial assemblage in a rice paddy field in Fujian, China. The Federation of European Materials Soc. *Microbiol Ecol.* 54, 131-140, 2005.
41. Colla, L.M., Reinehr, C.O., Reichert, C., Costa, J.A.V., Production of biomass and nutraceutical compounds by Spirulina platensis under different temperature and nitrogen regimes. *Bioresour. Technol.* 98 (7), 1489-1493, 2007.
42. Madhyastha, H.K., Vatsala, T.M., Pigment production in *Spirulina fussiformis* in different photophysical conditions. *Biomol. Eng.* 24(3), 301-305, 2007.
43. Sajilata, M.G., Singhal, R.S., Kamat, M.Y., Supercritical CO_2 extraction of γ-linolenic acid (GLA) from *Spirulina platensis* ARM 740 using response surface methodology. *J Food Eng.* 84, 321-26, 2008.
44. Ogbonda, K.H., Aminigo, R.E., Abu, G.O., Influence of temperature and pH on biomass production and protein biosynthesis in a putative *Spirulina* sp. *Bioresour. Technol.* 98, 2207-2211, 2007.
45. Pulz, O., Gross, W., Valuable products from biotechnology of microalgae. *Appl Microbiol. Biotechnol.* 65, 635-648, 2004.
46. Becker, E.W., Microalgae in human and animal nutrition. In: Richmond A (Ed). *Handbook of Microalgae Culture. Biotechnology and Applied Phycology.* Oxford: Blackwell Science. 2004.
47. Suleria, H.A.R., Osborne, S., Masci, P., Gobe, G., Marine-Based Nutraceuticals: An Innovative Trend in the Food and Supplement Industries. *Mar. Drugs.* 13, 6336-6351, 2015.
48. Galasso, C., Gentile, A., , Orefice, I., Ianora, A., Bruno, A., Noonan, D.M., Sansone, C., Albini, A., Brunet, C., Microalgal derivatives as potential nutraceutical and food supplements for human health: A focus on cancer prevention and interception. *Nutrients.*, 11, 1-22, 2019.

49. Mendes, A., Reis, A., Vasconcelos, R., Guerra, P., Lopes da Silva, T., Crypthecodinium cohnii with emphasis on DHA production: a review. *J Appl Phycol.* 21, 199-214, 2009.
50. Borowitzka, M.A., Carotenoid production using microorganisms. In: Cohen Z, Ratledge C (eds) *Single cell oils: microbial and algal oils.* AOCS Press, Urbana, pp. 225-240, 2010.
51. Singh, S., Kate, B.N., Banerjee, U.C., Bioactive compounds from cyanobacteria and microalgae: an overview. *Crit Rev Biotechnol* 25, 73-95, 2005.
52. Guil-Guerrero, J. L., Navarro-Juárez, R., López-Martínez, J. C., Campra-Madrid, P., Rebolloso-Fuentes, M.M., Functionnal properties of the biomass of three microalgal species. *J. Food Eng.* 65, 511-517, 2004.
53. Wells, M.L., Potin, P., Craigie, J.S., Raven, J.A., Merchant, S.S., Helliwell, K.E., Smith, A.G., Camire, M.E., Brawley, S.H., Algae as nutritional and functional food sources: revisiting our understanding. *J. Appl. Phycol.,* 29, 949-982, 2016.
54. Japelt, R.B., Jakobsen, J., Vitamin D in plants: A review of occurrence, analysis, and biosynthesis. *Front. Plant Sci.,* 4, 136, 2013.
55. Gruber, B.M., B-Group Vitamins: Chemoprevention? *Adv. Clin. Exp. Med.,* 25, 561-568, 2016.
56. Santiago-Morales, I.S., Trujillo-Valle, L., Márquez-Rocha, F.J., López Hernández, J.F., Tocopherols, phycocyanin and superoxide dismutase from microalgae as potential food antioxidants. *Appl. Food Biotechnol.* 5, 19-27, 2018.
57. Nishikawa, S., Hosokawa, M., Miyashita, K., Fucoxanthin promotes translocation and induction of glucose transporter 4 in skeletal muscles of diabetic/obese KK-A(y) mice. *Phytomedicine.* 19, 389-94, 2012.
58. Takahashi, K., Hosokawa, M., Kasajima, H., Hatanaka, K., Kudo, K., Shimoyama, N., Miyashita, K., Anticancer effects of fucoxanthin and fucoxanthinol on colorectal cancer cell lines and colorectal cancer tissues. *Oncol. Lett.* 10, 1463-1467, 2015.
59. Manirafasha, E., Ndikubwimana, T., Zeng, X., Lu, Y., Jing, K., Phycobiliprotein: Potential microalgae derived pharmaceutical and biological reagent. *Biochem. Eng. J.,* 109, 282-296, 2016.
60. Rizzello, C.G., Tagliazucchi, D., Babini, E., Rutella, G.S., Saa, D.L.T., Gianotti, A., Bioactive peptides from vegetable food matrices: Research trends and novel biotechnologies for synthesis and recovery. *J. Funct. Foods.,* 27, 549-569, 2016.
61. Hwang, E., Park, S.Y., Sun, Z.W., Shin, H.S., Lee, D.G. and Yi, T.H., The protective effects of fucosterol against skin damage in UVB-irradiated human dermal fibroblasts. *Mar. Biotechnol.,* 16, 361-370, 2014.
62. Liang, S., Liu, X., Chen, F., Chen, Z., Current microalgal health food R & D activities in China. *Hydrobiologia.,* 512, 45-48, 2004.
63. Bishop, M.A., West., Zubeck, M., Evaluation of microalgae for use as nutraceuticals and nutritional supplements. *Nutr. Food Sci.,* 2, 1-6, 2012.

64. Ishaq, A.G., Matias-Peralta, H.M., Basri, H., Bioactive compounds from green microalga-*Scenedesmus* and its potential applications: A brief review. *Pert. J. Trop. Agr. Sci.*, 39, 1-16, 2016.
65. Jensen, G.S., Ginsberg, D.I., Drapeau, C., Blue-green algae as an immuno-enhancer and biomodulator. *J. Nutraceu. Nutri.*, 3, 24-30, 2001.
66. Ibanez, E., Cifuentes, A., Benefits of using algae as natural sources of functional ingredients. *J Sci Food Agric.* 93 (4), 703-709, 2013.
67. Ryll *et al.*, Abstr. Eur Workshop, Microalgal Biotechnol. Germany pp. 56, 2003.
68. Yamaguchi, K., Recent advances in microalgal bioscience in Japan, with special reference to utilization of biomass and metabolites: a review. *J. Appl. Phycol.* 8, 487–502, 1997.
69. Soong, P., Production and development of *Chlorella* and *Spirulina* in Taiwan. In: Shelef G, Soeder CJ (eds) *Algae biomass*. Elsevier/North Holland Biomedical Press, Amsterdam, pp 97-113, 1980.
70. Borowitzka, L.J., Borowitzka, M.A., ß-Carotene (Provitamin A) production with algae. In: Vandamme EJ (ed) *Biotechnology of vitamins, pigments and growth factors*. Elsevier Applied Science, London, pp. 15-26, 1998.
71. Lorenz, R.T., Cysewski, G.R., Commercial potential for *Haematococcus* microalgae as a natural source of astaxanthin. *Trends Biotechnol.* 18, 160-167, 2000.
72. Pangestuti, R., Kim, S. K., Biological activities and health benefit effects of natural pigments derived from marine algae. *J. Funct. Foods* 3 (4), 255-266, 2011.
73. Soletto, D., Binaghi, L., Lodi, A., Carvalho, J. C. M., Converti, A., Batch and fed-batch cultivations of *Spirulina platensis* using ammonium sulphate and urea as nitrogen sources. *Aquaculture.* 243, 217-224, 2005.
74. Benedetti, S., Benvenuti, F., Pagliarani, S., Francogli, S., Scoglio, S., Canestrari, F., Antioxidant properties of a novel phycocyanin extract from the blue-green alga *Aphanizomenon flos-aquae*. *Life Sci.*, 75, 2353-2362, 2004.
75. Fernández-Sevilla, J., Acién Fernández, F., Molina Grima, E., Biotechnological production of lutein and its applications. *Appl Microbiol Biotechnol* 86, 27-40, 2010.
76. Wang, Y., Chen, T.,The biosynthetic pathway of carotenoids in the astaxanthin-producing green alga *Chlorella zofingiensis*. *World J Microbiol Biotechnol* 24, 2927-2932, 2008.
77. Hanagata, N., Secondary carotenoid accumulation in *Scenedemus komarekii* (Chlorophyceae, Chlorophyta). *J Phycol.* 35, 960-966, 1999.
78. Von Oppen-Bezalel, L., Shaish, A., Application of the colourless carotenoids, phytoene and phytofluene in cosmetics, wellness, nutrition, and theropcutics. In: Ben-Amotz A, Polle JEW, Subba Rao DV (eds) *The alga Dunaliella: biodiversity, physiology, genomics and biotechnology*. Science Publishers, Enfield, pp. 423-444, 2009.
79. Matsuura, H., Watanabe, M., Kaya, K., Echinenone production of a dark red-coloured strain of *Botryococcus braunii*. *J Appl Phycol.* 24, 973-977, 2012.

80. Kim, S.M., Jung, Y.J., Kwon, O.N., Cha, K.H., Um, B.H., Chung, D., Pan, C.H., A potential commercial source of fucoxanthin extracted from the microalga *Phaeodactylum tricornutum*. *Appl Biochem Biotechnol.* 166, 1843-1855, 2012.
81. Abidov, M., Ramazanov, Z., Seifulla, R., Grachev, S., The effects of xanthigen in the weight management of obese premenopausal women with non-alcoholic fatty liver disease and normal liver fat. *Diabetes. Obes. Metab.* 12 (1), 72-81, 2010.
82. Viskari, P. J., Colyer, C. L., Rapid extraction of phycobiliproteins from cultured cyanobacteria samples. *Anal. Biochem.*, 319, 263-271, 2003.
83. Glazer, A.N., Phycobiliproteins-a family of valuable, widely used fluorophores. *J Appl Phycol* 6, 105-112, 1994.
84. Hu, L., Huang, B., Zuo, M., Wei, H., Preparation of the phycoerythrin subunit liposome in a photodynamic experiment on liver cancer cells. *Acta Pharmacol Sin.* 29, 1539-1546, 2008.
85. Hu, Q., Industrial production of microalgal cell-mass and secondary products-major industrial species: *Arthrospira platensis*. In: Richmond A (ed) *Microalgal culture: biotechnology and applied phycology*. Blackwell Science, Oxford, pp. 264-272, 2004.
86. Metting, F. B., Biodiversity and application of microalgae. *J. Ind. Microbiol.* 17, 477-489, 1996.
87. Luiten, E. E. M., Akkerman, I., Koulman, A., Kamermans, P., Reith, H., Barbosa, M. J., Sipkema, D., Wijffels, R. H., Realizing the promises of marine biotechnology. *Biomol. Eng.* 20, 429-439, 2003.
88. Fernando, I. P. S., Nah, J. W., Jeon, Y. J., Potential anti-inflammatory natural products from marine algae. *Environ. Toxicol. Pharmacol.* 48, 22-30, 2016.
89. Volkman, J.V. Sterols in microorganisms. *Appl Microbiol Biotechnol.* 60, 495-506, 2003.
90. Robles Medina, A., Molina Grima, E., Giménez Giménez, A., Ibáñez González, M. J., Downstream processing of algal polyunsaturated fatty acids. *Biotechnol. Adv.* 16, 517-580, 1998.
91. Ward, O. P., Singh, A., Omega-3/6 fatty acids: alternative sources of production. *Process Biochem.* 40, 3627–3652, 2005.
92. Medina, A.R., Grima, E.M., Gimenez, A.G., Gonzalez, M.J.I., Downstream processing of algal polyunsaturated fatty acids. *Biotechnol Advances.* 16, 517-580, 1997.
93. Jiang, Y., Chen, F., Liang, S. Z., Production potential of docosahexaenoic acid by the heterotrophic marine dinoflagellate *Crypthecodinium cohnii*. *Process Biochem.* 34, 633-637, 1999.
94. Leu, S., Boussiba, S., Advances in the production of high-value products by microalgae. *Ind. Biotechnol.* 10 (3), 169-183, 2014.
95. Jha, D., Jain, V., Sharma, B., Kant, A., Garlapati, V. K.,Microalgae-based pharmaceuticals and nutraceuticals: An emerging field with immense market potential. *Chem Bio Eng Rev.* 4 (4), 257-272, 2017.

96. Chen, Y.C., Immobilized *Isochrysis galbana* for long-term storage and applications for feed and water quality control in clam (*Meretrix lusoria*) cultures. *J. Appl. Phycol.* 15, 439-444, 2003.
97. Certik, M., Shimizu, S., Biosynthesis and regulation of microbial polyunsaturated fatty acid production. *J. Biosci. Bioeng.* 87, 1-14, 1999.
98. Muller-Feuga, A., The role of microalgae in aquaculture: situation and trends. *J Appl Phycol.* 12, 527-534, 2000.
99. Le, T. M., Knulst, A. C., Rockmann, H., Anaphylaxis to Spirulina confirmed by skin prick test with ingredients of *Spirulina* tablets. *Food. Chem. Toxicol.* 74, 309-310, 2014.
100. Mouritsen, O. G., *Seaweeds: Edible, Available and Sustainable.* University of Chicago Press, Chicago, 2013.
101. Nicoletti, M., Microalgae nutraceuticals. *Foods.* 5, 54, 2016.
102. Rania, M.A., Hala, M.T., Antibacterial and antifungal activity of cynobacteria and green microalgae evaluation of medium components by Plackett-Burman design for antimicrobial activity of *Spirulina platensis. Global J. Biotechnol. Biochem.* 3(1), 22-31, 2008.
103. Michele Greque deMorais, Bruna da SilvaVaz, Etiele Greque de Morais, Jorge Alberto Vieira Costa. Biologically active metabolites synthesized by microalgae. *BioMed Res International.* 7-15, 2015.
104. Carvalho, R., Costa-Neves, A., Conserva, G. A. A., Biologically active compounds from cyano bacteria extracts: in vivo and in vitro aspects. *Rev. Bras. Farmacogan.* 23(3), 471-480, 2013.
105. Higuera-Ciapara, I., Felix-Valenzuela, L., Goycoolea, F.M., Astaxanthin: a review of its chemistry and applications. *Crit Rev Food Sci Nutr.* 46, 185-196, 2006.
106. Yuan, J.P., Peng, J., Yin, K., Wang, J.H., Potential health-promoting effects of astaxanthin: a high-value carotenoid mostly from microalgae. *Mol Nutr Food Res.* 55, 150-165, 2011.
107. Chu, W.L., Potential applications of antioxidant compounds from algae. *Curr Top Nutraceut Res.* 9, 83-98, 2011.
108. Mani, U.V., Iyer, U.M., Dhruv, S.A., Mani, I.U., Sharma, K.S., Therapeutic utility of Spirulina. In: Gershwin ME, Belay A, eds. *Spirulina in human nutrition and health.* Boca Raton: CRC Press. 71-99, 2008.
109. Dey, B., Lerner, D.L., Lusso, P., Boyd, M.R., Elder, J.H., Bergeri, E.A., Multiple antiviral activities of cyanovirin-N: blocking of human immunodeficiency virus Type 1 gp120 interaction with CD4 and coreceptor and inhibition of diverse enveloped viruses. *J. Virol.*, 74, 4562-4569, 2000.
110. Ayehunie, S., Belay, A., Baba, T.W., Ruprecht, R.M., Inhibition of HIV- 1 replication by an aqueous extract of *Spirulina platensis. J Aquir Immun Defic Syndr Hum Retrovirol.* 18, 7-12, 1998.
111. Boopathy, N.S., Kathiresan, K., Anticancer drugs from marine flora; an overview. *J. Oncol*, Article ID 214186: 18, 2010.

112. Schwartz, R.E., Hirsch, C.F., Sesin, D.F., Flor, J.E., Chartrain M, Fromtling RE. Pharmaceuticals from cultured algae. *J Ind Microbiol.* 5, 13-123, 1990.
113. Stevenson, C.S., Capper, E.A., Roshak, A.K., Marquez, B., Echman, C., Jackson, J.R., Mattern, M., The identification and characterization of the marine natural product scytonemin as a novel antiproliferative pharmacophore. *J. Pharmacol. Exp. Ther.* 303, 858-866, 2002.
114. Stolz, P., Obermayer, B., Manufacturing microalgae for skin care. *Cosmetics Toiletries.* 120, 99-106, 2005.
115. Dubey, S.K., Dubey, J., Mehra, S., Tiwari, P., Bishwas, A.J., Potential use of cyanobacterial species in bioremediation of industrial effluents. *Afr. J. Biotechnol.* 10, 1125-1132, 2011.
116. Omar, H.H., Algal decolorization and degradation of monoazo and diazo dyes. *Pak. J. Biol. Sci.* 11, 1310-1316, 2008.
117. Yan, H., Pan, G., Increase in biodegradation of dimethyl phthalate by *Closterium lunula* using inorganic carbon. *Chemosphere.* 55, 1281-1285, 2004.
118. Hallenbeck, P. C., Grogger, M., Mraz, M.,Veverka, D., The use of design of experiments and response surface methodology to optimize biomass and lipid production by the oleaginous marine green alga, *Nannochloropsis gaditana* in response to light intensity, inoculum size and CO_2. *Bioresour Technol.* 184, 161-168, 2015.
119. Pal, D., Khozin-Goldberg, I., Cohen, Z., Boussiba, S., The effect of light, salinity, and nitrogen availability on lipid production by *Nannochloropsis* sp. *Appl Microbiol Biotechnol.* 90 (4), 1429-1441, 2011.
120. Mandotra, S. K., Kumar, P., Suseela, M. R., Nayaka, S., Ramteke, P. W., Evaluation of fatty acid profile and biodiesel properties of microalga *Scenedesmus abundans* under the influence of phosphorus, pH and light intensities. *Bioresour Technol.* 201, 222-229, 2016.
121. Yeesang, C.,Cheirsilp, B., Effect of nitrogen, salt, and iron content in the growth medium and light intensity on lipid production by microalgae isolated from freshwater sources in Thailand. *Bioresour Technol.* 102(3), 3034 - 3040, 2011.
122. Sun, X., Cao, Y., Xu, H., Effect of nitrogen-starvation, light intensity and iron on triacylglyceride/carbohydrate production and fatty acid profile of *Neochloris oleoabundans* HK-129 by a two-stage process. *Bioresour Technol.* 155, 204-212, 2014.
123. Vitova, M., Bisova, K., Kawano, S., Zachleder, V., Accumulation of energy reserves in algae: from cell cycles to biotechnological applications. *Biotechnol. Adv.* 33(6), 1204-1218, 2015.
124. Khotimchenko, S.V., Yakovleva, I.M., Lipid composition of the red alga *Tichocarpus crinitus* exposed to different levels of photon irradiance. *Phytochemistry.* 66, 73-79, 2005.
125. Zhu, L. D., Hiltunen, E., Methylation of DACT2 accelerates esophageal cancer development by activating Wnt signaling. *Renew Sustain Energy Rev.* 54, 1285-1290, 2016.

126. Converti, A., Casazza, A.A., Ortiz, E.Y., Perego, P., Borghi, M.D., Effect of temperature and nitrogen concentration on the growth and lipid content of *Nannochloropsis oculata* and Chlorella vulgaris for biodiesel production, *Chemical Engineering and Processing: Process Intensification*, 48, 1146-1151, 2009.
127. Gardner, R. D., Lohman, E., Gerlach, R., Cooksey, K. E., Peyton, B.M., Comparison of CO2 and bicarbonate as inorganic carbon sources for triacylglycerol and starch accumulation in *Chlamydomonas reinhardtii*. *Biotechnol. Bioeng.* 110(1), 87-96, 2013.
128. Widjaja, A., Chien, C.C., Ju, Y.H., Study of increasing lipid production from fresh water microalgae Chlorella vulgaris. *J. Taiwan Inst. Chem. Engrs.* 40, 13-20, 2009.
129. Nakanishi, A., Aikawa, S., Ho, S.H., Development of lipid productivities under different CO2 conditions of marine microalgae *Chlamydomonas* sp. JSC4. *Bioresour Technol.* 152, 247-252, 2014.
130. Ota, M., Kato, Y., Watanabe, H., Fatty acid production from a highly CO_2 tolerant alga, *Chlorocuccum littorale*, in the presence of inorganic carbon and nitrate. *Bioresour Technol.* 100 (21), 5237-5242, 2009.
131. Ho, S.H., Chen, C.Y., Chang, J.S., Effect of light intensity and nitrogen starvation on CO2 fixation and lipid/carbohydrate production of an indigenous microalga *Scenedesmus obliquus* CNW-N. *Bioresour Technol.* 113, 244-252, 2012.
132. Aaronson, S., Effect of incubation temperature on the macromolecular and the lipid content of the phytoflagellate *Ochromonas danica*. *Journal of Phycologia*. 9, 111-113, 1973.
133. Boussiba, S., Vonshak, A., Cohen, Z., Avissar, Y., Richmond, A., Lipid and biomass production by halotolerant microalga *Nannochloroplasis salina*. *Biomass*. 12, 37-47, 1987.
134. Breuer, G., Lamers, P. P., Martens, D. E.,Draaisma, R. B., Wijffels, R. H., Effect of light intensity, pH, and temperature on triacylglycerol (TAG) accumulation induced by nitrogen starvation in *Scenedesmus obliquus*. *Bioresour. Technol.* 143, 1-9, 2013.
135. Sibi, G., Shetty, V., Mokashi, K., Enhanced lipid productivity approaches in microalgae as an alternate for fossil fuels - a review. *J. Energy Institute*. 89(3), 330- 334, 2016.
136. Li, H., Liu, Z., Zhang, Y., Li, B., Lu, H., Duan, N., Liu, M., Zhu, Z., Si, B., Conversion efficiency and oil quality of low-lipid high-protein and high-lipid low-protein microalgae via hydrothermal liquefaction. *Bioresour. Technol.* 154, 322-329, 2014.
137. Alabi, A. O., Tampier, M., Bibeau, E., Microalgae technologies and processes for biofuels/bioenergy production in British Columbia, Current Technology, Suitability and Barriers to Implementation. Final Report, The British Columbia Innovation Council. Seed Science Press, 2009.

138. Mandal, S., Mallick, N., Microalga *Scenedesmus obliquus* as a potential source for biodiesel production. *Appl Microbiol. Biotechnol.* 84(2), 281-291, 2009.
139. Sato, A., Matsumura, R.,Hoshino, N., Tsuzuki, M., Sato, N., Responsibility of regulatory gene expression and repressed protein synthesis for triacylglycerol accumulation on sulfur starvation in *Chlamydomonas reinhardtii. Front. in Plant Sci.* 5, 444, 2014.
140. Pribyl, P., Cepák, V., Zachleder, V., Production of lipids in 10 strains of Chlorella and Parachlorella, and enhanced lipid productivity in *Chlorella vulgaris. Appl Microbiol Biotechnol.* 94(2), 549-561, 2012.
141. Rodolfi, L., Chini Zittelli, G., Bassi, N., Microalgae for oil: Strain selection, induction of lipid synthesis and outdoor mass cultivation in a low-cost photobioreactor. *Biotechnol Bioeng* 102(1), 100-112, 2009.
142. Xu, N., Zhang, X.,Fan, X., Han, L., Zeng, C., Effect of nitrogen source and concentration on growth rate and fatty acid composition of *Ellipsoidion* sp. (Eustigmatophyta). *J. Appl. Phycol.* 13, 463-469, 2001.
143. Zhu, L. D., Li, Z. H., Hiltunen, E., Strategies for lipid production improvement in microalgae as a biodiesel feedstock. *BioMed Research International.* Article ID 8792548, 1-8, 2016.
144. Kirrolia, A., Bishnoia, N. R., Singh, N., Salinity as a factor affecting the physiological and biochemical traits of *Scenedesmus quadricauda. J. Algal Biomass Utilization.* 2, 28-34, 2011.
145. Takagi, M., Karseno, T., Yoshida, T., Effect of salt concentration on intracellular accumulation of lipids and triacylglyceride in marine microalgae *Dunaliella* cells. *J. Biosci. Bioeng.* 101 (3), 223-226, 2006.
146. Duan, X., Ren, G. Y., Liu, L. L., Zhu, W. X., Salt-induced osmotic stress for lipid overproduction in batch culture of Chlorella vulgaris. *Afr. J. Biotechnol.* 11(27), 7072-7078, 2012.
147. Ren, H.Y.,Liu, B.F., Kong, F., Zhao, L., Xie, G.J., Ren, N.Q., Enhanced lipid accumulation of green microalga *Scenedesmus* sp. by metal ions and EDTA addition. *Bioresour Technol.* 169, 763-767, 2014.
148. Yang, J.,Cao, J., Xing, G., Yuan, H., Lipid production combined with biosorption and bioaccumulation of cadmium, copper, manganese and zinc by oleaginous microalgae *Chlorella minutissima* UTEX2341. *Bioresour Technol.* 175, 537-544, 2015.
149. Hossain, A., Salleh, A., Boyce, A., Chowdhury, P., Naqiuddin, M., Biodiesel fuel production from microalgae as renewable energy. *Amer J. Biochem. Biotechnol.* 4, 250-254, 2008.
150. Johnson, M., Wen, Z., Production of biodiesel fuel from the microalga *Schizochytrium limacinum* by direct transesterification of algal biomass. *Energy Fuels.* 23(10), 294-306, 2009.
151. Xiaoli, L., Thomas Kiran, M., Ling, T., Rong, L., Archana, T., Gu, L., Optimization of growth conditions and fatty acid analysis for three freshwater diatom isolates. *Phycol Res.* 65(3), 177-187, 2017.

152. Balat, M., Potential alternatives to edible oils for biodiesel production-A review of current work. *Energ Convers Manage.* 52, 1479-92, 2011a.
153. Arshad, M., Adil, M., Sikandar, A., Hussain, T., Exploitation of meat industry by-products for biodiesel production: Pakistan's perspective. *Pak. J Life Soc Sci.* 12,120-5,2014a.
154. Huang, G. H., Chen, F.,Wei, D., Zhang, X.W., Chen, G., Biodiesel production by microalgal biotechnology. *Appl Energy*. 87, 38-46, 2010.
155. Tiwari, A., Pandey, A.,Cyanobacterial hydrogen production: A step towards clean environment. *Int. J. Hydrog. Energy.* 37, 139-150, 2012.
156. Ranga Rao, A., Ravishankar, G.A., Influence of CO_2 on growth and hydrocarbon production in *Botryococcus braunii. J. Microbiol. Biotechnol.* 17(3), 414-419, 2007.
157. Ghasemi, Y., Amini, S., Naseri, A., Najafabady, N., Mobasher, M., Dabbagh, F., Microalgae biofuel potentials. *Appl. Biochem. Microbiol.* 48(2), 126-144, 2012.
158. Hamelinck, C.N., van Hooijdonk, G., Faaij, A.P.C., Ethanol from lignocellulosic biomass: Techno-economic performance in short, middle and long-term. *Biomass & Bioenergy*. 28(4), 384-410, 2005.
159. Deng, M.D., Coleman, J.R., Ethanol synthesis by genetic engineering in cyanobacteria. *Appl. Environ Microbiol.* 65(2), 523-528, 1999.
160. Vijayaraghavan, K., Hemanathan, K., Biodiesel production from freshwater algae. *Energy Fuels* 23, 5448-5453, 2009.
161. Brennan, L., Owende, P., Biofuels from microalgae: a review of the technologies for production processing and extractions of biofuels and co-products. *Renew Sust Energ Rev.*, 14, 557-577, 2010.
162. Ugwu, C.U., Aoyagi, H., Uchiyama, H., Photobioreactors for mass cultivation of algae. *Bioresour Technol.* 99, 4021-4028, 2008.
163. Rawat, I., Ranjith Kumar R, Mutanda, T., Bux, F., Biodiesel from microalgae: A critical evaluation from laboratory to large scale production. *Appl. Energy.* 103, 444-467, 2013.
164. Shaikh Abdur, R., Saad Aldin, M.A., Mohammad Mozahar, H., Hugo, L., Biological CO_2 fixation with production of microalgae in wastewater: A review. *Renewable and Sustainable Energy Reviews*. 76, 379-390, 2017.
165. Schenk, P.M., Thomas-Hall, S.R., Stephens, E., Marx, U.C., Mussgnug, J.H., Posten, C., Second generation biofuels: High-efficiency microalgae for biodiesel production. *Bioenergy Res.* 1, 20-43, 2008.v

Index

www.ingramcontent.com/pod-product-compliance
Lightning Source LLC
LaVergne TN
LVHW011528200926
848484LV00001B/196